CENTRIFUGAL SEPARATIONS
IN MOLECULAR AND CELL BIOLOGY

Centrifugal Separations in Molecular and Cell Biology

Edited by
G.D. BIRNIE
The Beatson Institute for Cancer Research, Glasgow

and

D. RICKWOOD
The University of Essex, Colchester

BUTTERWORTHS
LONDON – BOSTON
Sydney – Wellington – Durban – Toronto

The Butterworth Group

United Kingdom   Butterworth & Co (Publishers) Ltd
London   88 Kingsway, WC2B 6AB

Australia   Butterworths Pty Ltd
Sydney   586 Pacific Highway, Chatswood, NSW 2067
                  Also at Melbourne, Brisbane, Adelaide and Perth

South Africa   Butterworth & Co (South Africa) (Pty) Ltd
Durban   152—154 Gale Street

New Zealand   Butterworths of New Zealand Ltd
Wellington   T & W Young Building, CPO Box 472
                  77—85 Customhouse Quay

Canada   Butterworth & Co (Canada) Ltd
Toronto   2265 Midland Avenue
                  Scarborough, Ontario, M1P 4S1

USA   Butterworth (Publishers) Inc
Boston   19 Cummings Park, Woburn, Massachusetts 01801

First published 1978

ISBN 0 408 70803 4

©Butterworth & Co (Publishers) Ltd 1978

British Library Cataloguing in Publication Data

Centrifugal separations in molecular and cell biology
    1. Centrifuges 2. Molecular biology — Technique
    I. Birnie, George David II. Rickwood, David
    574.8'8'028   QH324.9.C4   78-40042

    ISBN 0-408-70803-4

Typeset by Scribe Design, Medway, Kent
Printed by W & J Mackay Ltd., Chatham

Preface

The centrifuge has played an essential, although not always acknowledged, role in almost every advance in molecular and cell biology, and it is now one of the most basic and valuable pieces of equipment available in biology laboratories of all kinds and sizes. Over the years the laboratory centrifuge, particularly the high-speed version, the ultracentrifuge, has undergone extensive development and modern machines are highly sophisticated and versatile instruments. However, despite daily use in many laboratories, the full potential of preparative centrifuges is rarely realized. This is because the proper application of many centrifugation techniques requires considerable expertise which is largely passed on by word of mouth, since many of the pertinent details are widely scattered in the literature. It was the realization that laboratory workers as well as students would find it beneficial to have these details drawn together in the one volume which prompted us to initiate the preparation of this book. The subject of the book is the application of modern centrifugation technology in molecular and cell biology. Its purpose is to present a detailed discussion of all aspects of the methodologies for the separation and fractionation of biological particles by centrifugation on both the preparative and analytical scales and, most important, to present this in such a form that readers with little or no previous experience of the techniques can easily make effective use of the most sophisticated of these methods.

The major emphasis of this book is on the practical aspects of the subject. Thus, a great deal of space has been devoted to describing how the various types of centrifugal separations are actually done, why they are best done in particular ways, the relative merits of different types of centrifugation techniques and how these vary according to circumstances, and the limitations of the methods and the ways in which these limitations may be minimized. Much space has also been given to the interpretation of the data obtained by the various centrifugation methods, and many examples of separations are described to illustrate the points made. In addition, the mathematics of basic sedimentation theory has been presented in a very simplified form, with many worked examples in illustration, to show non-mathematicians how they can easily apply the conclusions drawn from this theory to data from their own experiments and, thus, make more effective use of the equipment at their disposal. In many ways, therefore, the contents of this book complement those of its two predecessors, *Subcellular Components* (1972) and *Subnuclear Components* (1976), also published by Butterworths, since most of the methods described in these depend on one or more centrifugation techniques.

This book is multi-authored because centrifugation is such a wide field that it is difficult for any one author to give an equally authoritative, in-depth description of every one of its many aspects. We are indebted to the authors for having given so generously of their time and experience in preparing their chapters; we are particularly grateful to them for encouraging us to assemble this book, and for their patience and co-operation with our attempts to present a reasonably integrated overview of the whole field. We also thank the publishers for their help, and the personnel of the various centrifuge manufacturers for answering our questions promptly and in full detail.

G.D. BIRNIE and D. RICKWOOD

Contributors

G.D. Birnie
The Beatson Institute for Cancer Research, Wolfson Laboratory for Molecular Pathology, Garscube Estate, Switchback Road, Glasgow G61

Ailsa M. Campbell
Department of Biochemistry, The University of Glasgow, Glasgow G12 8QQ

Robert Eason
Department of Biochemistry, The University of Glasgow, Glasgow G12 8QQ

Lars Funding
Department of Clinical Chemistry, Vejle Hospital, Vejle, Denmark

John M. Graham
Department of Biochemistry, St George's Hospital Medical School, Blackshaw Road, Tooting, London SW17 0QT

Niels Peter H. Møller
Institute of Medical Biochemistry, The University of Aarhus, Aarhus, Denmark

J. Molloy
MSE Scientific Instruments Ltd, Manor Royal, Crawley RH10 2QQ, Sussex

D. Rickwood
Department of Biology, The University of Essex, Wivenhoe Park, Colchester CO4 3SQ, Essex

David Ridge
Department of Biochemistry, University College London, Gower Street, London WC1E 6BT

S.P. Spragg
Department of Chemistry, The University of Birmingham, Birmingham B15 2TT

Jens Steensgaard
Institute of Medical Biochemistry, The University of Aarhus, Aarhus, Denmark

Contents

1 Introduction: Principles and Practices of Centrifugation

D. RICKWOOD
Department of Biology, The University of Essex
G.D. BIRNIE
The Beatson Institute for Cancer Research, Wolfson Laboratory for Molecular Pathology, Glasgow

This book is designed both as a complete guide for the novice and as an aid to the worker who has already had some experience of centrifugal techniques. This introductory chapter discusses the various techniques which are available to the researcher and describes the layout of the book, thus directing the reader to the relevant chapters which provide the information necessary to carry out each type of fractionation.

BASIC CONCEPTS OF SEDIMENTATION THEORY

Essentially, a centrifuge is a device for separating particles from a solution. In biology, the particles are usually cells, subcellular organelles, or large molecules – all of which are called 'particles' to simplify the terminology. The physical parameters which determine the extent of fractionation apply equally to such diverse particles as macromolecules and cells, although the nature of the particles (for example, their lability, sensitivity to osmotic pressure, etc.) may place restraints on the centrifugation conditions that can be used. The following is a simplified introduction to some of the basic parameters which govern the sedimentation and separation of particles in a centrifugal field. A detailed analysis of the theoretical aspects of centrifugal separations is given in Chapter 2.

Some of the basic principles of the sedimentation theory originate from Stokes's law. If the sedimentation of a sphere in a gravitational field is considered it can be shown that, as the velocity of a spherical particle reaches a constant value, the net force on the particle is equal to the force resisting its motion through the liquid. This resisting force is called frictional or drag force. From Stokes's law it can be calculated that the sedimentation rate, v, of a particle is given by

$$v = \frac{d^2(\rho_p - \rho_m)}{18\mu} \times g$$

From this equation, it can be seen that:

1. The sedimentation rate of a given particle is proportional to the square of the diameter, d, of the particle.

1

2. The sedimentation rate is proportional to the difference between the density of the particle and the density of the liquid medium, $(\rho_p - \rho_m)$.
3. The sedimentation rate is zero when the density of the particle is equal to the density of the liquid medium.
4. The sedimentation rate decreases as the viscosity, μ, of the liquid medium increases.
5. The sedimentation rate increases as the force field, g, increases.

The force field relative to the earth's gravitational field (RCF) exerted during centrifugation is defined by the equation

$$RCF = \frac{\omega^2 r}{980}$$

where r is the distance between the particle and the centre of rotation in cm; the rotor speed ω in rad/sec can be calculated from the equation

$$\omega = \text{rev/min} \times \frac{2\pi}{60} = \text{rev/min} \times 0.104\,72$$

The sedimentation velocity per unit of centrifugal force is called the sedimentation coefficient, s:

$$s = \frac{1}{\omega^2 r} \times \frac{dr}{dt}$$

where dr/dt is the rate of movement of the particle in cm/sec. Sedimentation coefficients are usually expressed in svedbergs (S), equivalent to 10^{-13} sec. Thus, a particle whose sedimentation coefficient is measured at 10^{-12} sec, i.e. 10×10^{-13} sec, is said to have a sedimentation coefficient of 10 S.

The mathematical bases of sedimentation theory (both rate-zonal and isopycnic) have concerned many authors since the methods were first introduced, and a detailed analysis of this aspect of centrifugation is described in Chapter 2. Many biologists consider that the mathematics included in most discussions of sedimentation theory is esoteric in the extreme, and constitutes a considerable disincentive to their attempting to understand the basic parameters governing fractionations in a centrifugal field. To help counteract this feeling, Chapter 2 is so organized that the basic theories are presented in the text of the chapter with the main mathematical equations involved simply being stated, whereas the detailed derivations of the equations are grouped together in the Appendix to Chapter 2. The use of these equations for the interpretation of data from the analytical ultracentrifuge is explained in Chapter 8, which also gives some indication of the accuracy and reliability of the theory in practical situations. In fact, a considerable amount of valuable quantitative information can be obtained by applying some of these equations to data from experiments done in preparative ultracentrifuges, and the theme is extended in Chapters 5 and 6, which show how these quantitative data can easily be obtained by the use of no more than a little simple arithmetic.

CENTRIFUGATION METHODS

There are three main types of centrifugal fractionation, namely, (*i*) differential pelleting (differential centrifugation); (*ii*) rate-zonal density-gradient sedimentation; and (*iii*) isopycnic density-gradient sedimentation.

Of these techniques, differential pelleting is the method most commonly used for fractionating material according to size. In this method, the material to be fractionated is initially distributed uniformly throughout the sample solution, which is the sole occupant of the centrifuge tube (*Figure 1.1a*). After centrifugation the pellet is enriched in the larger particles of the mixture (*Figure 1.1b*). However, the pellet obtained always consists of a mixture of the different species of particle, and it is only the most slowly sedimenting component of the mixture that remains in the supernatant liquid which can be purified by a single centrifugation. The amount of contamination in the pellet can be reduced by washing it (that is, by resuspending and recentrifuging), but this inevitably

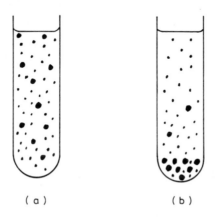

(a)   (b)

Figure 1.1   Effect of centrifuging a suspension of heterogeneous particles

reduces the yields obtained. Some improvement can be made by, for example, sedimenting the particles through a pad of dense sucrose, although such methods are not universally applicable. An example of a general scheme which is the basis of many used to fractionate cells into their components is outlined in *Chart 1.1*. However, it is not possible to discuss such general schemes in any detail. The actual method which should be used to isolate any particular species of particle depends on a host of independently variable factors, including the tissue from which the particles are to be isolated, the purpose for which they are required, the precise composition of the medium in which the cells are homogenized, and the interaction of other components of the homogenate with the particles of interest. No general scheme can make allowances for the variation in all of these factors, and it is, therefore, much more satisfactory to look at the problems involved from the point of view of each individual species of particle. This is beyond the scope of a book of this kind, and reference should be made to publications such as *Subcellular Components: Preparation and Fractionation* (2nd edn) and *Subnuclear Components: Preparation and Fractionation*, published by Butterworths in 1972 and 1976, respectively, which deal with the problems involved in isolating the various components of eukaryotic cells from a variety of tissues. The simplicity of differential pelleting methods makes them very attractive for a large number of fractionations on a preparative scale. Moreover, the methods have the advantage that they can be performed with fixed-angle rotors which generally have a higher capacity than the swing-out type.

Chart 1.1   FRACTIONATION OF CELL COMPONENTS

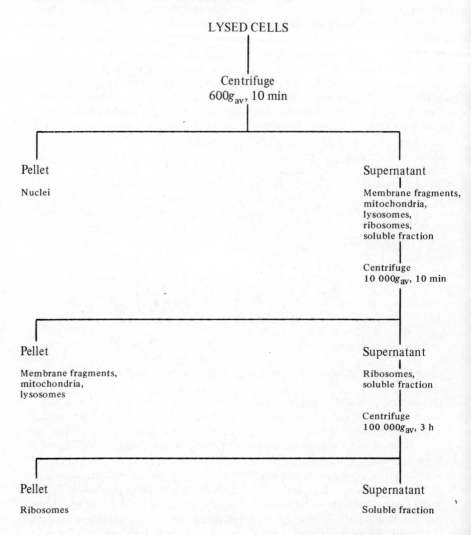

LYSED CELLS

Centrifuge
$600g_{av}$, 10 min

Pellet

Nuclei

Supernatant

Membrane fragments,
mitochondria,
lysosomes,
ribosomes,
soluble fraction

Centrifuge
$10\ 000g_{av}$, 10 min

Pellet

Membrane fragments,
mitochondria,
lysosomes

Supernatant

Ribosomes,
soluble fraction

Centrifuge
$100\ 000g_{av}$, 3 h

Pellet

Ribosomes

Supernatant

Soluble fraction

However, sometimes the material is not cleanly pelleted at the bottom of the tube in a fixed-angle rotor, but is smeared down the side of the tube. This is the result either of too rapid a rate of acceleration of the rotor (especially in the case of very large particles), or of too high a concentration of material in the tube.

The efficiency of the fractionation of particles according to size and shape can be improved markedly by using rate-zonal centrifugation through a density gradient, although this does mean that the sample capacity of each tube is greatly reduced. This technique involves layering the sample on to the top of a liquid column which is stabilized by a gradient of an inert solute, commonly sucrose. It is important that the maximum density of the gradient is less than the buoyant density of the particles and that the gradients are centrifuged in swing-out or zonal rotors to minimize wall effects (see p. 63).

When centrifuged, the particles move down the gradient at a rate which depends on the parameters given in Stokes's equation. Since one frequently only wants to separate the different particles rather than measure their sedimentation coefficients accurately, Chapter 3 discusses the general and practical aspects of such separations with the emphasis on the use of swing-out rotors. On the other hand, if very large amounts of sample are to be fractionated it is often more convenient to use a zonal rotor. Moreover, the resolution of a mixture into its components is significantly better in zonal rotors. Details of the design and use of these rotors are discussed in Chapter 4. Further, if care is taken these preparative techniques can be used analytically. Using the methods described in Chapter 5, the data obtained from rate-zonal sedimentation experiments in either tubes or zonal rotors will yield accurate estimates of the sedimentation coefficients of particles.

The third method for separating particles is isopycnic sedimentation in a gradient whose maximum density exceeds that of the particles. This is an equilibrium technique in which particles are separated on the basis of their buoyant densities, independently of the time of centrifugation and of the size and shape of the particles, although these parameters do determine the rate at which equilibrium is reached and the width of the bands formed at equilibrium. It is important to realize that the effective buoyant density of any particle is a function of the actual density of the particle (determined by its partial specific volume) and its degree of hydration. For example, the density of non-hydrated DNA is close to 2.0 g/cm^3, but its observed buoyant density can vary from 1.7 g/cm^3 to 1.1 g/cm^3, depending on the water activity of the gradient medium. For instance, the density of sodium DNA is 2.0 g/cm^3; in NaI the observed buoyant density of DNA is 1.52 g/cm^3, whereas in metrizamide it is 1.12 g/cm^3, corresponding to a hydration of 8 and 68 moles of water per mole of nucleotide, respectively. Isopycnic centrifugation in the salts of alkaline metals (for example, CsCl and NaI) has been widely used for the separation, fractionation and purification of macromolecules and nucleoproteins; this aspect is discussed in Chapter 6. On the other hand, isopycnic sucrose gradients have been used widely for the fractionation of such cell organelles as mitochondria and nuclei, whereas the more osmotically inert polysaccharides (for example, Ficoll) have been used for the separation of cells. Moreover, a number of other non-ionic density-gradient media have recently been introduced and comparisons between the established and new non-ionic gradient media are discussed in Chapter 7.

The techniques mentioned so far can be used for fractionating particles on either a preparative or an analytical scale. However, in order to obtain really accurate quantitative data it is necessary to use the purpose-built analytical ultracentrifuges. Using these centrifuges it is possible to make measurements on a number of different samples while they are actually being sedimented, the times required for such measurements usually being quite short. The use of the analytical centrifuge and its associated specialized techniques are discussed in detail in Chapter 8.

DESIGNING CENTRIFUGATION EXPERIMENTS

Most centrifugation experiments are generally adaptations of fractionation procedures published by other workers studying similar types of particles. The most common difficulty is that they have used a different make or type of centrifuge or centrifuge rotor. It is not possible to give any sensible general advice as to the relative merits and demerits of different types and makes of centrifuge since the considerations which influence any one person's choice from among the models available embrace such diverse factors as the purposes for which the centrifuge is required and the skills of its users, and the location of the nearest centrifuge engineer, his competence, and the ready availability of spare parts. More serious and general problems arise when the other workers whose method is being followed have specified the type of rotor used and its speed, but not the relative centrifugal force (or other details such as tube angle) which is required to obtain the desired separation. In order to reproduce the experimental conditions when a different type or make of rotor is to be used, it is essential to know the precise dimensions and characteristics of each rotor, and for this purpose Chapter 9 provides tables of data from which direct comparisons of most commercially available rotors can be made. The other important factor to be considered is the choice of centrifuge tube, since some types of tube fail when exposed to particular solvents, extremes of pH, or large density differences. The suitability of each type of tube for a variety of situations is also discussed in Chapter 9.

CARE OF CENTRIFUGES AND ROTORS

The investment of buying an ultracentrifuge and rotors is similar to buying a Rolls-Royce car, and both should be held in the same high esteem and treated accordingly. Thus, for the complete novice with no experience of using a particular centrifuge or a particular rotor it is very important that he or she should read the manufacturers' instructions and, if possible, consult someone who can advise on practical details. For example, at first sight some rotors appear compatible with a particular model of centrifuge, but in reality they must not be used in that model because they are too heavy or, in the case of swing-out rotors, the bowl is too small! The second important factor is the balancing of tubes since, although the flexible drive of an ultracentrifuge is designed to withstand minor imbalances in the rotor, running under these conditions causes the centrifuge to deteriorate faster than normal. It is also important to contain the sample during centrifugation, particularly as the samples are often biohazardous or radioactive. Also, the rotors of all ultracentrifuges run in a high vacuum to minimize heating due to air resistance and so it is especially important to isolate the sample, to prevent evaporation, by capping the tubes securely. Another problem is that of spillage of sample material into the rotor buckets. Many solutions, particularly alkaline and salt solutions, rapidly corrode aluminium and, in some cases, even titanium rotors. Rotors *must always* be rinsed thoroughly with distilled water and left to drain after use. A more detailed account of rotor care is given in Chapter 9, whereas the special problems associated with the maintenance of zonal rotors are discussed in Chapter 4.

2 The Bases of Centrifugal Separations

S.P. SPRAGG
Department of Chemistry, The University of Birmingham

The origin of zonal centrifugation may be hidden in scientific literature but Brakke (1951) and Anderson (1955) are credited with being the first workers to demonstrate the practical advantages of the method. Since then, views have hardened and opinions divided between which of the many advocated procedures give the best results. Even terms have been disputed, but for this discussion 'zone' and 'band' are taken as being synonymous terms. In general, two separate types of zonal experiment have been applied to separating particles: rate-zonal and isopycnic equilibrium. In rate-zonal experiments the zone migrates with the field and its width increases with time, whereas in equilibrium experiments the zone settles at its isopycnic density (isodensity is considered to be synonymous with isopycnic): hence, its shape and position are invariant with time. Both methods differ from the widely applied differential procedure, in which a plateau of concentration is maintained throughout the experiment. In zonal experiments the concentrations both of the supporting gradient and the zonal constituents are never constant with respect to radius once the experiment has started. The introduction of the inert gradient into the experiment presents problems of resolution which are not normally encountered in differential experiments. In this chapter an outline is given of the theoretical treatments of some of the problems encountered in zonal experiments, and emphasis has been placed on conditions which affect resolution of zones when separating a mixture. Many of the factors which affected shape and migration of zones were understood by 1967, and Schumaker (1967) reviewed this work giving useful solutions to the transport equations for zones. Details of these solutions are not repeated here. Instead, an attempt is made to collect together those features which affect the shape and positioning of zones in both velocity and equilibrium experiments.

The major equations are given in the main text, whereas their derivation is given in the Appendix (pp. 22–30). A list of the symbols used and their definitions are given in *Table 2.1*.

COMPARISON OF ZONAL CENTRIFUGATION WITH COMPLEMENTARY TECHNIQUES

When Svedberg first introduced the use of the ultracentrifuge to biochemistry the major problem in physical biochemistry was defining and detecting heterogeneity of macromolecules. Since those early days, analytical techniques more powerful than centrifugation have been developed for detecting heterogeneity, leaving the analytical centrifuge for measuring absolute molecular weights and

Table 2.1 SYMBOLS AND DEFINITIONS

Symbol	Definition
c	concentration (g/cm^3), using subscript where necessary to indicate substance
c_l	loading concentration
c_0	concentration at zero time
D	diffusion coefficient
h	thickness of zone perpendicular to the radius
J	rate of flow
L	length of cylindrical rotor
M	molecular weight
m	mass of macromolecule, or when used as subscript indicates macromolecule
η	viscosity
ϕ	sectorial angle
ω^2	acceleration of rotor ($\pi/30 \times$ rev/min)^2
R	gas constant
r	radius
$\bar{r}$	average radius of zone
r_{max}	radial position of maximum concentration of zone
r_b	radius of bottom of column of liquid
r_0	radius of meniscus
ρ	density; when bearing subscripts it refers to solvent (s), water (w) and total (T)
s	sedimentation coefficient, constant or as measured according to subscript
se( )	standard error of term within the brackets
σ	width of zone
T	temperature (K)
t	time in seconds
θ	suffix defining radius of centre of zone at isopycnic equilibria
$\bar{v}$	partial specific volume with subscripts as for ρ
V_i	variance of ith model
Z_n	nth moment

preparative machines for large-scale separation of particles. The resolution of separated components achieved with the centrifuge depends on differences between masses, densities and frictional coefficients of the particles (the frictional coefficients are phenomenological measurements of their chemical activities in solution). Comparing these parameters with the operational bases of other common analytical procedures shows that centrifugation is the only method which depends primarily on the masses of the components of a mixture for separation (*Table 2.2*).

Electrophoresis is complementary to centrifugation, and depends on differences in charge, density and frictional factors for separation. The resolution of separated components in a mixture can be higher than for centrifugation; unfortunately, since charge and biological activity of particles are dependent on pH, the range

Table 2.2 THE MAJOR PARAMETERS OF PARTICLES WHICH AFFECT SEPARATION BY SOME OF THE COMMONLY USED METHODS

Method	Operational properties
Centrifuge	Mass, density and frictional factor
Gel chromatography	Frictional factor and adsorption
Ionic chromatography and gel electrophoresis	Frictional factor, charge
Isoelectric focusing	Charge

of conditions that can be applied while still retaining activity is limited. This is particularly true for particles constructed from aggregates of molecules, where changing pH or ionic strength dissociates the particles into inactive units. In this case centrifugation and gel chromatography are possibly the only tools available for separating mixtures, and since the former can be applied to large-scale separations of mixtures in almost any solvent, it has become ubiquitous in all biochemical laboratories. Part of the advantage is lost in zonal experiments where the particle is subjected to relatively high concentrations of small molecules forming the gradient. These small molecules can react with the particle producing inactivity, but the major undesirable effect is to reduce resolution below the expected values in mixtures of particles and macromolecules. For most of this discussion it has been assumed that there is no chemical interaction between the gradient and the zonal constituents.

Techniques which depend on a gel matrix for support and separation are usually cheaper to use than centrifuges and require less expertise to achieve good results. However, the limits on the size and quantity of particles which can be effectively separated restrict the useful range of these methods. The capacity of gels is small compared with that of most centrifuge rotors and maximum resolution can be achieved only if the total mass of macromolecules is kept low. Boundaries become distorted if the width of columns is enlarged to accommodate large masses of solutes, and usually it is better to use a cascade of columns rather than increase the diameter of one column beyond normal laboratory limits. The mass which can be separated successfully in zonal experiments is also limited but, since this is determined partly by the gradient, the stable mass in a zone can be raised to relatively high levels by steepening the gradient. This makes zonal centrifugation an attractive method for separating and analysing mixtures of particles and macromolecules.

Zonal experiments are carried out in either tubes or zonal rotors (their use is summarized by Anderson, 1966a). The most obvious difference between them is their relative capacities. For example, a BXIV rotor (Anderson, 1966b) can hold as much as 30 times the mass of sample as a tube 4 cm in diameter loaded at 4 cm from the centre of rotation. Some of the simplicity of the zonal method is lost when zonal rotors are used, because they must be loaded and unloaded while rotating. A less obvious advantage of zonal rotors, but one which is important when considering resolution, arises from their construction – convection occurs in tubes as a result of accumulation of material on the walls leading to density inversions which broaden the zones. This cannot occur in zonal rotors.

Differential *versus* Zonal Centrifugation

A differential experiment may be defined as one in which a plateau of concentration of macromolecule must be preserved throughout the experiment. This differs from zonal experiments, where no plateau exists once the experiment has started. On comparing results from these two types of experiment (*Figure 2.1*) it can be seen that in zonal experiments a complete separation of the components is theoretically possible, whereas in differential experiments only the slowest moving component can be purified (see Chapter 1). Resolution of separated components is determined by the extent to which the zones broaden during the experiment (broadening by diffusion of solutes is proportional to the

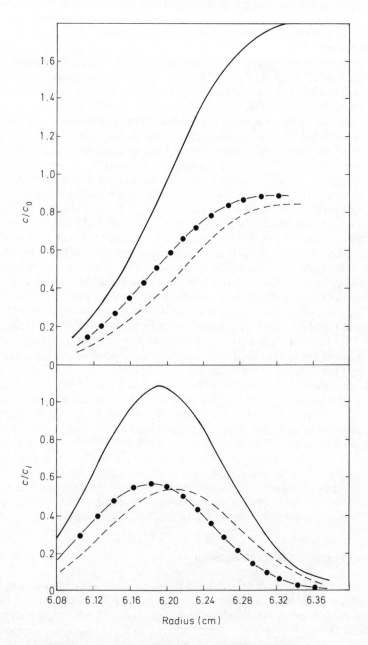

Figure 2.1 Profiles of two boundaries formed from molecules having s_{20w} of 7.5 × 10^{-13} sec (●——●) and 8.5 × 10^{-13} sec (-----) and having the same molecular weight (10^5 daltons) and partial specific volume (0.75 cm^3/g). The profiles were calculated for boundaries starting at 6.0 cm after 3000 sec at 35 000 rev/min. The zone was assumed to be 0.1 cm thick at the start. The continuous line is the total relative concentration and equivalent to the one observed in an experiment

square root of time of the experiment). The experiment simulated in *Figure 2.1* was for two molecules having s-rates of 7.5×10^{-13} sec and 8.5×10^{-13} sec (see later for definition of s) and for observations made 3000 sec after the start of the run at 35 000 rev/min. In these calculations it was assumed that the molecules were ideal, showing no concentration dependence for sedimentation and diffusion. Under these conditions the equation of Gehatia and Katchalsky (1959) as modified by Vinograd and Bruner (1966) is applicable for zones, and the equation of Fujita and MacCosham (1959) was used for the differential boundaries. If zonal calculations had been made for an isopycnic experiment, then the degree of separation would have been negligible, since the same partial specific volumes were chosen for this simulation. In this case a weight-average diffusion coefficient would define the width of the zone.

THEORY OF RATE-ZONAL EXPERIMENTS

Rate-zonal experiments are designed to separate the components of a mixture of particles on the basis of their relative velocities. Since centrifugal force increases with increasing radius absolute velocities also increase and it is normal practice to define a coefficient, s, which has the dimensions of time but is independent of radius. Physically, s is the relative velocity of the particle when subjected to unit acceleration. Before this coefficient can be calculated it is necessary to locate and measure the radial position of the centre of the zone which is independent of diffusion. The location of this co-ordinate is discussed in the following sections.

Sedimentation Coefficient

Under the influence of a gravitational field a zone moves at a velocity determined by a balance between the local force and the frictional resistance it meets. This resistance arises from two sources: (*i*) the viscosity and density of the solvent, which in differential centrifugation is considered to be invariant with radius, and (*ii*) diffusion of particles into regions of low concentration. Diffusion does not affect the position of the centre of mass of the zone, and measurement of the temporal change of this radial co-ordinate (often called the centroid of the boundary) is required in the calculations of s (defined as equivalent to $d\ln r/dt\,\omega^2$). Since s is dependent on the nature of the solvent, a normalized coefficient ($s_{20,w}$) is defined as the coefficient the particle would have in water at 20 °C. $s_{20,w}$ can be calculated from eqn (2.1) if the densities and viscosities of the solution and water at 20 °C, and temperature of the experiment, are known [using eqn (2.1) assumes $\bar{v}$ is independent of solvent and temperature] :

$$s_{20,w} = s \left[\frac{1 - \bar{v}\rho_{20,w}}{1 - \bar{v}\rho_{T,s}} \right] \frac{\eta_{T,s}}{\eta_{20,w}} \tag{2.1}$$

Even after this correction the $s_{20,w}$ is not constant over all experiments because its value increases with decreasing concentration of the particles, eqns (2.2a) and (2.2b):

$$s_{20,w} = s_{20,w}^{\circ}(1 - gc) \qquad (2.2a)$$

$$s_{20,w} = \frac{s_{20,w}^{\circ}}{1 + kc} \qquad (2.2b)$$

where g and k are constants having dimensions of reciprocal concentration and whose values are determined mainly by the shape of the particle (see p. 279). Thus, in any centrifugal experiment a weight average $s_{20,w}$ can be estimated provided the change in the radial position of the centroid of the zone with time is known.

Location of the Centroid

Often, the peak of the zone is equated with the centroid, but as is well known for differential experiments this is true only for symmetrical boundaries which in turn can only occur in the absence of a centrifugal field. Normally, the centroid must be calculated using second-moment analysis involving summations of concentration and radii:

$$(\bar{r})^n = \frac{\displaystyle\int_{r_a}^{r_b} c\, r^n\, dr}{\displaystyle\int_{r_a}^{r_b} c\, dr} \qquad (2.3)$$

The second moment is required to calculate the centroid ($\bar{r}$) for sector-shaped vessels, $n = 2$, eqn (2.3). The integration limits, r_a and r_b, correspond to radii where the concentrations are zero, but are generally equal to the radii of meniscus and bottom of the vessel. Practically, the integrals in eqn (2.3) can be evaluated by a numerical procedure (for example, Simpson's procedure) from concentrations measured at equal intervals across the zone.

It is instructive to calculate the level of errors introduced by using the maximum of the zone rather than the centroid. This can be calculated from eqn (2.4) (Vinograd and Bruner, 1966) which relates the true s with that calculated from the maximum (s_{max}) for approximately symmetrical zones (see Appendix for derivation):

$$s_{max} = s \left[1 - \frac{Dt \exp(s\omega^2 t)}{r_{max}^2}\right] \qquad (2.4)$$

Equation (2.4) shows that the error is determined mainly by the magnitude of the diffusion coefficient [$\exp(s\omega^2 t) \simeq 1$], and since D decreases as the macromolecular weight increases, the error decreases with increasing molecular weight. Errors (with $s_{max} < s$) in the range of 0.4% for ribonuclease to 0.005% for ϕX virus were quoted by Vinograd and Bruner (1966). Considerably larger errors occur if the radial position of the maximum is used in place of the centroid for asymmetrical zones. Asymmetry can be caused by either enhanced concentration dependency of s (for example, in the centrifuging of nucleic acids and fibrous proteins) or heterogeneity. Hence, in order to estimate true weight average $s_{20,w}$ the centroid and not the peak must be employed, and this requires second-moment calculations.

Resolution in Velocity Experiments

Resolution is a difficult parameter to define, since its value is determined by the relative sedimentation and diffusion coefficients of the particles (an example of incomplete resolution is shown in *Figure 2.1*). Both parameters are reduced by increased viscosity of the supporting gradient, but only sedimentation is affected by changing density. These properties of the solvent are ill-defined variables, making the derivation of explicit solutions to transport equations involving relative coefficients rather than $s_{20,w}^{\circ}$ difficult. Normally, it is possible to simulate the experiment by the introduction of adjustable parameters into simplified equations such as that of Gehatia and Katchalsky (1959), or by using numerical procedures as discussed briefly on p. 21. For the present, it can be assumed that the limiting factor on resolution is the extent to which the zone broadens during the experiment, and this in turn depends on the length of time it takes to complete the run.

Zonal Broadening

Many descriptions of zonal broadening follow classical theories of sedimentation and assume that diffusion and heterogeneity are the sole cause of broadening, but this treatment is inadequate for zonal experiments. The zone is supported by the positive gradient and a stable zone can be produced only if the zonal density at any position does not exceed that of the density of the gradient beneath the zone. The density of the zone is equal to the sum of densities of the solvent (including buffers and gradient material) and of particles. Thus a stable zone can be defined by

$$\rho_{T}^{(r)} \leqslant \rho_{s}^{(r+dr)} \tag{2.5}$$

Anderson (1955) first realized that density instabilities reduced resolution in zonal experiments, while Svensson, Hagdahl and Lerner (1957) were the first workers to formulate a relationship to define a stable zone. Later, Brakke (1964) produced detailed experimental demonstrations of instability in intact zones. These early considerations have led to more exact studies based on the relative diffusions of the components in the zone; eqn (2.6) related ρ_T with the densities of the components (see Appendix for derivation):

$$\rho_{T}^{(r)} = \rho_{w} + c_{s}^{(r)}(1 - \rho_{w}\bar{v}_{s}) + c_{m}^{(r)}(1 - \rho_{w}\bar{v}_{m}) \tag{2.6}$$

Stability requires that $d\rho_{T}/dr \geqslant 0$; hence the limit of stability is given by eqn (2.7) (see Appendix for derivation):

$$\left(\frac{\Delta\rho_{s}}{\Delta\rho_{m}}\right)_{r} = \frac{(\rho_{s}^{(1)} - \rho_{s}^{(2)})}{c_{m}(1 - \bar{v}_{m}\rho_{s}^{(2)})} \leqslant -\left(\frac{D_{s}}{D_{m}}\right)^{n} \tag{2.7}$$

Equation (2.7) was given by Schumaker and Halsall (1971) in which $\rho_s^{(1)}$ is the initial density of the gradient in the zone while $\rho^{(2)}$ is that of the underlay. From theoretical considerations the index n has been assigned values ranging from -1.5 (Svensson, Hagdahl and Lerner, 1957) to -2.5 for thin overlays (Sartory, 1969). Berman (1966) examined stability caused by density instabilities at the start of the experiment and produced eqn (2.8) (see Appendix for derivation):

$$m_{max} = \frac{\Pi L \rho_m (d\rho_s/dr)\bar{r}(\Delta r)^2}{\rho_m - \rho_s} \tag{2.8}$$

This equation is simpler to apply than eqn (2.7) but, as shown later, it is inaccurate.

Diffusion-induced convection has been applied in hydrology to explain the presence of zones of changing salinity in oceans. These rest on gradients of density produced by differences in temperature (see, for example, the work of Turner and his colleagues, 1968, 1969 and 1970). In the ocean, fingers of high salinity move out from higher concentrations into regions of lower concentrations, so broadening the zones at rates greater than would be predicted by diffusion. In the centrifuge, the same process produces droplets from zones containing particles but not macromolecules (Nason *et al.*, 1969). The fingers move into regions of higher centrifugal field, forcing them to move faster than the main zone until eventually they form discrete drops. In zones formed from macromolecules droplets do not form, since the rates of diffusion are sufficiently large that discrete fingers cannot form. Instead, the zone broadens at a rate which is not determined solely by the diffusion of the macromolecule, but by the relative rates of diffusion, eqn (2.7).

Practically, it would be useful if a realistic value could be assigned to n in eqn (2.7); then calculations could be made for predicting conditions which will reduce the instability and hence give tighter zones. Careful experiments by Schumaker and Halsall (1971) with proteins suggest that $n \simeq -1$; however, the experiments presented by Spragg and Rankin (1967) with phage T3 on a gradient of sucrose $(D_s/D_m \simeq 68.7)$ show that the zone was stable when $n = -2.9$ and unstable for $n = -1.5$. Sartory (1969) showed that the value for n changes with thickness of both the zone and the overlay, and so it is not surprising to find that the experimental values for n vary between experiments. Nevertheless, the results demonstrate that the mass of a macromolecule or particle carried by a stable zone is less than the value calculated by eqn (2.8). The error is sufficient to reduce the load to as little as 10% of that calculated for molecules of the size of bovine serum albumin. With larger particles in thin zones and reasonable overlays (50 ml in a BXIV rotor), eqn (2.8) is a useful guide to calculating the maximum load that can be floated on a gradient while maintaining the expected width (see also pp. 129–133). Using macromolecules, the only way to conduct experiments and maintain narrow zones with high concentrations is to steepen the supporting gradient. Then it is possible to reduce diffusion by utilizing the increased viscosity in the direction of the centrifugal field (Spragg, Morrod and Rankin, 1969).

All theory and experimental work on instabilities has utilized zonal rotors. Instability occurs in tubes even without overlays, but here broadening will be larger than in rotors because the sedimenting particles accumulate on the walls to produce density inversions. The simplest guide when designing experiments for tubes is to assume at least a 90% reduction on the estimated load calculated from eqn (2.8).

The diffusion coefficients of biological particles surrounded by a semi-permeable membrane are complex parameters compounded from diffusion of exchangeable solutes and the overall coefficients of the intact particles. The exchange with the gradient means its density will increase with increasing osmotic pressure. The theoretical treatments discussed previously assume constant diffusion coefficients and densities for the particle. Although it is possible to extend the model to situations involving varying parameters, it is doubtful whether the results would be sufficiently general to provide usable guides for experiments. Often, the exchange between gradient and particle can be reduced by constructing the gradient from large inert molecules such as Ficoll. Equation (2.7) shows that increasing the molecular weight would reduce convective instabilities $(D_s/D_m \rightarrow 1)$, so allowing increased loading of the initial zone. This gain would be sacrificed by increased experimental time caused by the considerably higher viscosity of Ficoll compared with water. Interaction between gradient and zonal constituent would reduce resolution in rate-zonal experiments, but its practical importance only becomes significant in isopycnic experiments.

Practical studies on the homogeneity of separated zones by testing for symmetry of zone or comparing the maximum of the peak with the calculated centroid, eqn (2.3), can be considerably in error for unstable zones. It would seem initially that asymmetrical zones could be caused only by heterogeneity, but the previous discussion shows that density instabilities produce similar shapes. However, in carefully designed experiments asymmetry is caused either by heterogeneity or enhanced concentration dependency of $s_{20,w}$. Conversely, symmetrical zones do not prove homogeneity because, as Cox (1966) showed, a disturbed gaussian peak corrects itself with time, and as seen in *Figure 2.1* the sensitivity of zones to heterogeneity is poor. Readjustment may cause small zones to be engulfed, leaving a symmetric heterogeneous zone at the end of the experiment. These uncertainties mean that statements concerning purity based on results from zonal experiments are not reliable unless consistent results are obtained from experiments made with varying concentrations.

With most zonal experiments the zones are located after unloading by spectrophotometric measurements or, in experiments with analytical ultracentrifuges, by combining refractometric measurements with light absorption. Thus, generally only concentrations of the solute that are greater than 0.1 mg/ml can be studied by these methods. This is considerably greater than concentrations normally found in biological materials and is also greater than the levels used for testing activity. Cohen, Giraud and Messiah (1967) reduced these difficulties by enlisting the biological activity to detect active material in moving zones. They filled a cell with gradient containing a substrate whose concentration could be determined spectrophotometrically and which changed absorbance during the reaction. The zone containing the enzyme was layered on to the gradient and the migration of this zone detected by following the boundary formed between the absorbing substrate and non-absorbing products. Provided the enzyme does not convert all the substrate before the zone has moved to a new position and the rate is proportional to concentration, it is then possible to follow the migration of the active component. The application of this novel and powerful procedure has been limited mainly to experiments with the analytical ultracentrifuge (see p. 280), but there is nothing to prevent it being extended to tubes or zonal rotors; then, the estimation of the converted substrate would be made at the end of the experiment.

ISOPYCNIC ZONAL EXPERIMENTS

In contrast with rate-zonal experiments where separation is achieved by differences in mass, isopycnic experiments are designed to separate particles by differences in density (Meselson, Stahl and Vinograd, 1957). The species reach true chemical and physical equilibrium to give a gaussian zone with the peak centred at the density for the macromolecule or particle. The experiment may be started either by mixing an inert solute of high density with the macromolecule or particle to give a uniform solution of both at the beginning of the experiment, or by layering a zone on a preformed gradient. In both methods, the gradient of inert material must include the range of densities at which the particles are expected to band.

Time to Reach Equilibrium

In general, preparative experiments use columns of liquid extending over relatively large radial ordinates (1–4 cm are commonly employed), but the time to reach equilibrium is inversely proportional to diffusion and directly proportional to the square of the length of the column (Fujita, 1961). Hence, with small molecules prolonged runs are necessary before equilibrium is achieved (5–7 days). It would be useful if formulae could be given for calculating the time to reach equilibrium. Baldwin and Shooter (1963) considered this problem and showed that eqn (2.9) is the approximate relationship between time for a molecule at r to reach r_θ :

$$\frac{r_\theta - r}{r} = \frac{r_\theta - r_0}{r_0} \exp\left(\omega^2 r_\theta \frac{ds}{dr} t\right) \tag{2.9}$$

Viscosity and density of the supporting gradient resist movement of the molecule, and in the simplest case where a difference in density between the macromolecule and the gradient is the predominant factor, the change in velocity can be represented by ds/dr, eqn (2.9). $ds/dr = (ds/d\rho_T) \times (d\rho_T/dr)$, whereas $ds/d\rho_T = -Kv$, where $K = MD/RT$; $d\rho_T/dr$ is the linear gradient in the container. The other variables will be known approximately at the start of the experiment, so if a value could be assigned to $(r_\theta - r)/r$ then t, the time to reach approximate equilibrium, could be calculated. True equilibrium can only be reached after infinite time when $(r_\theta - r)/r = 0$ but, for practical purposes, Baldwin and Shooter (1963) suggest a value of 0.001. Using eqn (2.9) the approximate time to reach equilibrium for a molecule of 10^5 daltons, $D = 10^{-7}$ cm^2/sec and $\bar{v} = 0.75$ cm^3/g which sediments in a gradient of 10^{-4} g/cm^4 with an angular acceleration of 10^7 sec^{-2}, is approximately 6 days if the initial and final distances were 2 cm and 5 cm, respectively. A significant reduction in this time could be achieved by either decreasing ds/dr (that is, increasing $d\rho/dr$) or increasing the speed of the rotor.

Resolution of Isopycnic Zones

As with velocity experiments, the resolution between components at equilibrium is defined by the widths and overlaps of the zones. These widths are normally defined by their bandwidths (width at two-thirds their height). The radial position of the maximum of the zone is the isopycnic value of the macromolecule or particle and is described by eqn (2.10) (see Appendix for derivation):

$$r_\theta = \frac{RT}{M\bar{v}\omega^2 (d\rho/dr)_{r_\theta}} \times \frac{1}{\sigma^2} \qquad (2.10)$$

Hence, the radial separation of two molecules at equilibrium becomes

$$(r_\theta^{(2)} - r_\theta^{(1)}) = \frac{RT}{(d\rho/dr)\omega^2} \left[\frac{1}{M_2 \bar{v}_2} \frac{1}{\sigma_2^2} - \frac{1}{M_1 \bar{v}_1 \sigma_1^2} \right] \qquad (2.10a)$$

when it can be seen that radial resolution is proportional to the reciprocal of $(d\rho/dr)\omega^2$. Thus, the steeper the gradient the nearer the centres of the two zones at equilibrium. Decreasing speed pushes them further apart but increases the time to reach equilibrium ($t_{eq} \propto 1/\omega^4$). For this reason it is better to use shallow gradients, but this can lead to more overlap of lower concentrations of the zones because $\sigma^2 \propto 1/(d\rho/dr)$. It is shown in the Appendix that the mass contained by a zone is proportional to σ^2 and inversely proportional to $(d\rho/dr)$; hence larger masses supported on shallower gradients lead to increased widths which reduces resolution of components. It would be difficult to produce optimum conditions to satisfy these interdependent yet conflicting parameters. Experimenters must decide which is the important requirement of their experiment — partially separating large masses or completely resolving small masses — before designing the gradient.

Interaction Between Solutes in Isopycnic Experiments

The results of Birnie, Rickwood and Hell (1973) and Rickwood *et al.* (1974) have confirmed a suspicion that solutes in the gradient are often not inert. In their work with proteins equilibrating on gradients formed from metrizamide [2-(3-acetamido-5-*N*-methylacetamido-2,4,6-tri-iodobenzamido)-2-deoxy-D-glucose] multiple isopycnic zones were produced. When these zones were isolated and re-run, the pattern was repeated showing that the system was in equilibrium. Metrizamide has a density greater than proteins, and when the protein combines with metrizamide to form a complex, the result is several species having densities determined by the proportion of metrizamide to protein. Skerrett (1975) examined this situation theoretically and showed that, for the protein, both the measured weight-average molecular weight and its partial specific volume would be considerably in error when interaction occurs (some of the relationships are summarized in the Appendix). If the rates of reaction are fast relative to sedimentation, then symmetrical zones are produced which are broadened and give erroneous molecular weights (or densities) if interaction were not considered. Slow reactions lead to multiple zones of the type reported by Birnie, Rickwood and Hell (1973) and Rickwood *et al.* (1974). These reactions between the macro-molecule and the solutes in the gradient must not be confused with interactions between macromolecules themselves (Howlett and Jeffrey, 1973). Density changes are small during these associative reactions and only become important when compressibility of the complex is large, making it possible for pressures generated in the centrifuge to reverse the association.

STATISTICAL ESTIMATION OF MOLECULAR PARAMETERS

Often, an experiment is largely designed to separate components in a mixture, but since the method used to detect zones also gives information which can be used to calculate molecular parameters, this additional data can be valuable when identifying particles. In addition to measuring the radial co-ordinates of the zones, records of density (which can be calculated from concentration) of the solutes in the gradient are required.

s-Values

Historically, the method of Martin and Ames (1961) was the first procedure described for calculating $s_{20,w}$ and this was expanded by Bishop (1966). She calculated an *s*-value for each fraction collected at the end of an experiment from zonal rotors. Then, knowing temperature, speed and type of rotor, she calculated $s_{20,w}$ for each fraction. The densities and viscosities of the fractions containing sucrose were calculated from empirical relationships developed by Barber (1966). The relationship between $s_{20,w}$ and other variables is given in eqn (2.11), derived from eqn (2.1):

$$s_{20,w} = \frac{1}{\omega^2 t} \int_{r_q}^{r_i} \frac{(\rho_m - \rho_{20,w})\eta_{r,s} dr}{(\rho_m - \rho_{s,r_1})\eta_{20,w} r} \tag{2.11}$$

which, for finite fractions and constant densities for particles, gives

$$s_{20,w} = \frac{1}{\omega^2 t} \frac{(\rho_m - \rho_{20,w})}{\eta_{20,w}} \sum_{r_q}^{r_i} \frac{2\eta_{r,s,f_i}(r_i - r_{i-1})}{(\rho_m - \rho_{r,s,f_i})(r_i + r_{i-1})} \tag{2.12}$$

where i denotes the specific fraction f_i, $(r_i - r_{i-1})$ is width of fraction and r_q is the radius of the centre of mass of the zone. The summation is calculated over all samples starting at the loading radius, and radial ordinates r_i are calculated from the volumes of samples up to f_i (including overlay), provided the geometry and dimensions of the vessel are known. Schumaker (1967) gives formulae for calculating volumes of tubes, and Bishop (1966) formulae for zonal rotors. These calculations require no statistics; instead the programme produces a series of tabulated values of $s_{20,w}$ for a particle if it had moved to the position occupied by the fraction. The experimenter must compare this table with the spectrophotometric results to find a $s_{20,w}$ for the species of interest (see also Chapter 4).

Schumaker and Halsall (1969) produced a more direct method for calculating $s_{20,w}$ for situations where a few different gradients are used. The method depends on rearranging eqn (2.11) to give eqn (2.13):

$$s_{20,w}\omega^2 t = \frac{(1 - \bar{v}_m\rho_{20,w})}{\eta_{20,w}} \int_{\ln r_0}^{\ln r} \frac{\eta_{r,s}}{(1 - \bar{v}_m\rho_s)} \, d\ln r \qquad (2.13)$$

In this equation the right-hand side represents corrections for density and viscosity which must be estimated. Schumaker and Halsall (1969) used a graphical procedure to evaluate the integral by plotting $\eta_s/(1 - \bar{v}_m\rho_s)$ against $\ln r$ and then estimating the area of the curve. This area is proportional to $\omega^2 t \times s_{20,w}$ and the proportionality constant is $(1 - \bar{v}_m\rho_{20,w})/\eta_{20,w}$. An alternative to the graphical procedure would be to fit the viscosity and density results to a curvilinear polynomial (see Appendix) which can then be integrated using simple rules for integration. The coefficients of the integral can then be used to evaluate the integral between the limits of any particular experiment (that is, r_0 and r), making it relatively easy to simplify the calculations.

Diffusion coefficients can be estimated from records of zonal experiments, and Halsall and Schumaker (1970) showed that relatively accurate estimates are possible with this technique if low concentrations of proteins are examined. Equation (2.25) is the mathematical model used to analyse the results, and for a symmetrical zone at low speeds the maximum of the peak is approximately equal to the centroid of the zone. Equation (2.25) can be rearranged to give a linear relationship containing $\ln[(r_0/r)^{1/2}/c_{r,t}]$. This can be plotted against $(r_{max} - r)^2$ to give a straight line, the slope of which equals $1/(4D\,a\,t)$. A least squares regression analysis will be required to estimate the slope, and the calculations must include a weighting coefficient to allow for distortion of errors caused by using the logarithm of reciprocal concentration instead of true concentration. This weighting adjusts for distortion in errors after conversion for the case where precision with which $c_{r,t}$ can be measured varies over the zone. This weight equals $(c_{r,t}^2)^{-1}$. Errors calculated by the statistical analysis estimate the precision of the estimate, but the overall accuracy with which D is estimated will be affected both by stirring produced when introducing the zone into the rotor and by the accuracy with which the time can be estimated. Hence, care is required when loading and runs must be carried out for relatively long times (see Chapters 4 and 5).

$M\bar{v}$ can be calculated from isopycnic experiments by rearranging eqn (2.45) to give

$$\ln(c_\theta/c_r) = \frac{\omega^2 r_\theta M\bar{v}(d\rho_s/dr)(r - r_\theta)^2}{2RT} \qquad (2.14)$$

Thus plotting $\ln c_r$ against $-(r - r_\theta)^2$ gives a straight line whose intercept is $\ln c_\theta$ and slope $[\omega^2 r_\theta M\bar{v}(d\rho_s/dr)]/2RT$. Thus, knowing r_θ and assuming $d\rho_s/dr$ is constant, Mv can be calculated. If $d\rho/dr$ is not constant across the zone, then it must be measured at each radius and $-(r - r_\theta)^2(d\rho_s/dr)_r$ used for the abscissae. The estimation of these parameters requires statistical methods and again it is wise to include a weighting factor to allow for the conversion of c_r to $\ln c_r$. This time it is c_r^2. Converting $M\bar{v}$ into M or $\bar{v}$ requires that the other parameter be known, and since many experiments are carried out in strong salt solutions (for example, CsCl), the precautions discussed by Cassassa and Eisenberg (1964) must be observed when assuming a value for $\bar{v}$.

Statistical Procedures

The procedures outlined in the foregoing discussion are for a single species of particle or macromolecule, but it is often necessary to test whether results are consistent with homogeneous systems. Heterogeneity will distort the linear relationships and a quadratic equation will fit the record significantly better than a straight line, so tests must be made for linearity. Formulae for fitting quadratic and straight lines are given later and come from the paper of Trautman, Spragg and Halsall (1969). These equations were formulated for desk calculators, and more general concepts suitable for digital computers are outlined in the Appendix.

The problem is to estimate parameters a, b and c, eqn (2.15), from a set of data consisting of n pairs of ordinates (y_i) and abscissae (x_i):

$$y = a + bx + cx^2 \tag{2.15}$$

The coefficients can be calculated from the following formulae:

$$a = \bar{y} - b\bar{x} - c\bar{x}^2$$

$$c = (S_{22}S_{1y} - S_{12}S_{2y})/d$$

$$b = (S_{11}S_{2y} - S_{12}S_{1y})/d$$

In these equations $S_{11} = \Sigma wx^2 - (\Sigma wx)^2/\Sigma w$, $S_{22} = \Sigma wx^4 - (\Sigma wx^2)^2/\Sigma w$, $S_{12} = \Sigma wx^3 - (\Sigma wx)(\Sigma wx^2)/\Sigma w$, $S_{1y} = \Sigma wxy - (\Sigma wx)(\Sigma wy)/\Sigma w$, $S_{2y} = \Sigma wx^2y - (\Sigma wx^2)(\Sigma wy)/\Sigma w$, and $d = S_{11}S_{22} - S_{12}^2$. The weighting coefficient, w, is unity when no weighting is required, whereas terms with a bar above are means of that variable. The standard errors can be computed from the sums using the variance, V_c, equals sse $\times (n-3)^{-1}$, where sse $= \Sigma w(y - y)^2 = S_{yy} - bS_{1y} - cS_{2y}$; $S_{yy} = \Sigma wy^2 - [(\Sigma wy)^2/\Sigma w]$, so giving for se($a$) = $\{ [1/\Sigma w + (S_{22}\bar{x}^2 - 2S_{12}\bar{x}^3 + S_{11}\bar{x}^4)/d] V_c\}^{1/2}$, se($b$) = $(S_{22}V_c/d)^{1/2}$ and se(c) = $(S_{11}V_c/d)^{1/2}$. The tedious part of this procedure is in the computation of the sums of squares (S_{11}); the subsequent calculations are routine. The quadratic coefficient (c) can be tested for significance by computing Student's 't',

$$\text{Student's } 't' = c/\text{se}(c) \tag{2.16}$$

and comparing the value with tables of Student's 't' for 95% significance. If the results do not fit a curve, then in eqn (2.15) c is probably zero, so they fit a straight line and the coefficients a and b must be recalculated from the following equations:

$$a = \bar{y} - b\bar{x}$$

$$b = S_{1y}/S_{11}$$

whereas $se(a) = [(\Sigma wx^2/\Sigma w)V_s]^{1/2}$ and $se(b) = (V_s/S_{11})^{1/2}$, where $V_s = S_{yy}/(n-2)$. These sums have already been calculated for the quadratic equation, making the second fitting easier. The least square principle can be extended to fit non-linear mathematical models to experimental observations (see, for example, McCallum and Spragg, 1972), but the procedure is beyond the scope of this review.

SIMULATION OF ZONAL EXPERIMENTS

Normally, trial experiments must be run in order to define optimal conditions for separation. This is both tedious and expensive in materials when large rotors are used, and it would be useful if experiments could be simulated on a computer before the experiment. Possibly the most important variables in any experiment are gradients of density and viscosity, because these determine the mass of material which can be loaded and indicate the $\omega^2 t$ necessary to complete the separation. As in *Figure 2.1*, the shape of a zone can be calculated using the Katchalsky–Gehatia equation, eqn (2.24), but this makes no allowance for changing concentration of supporting material. A more realistic approach is to employ numerical procedures for integrating the transport equations. Only then is it possible to include changing values for densities and viscosities in the calculations. It is beyond the scope of this chapter to examine these methods in detail, and readers are referred to the work of Cox (1971) and Dishon, Weiss and Yphantis (1966) for numerical procedures found to be successful in analysing equations of the ultracentrifuge. These methods have been applied to cases of mechanical equilibrium and differential experiments in the ultracentrifuge, but the methods are equally applicable to zonal experiments. A procedure described by Sartory, Halsall and Breillat (1976) is particularly suitable for testing isopycnic experiments. In this case it is necessary to compute the redistribution of the supporting materials during the experiment. Realistic results are only achieved if allowance is made for the presence of relatively high concentrations of these materials. The result of the calculations is a gradient with zones containing the macromolecule or particle distributed along the radius. From this, it is possible to decide whether the proposed experiment will be successful.

Since these numerical methods are specialized and, hence, not widely applied, attempts have been made to prescribe idealized gradients for differing situations. These are based generally on rearrangements of eqns (2.1) and (2.2), by integrating them for changing densities and viscosities with increasing radius. For example, there is the isokinetic gradient of Noll (1967) in which particles move at constant speed, and the equivolumetric gradient of Pollack and Price (1971) in which the zone sweeps through equal volumes of gradient in equal times (see Chapter 4). These methods are useful and no doubt widely applied by experimenters when finding and testing experimental gradients.

APPENDIX

Relationships Between s and Radial Co-ordinates

In a sector-shaped vessel the nth moment of the mass of material in a zone is given by

$$Z_n = \phi h \int_{r_o}^{r_b} r^n c \, dr \tag{2.17}$$

From eqn (2.17) a series of moments (Z_n) of order n can be defined as $Z_n = m_n/m_0$, where m_{sub} = integral with r^{sub}: equivalent to eqn (2.3). Vinograd and Bruner (1966) used temporal derivatives of eqn (2.17), eqn (2.18), to define the velocity of the centre of gravity of the zone:

$$\frac{dZ_n}{dt} = \phi h \int_{r_o}^{r_b} r^{(n+1)} \left(\frac{\partial c}{\partial t}\right)_r dr \tag{2.18}$$

Substituting into eqn (2.18) for $(\partial c/\partial t)_r$ from Fick's law of diffusion, eqn (2.19), and integrating by parts knowing that $J = 0$ at r_0 and r_b gives eqn (2.20):

$$\left(\frac{\partial c}{\partial t}\right)_r = -\frac{1}{r}\left(\frac{\partial rJ}{\partial r}\right)_t \tag{2.19}$$

$$\frac{dZ_n}{dt} = \phi h n \int_{r_o}^{r_b} r^n J \, dr \tag{2.20}$$

where

$$J = -D(\partial c/\partial r)_t + s\omega^2 rc \tag{2.21}$$

assuming D and s are independent of concentration. The centre of mass of a zone in a sector-shaped vessel is defined by the second moment (Schachman, 1959), hence

$$\frac{dZ_n}{dt} = \frac{\phi h n}{m_0}\left\{\omega^2 s \int_{r_o}^{r_b} r^n c \, dr - D \int_{r_o}^{r_b} r^n \frac{\partial c}{\partial r_t} dr\right\} \tag{2.22}$$

Again, integrating the right-hand side by parts knowing $c_{r_o} = c_{r_b} = 0$ gives

$$\frac{dZ_n}{dt} = ns\omega^2 Z_n - n^2 D Z_{(n-2)} \tag{2.23}$$

With $n = 2$, eqn (2.23) gives the unweighted or unbiased definition of s ($d\ln(Z_2)^{1/2}/dt = s\omega^2$). The first moment ($n = 1$) defines s for rectangular co-ordinates, making $d\ln(Z_1)/dt = s\omega^2 + (D/Z_1^2)$. Thus, $s_{Z_1} = s_{Z_2}(1 + D/s\omega^2 Z_1^2)$ showing that s calculated from the first moment is greater than that from the second and correct moment.

An error can be calculated for the occasion where s is calculated from the migration of the maximum value for the peak, which in turn is located by finding the expression for dc/dr and putting this equal to zero. Gehatia and Katchalski (1959) derived the integral expression for sedimentation of a very thin lamella in cylindrical vessels which, when modified by Vinograd and Bruner (1966), becomes

$$c_{r,t} = \frac{c_l r_0 \delta}{\sigma^2} \exp\left[-\frac{(r^2 + r_g^2)}{2\sigma^2}\right] I_0\left[\frac{rr_g}{\sigma^2}\right] \tag{2.24}$$

$I_0(F)$ is the zero order Bessel function which can be approximated as

$$I_0(F) = \sum_{l=0}^{\infty} [(F/2)^{2l}/l!l!] = [e^F/(2\pi F)^{1/2}](1 + 1/8F + 9/128F^2 + \ldots),$$

r_0 is the position of the zone at zero time, and c_l is the concentration of the sample. The centre of sedimentation, r_g (Schumaker and Rosenbloom, 1965) is defined as an averaged centroid calculated from $\ln r$ and usually shown as $<\ln r>$ given by $<\ln r> - <\ln r>^0 = s\omega^2(t - t_0)$, where $<\ln r>$ is equivalent to r_g [and equal to $r_0 \exp(s\omega^2 t)$]. σ is the width of the zone at time t and equals $2(Dat)^{1/2}$ with $a = [\exp(2s\omega^2 t) - 1]/2s\omega^2 t$. Reducing eqn (2.24) to include only the first term of the Bessel function (the series converges rapidly) gives eqn (2.25) (equation 54 in Vinograd and Bruner, 1966, δ = width of zone at zero time):

$$c_{r,t} = \frac{c_l \delta (r_0/r_g)}{2\pi^{1/2}} \frac{r_g^{1/2}}{r} \exp\left[-\frac{(r - r_g)^2}{2\sigma^2}\right] \tag{2.25}$$

Differentiating eqn (2.24) with respect to time and equating to zero gives the expression $r_g = -\sigma^2/2r_{max}$. Following the method employed to derive eqn (2.22) and substituting for σ^2 and r_g to introduce r_{max}, then the approximate expression given in eqn (2.4) is for s calculated from the maximum. Thus r_{max} is shifted back relative to the centre of gravity by asymmetric diffusion induced by the centrifugal field. It is possible, by following through a similar logic for a rectangular cell using the simpler solution for the transport equation, eqn (2.26), where $x_g \equiv r_g$, to show that $x_{max} = x_g$, showing that no asymmetry occurs in rectangular cells:

$$c_{(x,t)} = \frac{c_l \delta}{(2\pi)^{1/2}\sigma} \exp\left[-\frac{(x - x_g)^2}{2\sigma^2}\right] \tag{2.26}$$

Convective Instabilities in Zones

In the typical 'wedge' zone inserted between the overlay and gradient, and where the concentration of the gradient solute continues to decrease through the zone into the overlay, the small molecule diffuses faster into the zone from the overlay than the macromolecule moves into the overlay. The result is an increase in density of the zone, causing it to fall and broaden. This process can be described mathematically by the following approximate procedure.

The density of a solution containing three solutes each having constant densities in the solution is given by

$$\rho_T = \frac{Y_m}{Y_m + Y_w + Y_s}\rho_m + \frac{Y_w}{Y_m + Y_w + Y_s}\rho_w + \frac{Y_s}{Y_m + Y_w + Y_s}\rho_s$$

$$(2.27)$$

where Y_i is weight fraction of component i. Since $c_m/\rho_m + c_w/\rho_w + c_s/\rho_s = 1$ at all positions, then $c_w = \rho_w - c_m\bar{v}_m\rho_w - c_s\bar{v}_s\rho_w$, in which $1/\rho_i$ has been replaced by $\bar{v}_i$. Thus eqn (2.27) becomes

$$\rho_T^{(r)} = \rho_w + c_s^{(r)}(1 - \bar{v}_s\rho_w) + c_m^{(r)}(1 - \bar{v}_m\rho_w)$$

$$(2.28)$$

For stability $d\rho_T/dr \geqslant 0$, hence differentiating eqn (2.28) with respect to radius and assuming no sedimentation for the moment gives

$$\left(\frac{d\rho_T}{dx}\right)_t = \left(\frac{d\rho_s}{dx}\right)_t (1 - \bar{v}_s\rho_w) + \left(\frac{d\rho_m}{dx}\right)_t (1 - \bar{v}_m\rho_w)$$

$$(2.29)$$

From Fick's first law, eqn (2.19), $(dc_i/dx)_t$ can be replaced by $-J_i/D_i$, which when assuming perfect differentials gives

$$\frac{d\rho_r}{dx} = -\frac{J_s}{D_s}(1 - \bar{v}_s\rho_w) - \frac{J_m}{D_m}(1 - \bar{v}_m\rho_w)$$

$$(2.30)$$

Equating eqn (2.30) to zero gives eqn (2.31), the expression for minimum stability, which states that when the left-hand side is less than the right-hand side, convection occurs:

$$\frac{J_s}{J_m} = -\frac{D_s(1 - \bar{v}_m\rho_w)}{D_m(1 - \bar{v}_s\rho_w)}$$

$$(2.31)$$

J_s/J_m can be replaced by $\Delta m_s/\Delta m_m$ when considering the relative transport of masses.

Equation (2.31) can be expanded to include sedimentation and becomes eqn (2.32), which shows that more material can be carried in a rapidly sedimenting zone than in a stationary one, eqn (2.31), because the velocity caused by sedimentation is greater than that by diffusion:

$$\frac{J_m - c_m s_m\omega^2 r}{J_s - c_s s_s\omega^2 r} = -\frac{D_m(1 - \bar{v}_s\rho_w)}{D_s(1 - \bar{v}_m\rho_w)}$$

$$(2.32)$$

If densities are retained in eqn (2.28) rather than partial specific volumes, this gives eqn (2.33) which describes the limits of stability:

$$\frac{dc_s}{dr}\left(1 - \frac{\rho_w}{\rho_s}\right) = -\frac{dc_m}{dr}\left(1 - \frac{\rho_w}{\rho_m}\right) \tag{2.33}$$

But $[1 - (\rho_w/\rho_s)] = -(D_s/D_m)(1 - \rho_w/\rho_m)$ and $(dc_s/dr)(1 - \rho_w/\rho_s) = d\rho_s/dr$: therefore, substituting in eqn (2.33) gives eqn (2.34), which, when rearranged, gives eqn (2.7) (Schumaker and Halsall, 1969):

$$\frac{d\rho_s}{dr} = -\frac{D_s}{D_m}\left(1 - \frac{\rho_w}{\rho_m}\right) \tag{2.34}$$

These idealized equations assume that mass transfer is determined by diffusion and sedimentation, but vibrations will increase the rates of transfer. To allow for this uncertainty, an exponent n is included in the stability equations, eqn (2.7). Sartory (1969) applied perturbation theory in an attempt to simulate external stirring and calculate the time required for a zone to become unstable. Depending on the thickness of the overlay he found that n varies between -2.5 for thin, and -1.5 for thick, overlays.

A simpler equation which does not include temporal changes was derived by Berman (1966) for conditions of stability. He started from the conservation of mass criterion in a stable band when eqn (2.35) is true:

$$\int_{r_1}^{r_1'} \frac{f\eta}{\overline{v}(\rho_m - \rho_s)}\frac{dr}{r} = \int_{r_2}^{r_2'} \frac{f\eta}{\overline{v}(\rho_m - \rho_s)}\frac{dr}{r} \tag{2.35}$$

where $(r_1' - r_1)$ and $(r_2' - r_2)$ are the widths of the zone at two radial positions $\overline{r}_1$ and $\overline{r}_2$ $(f = $ frictional coefficient). The total mass in the zone is given by

$$m_{max} = 2\pi L \int_{r_i}^{r_i'} \rho_m c_m \overline{v} r\, dr \tag{2.36}$$

and hence for a binary mixture the local stability from eqn (2.28) becomes

$$\rho_s(r_j) \geqslant \rho_s(r_j) + c_s(r_j)\overline{v}(\rho_m - \rho_s) \tag{2.37}$$

Substituting eqn (2.37) into eqn (2.36) provides the relationship relating the maximum mass that a stable zone can carry to the density of the gradient:

$$m_{max} = 2\pi L \int_{r_i}^{r_j} \rho_m \frac{(\rho_s^{(r_j)} - \rho_s^{(r)})}{(\rho_m - \rho_s^{(r_j)})} r\, dr \tag{2.38}$$

Equations (2.35) and (2.38) can be integrated, provided the widths are small relative to the radius (thin zones). Thus, the limits of integration can be changed by the following logic:

$$0 = \int_{r_1}^{r_1'} - \int_{r_2}^{r_2'} = \int_0^{r_1'} - \int_0^{r_1} - \int_0^{r_2'} + \int_0^{r_2} = \int_{r_1}^{r_2} - \int_{r_1'}^{r_2'}$$

which, when applied to eqn (2.35) and replacing primed variables by averages over the initial thin zone, give the approximate equation

$$\frac{f\eta(r_2 - r_1)}{\bar{v}(\rho_m - \rho_s)\bar{r}} = \frac{f\eta'(r_2' - r_1')}{\bar{v}(\rho_m - \rho_s')\bar{r}'} \tag{2.39}$$

From this, the ratio of final to initial bandwidths is given as

$$\frac{\Delta r'}{\Delta r} = f\bar{v} \frac{\rho_m' - \rho_s'}{\rho_m - \rho_s} \frac{\eta'\bar{r}'}{\eta'\bar{r}} \tag{2.40}$$

The density $\rho_s(r)$ can be expanded in a Taylor series around r and neglecting high-order terms converts eqn (2.38) to eqn (2.8).

Isopycnic Equilibrium

Van Holde (1971) derived the relationship between M, σ_{eq}, $d\rho/dr$ and the centrifuge parameters by the following logic.

For physical equilibrium in a centrifuge where $J = 0$,

$$\frac{dc_m}{dr} \frac{1}{c_m} = \frac{\omega^2 r_\theta M(1 - \bar{v}_m \rho_s)}{RT} \tag{2.41}$$

which, assuming a linear constant gradient for the density of the form

$$\rho_s(r) = \frac{1}{\bar{v}_m} + (r - r_\theta) \frac{d\rho_s}{dr}$$

gives

$$\frac{dc_m}{dr} \frac{1}{c_m} = \frac{\omega^2 r_\theta M}{RT} \left\{ 1 - \bar{v}_m \left[\frac{1}{\bar{v}_m} + (r - r_\theta) \frac{d\rho_s}{dr} \right] \right\} \tag{2.42}$$

$$= \frac{-\omega^2 r_\theta M}{RT} \bar{v}(r - r_\theta) \frac{d\rho_s}{dr} \tag{2.43}$$

For small deviations from equilibrium, then $r \simeq r_\theta$ and $(r - r_\theta)dr = \frac{1}{2}d(r - r_\theta)^2$ and eqn (2.43) becomes

$$d\ln c = \frac{-\omega^2 r_\theta M \bar{v}}{2RT} \left(\frac{d\rho_s}{dr} \right)_{r_\theta} d(r - r_\theta)^2 \tag{2.44}$$

Equation (2.44) can be integrated to give

$$c_r = c_{r_\theta} \exp\left[-\frac{\omega^2 r_\theta M\bar{v}\left(\frac{d\rho_s}{dr}\right)_{r_\theta}(r-r_\theta)^2}{2RT}\right]$$
(2.45)

By analogy with the definition of probability where

$$dP = \frac{1}{\sigma(2\pi)^{1/2}} \exp\left[-\frac{(x-\bar{x})^2}{2\sigma^2}\right]dx$$

eqn (2.45) becomes

$$\sigma_{eq} = \left(\frac{RT}{\omega^2 r_\theta M\bar{v}\left(\frac{d\rho_s}{dr}\right)_{r_\theta}}\right)^{1/2}$$
(2.46)

and σ_{eq}^2 equals the half-width of a gaussian peak. The shape of the zone is determined by the net velocity of the molecules in the gradient. Putting the velocity $U = sG$, where $U \equiv dx/dt$ and G is the centrifugal field, and assuming $s = \bar{r}(\rho_s)$ and with no concentration dependency, then Baldwin and Shooter (1963) define s at any radius by a Taylor series,

$$s = s_\theta + \frac{ds}{d\rho}(\rho - \rho_\theta) + \frac{1}{2}\frac{d^2s}{d\rho^2}(\rho - \rho_\theta)^2 + \ldots$$
(2.47)

Since $s = 0$ when $\rho = \rho_\theta$, therefore $s = \frac{ds}{d\rho}(\rho - \rho_\theta)$.

From the definition of a sedimentation coefficient ($s\omega^2 r = dr/dt$), and substituting for s in eqn (2.47), gives

$$\frac{d(r-r_\theta)}{dt} = \omega^2 \frac{ds}{dr}(r-r_\theta)[r_\theta + (r-r_\theta)]$$
(2.48)

which, when integrated, gives

$$\ln\left(\frac{r_\theta - r}{r}\right) = \left(\omega^2 r_\theta \frac{ds}{dr}\right)t + \ln\left(\frac{r_\theta - r_0}{r_0}\right)$$
(2.49)

or

$$\frac{r_\theta - r}{r} = \frac{r_\theta - r_0}{r_0} \exp\left[\omega^2 r_\theta \frac{ds}{dr} t\right]$$
(2.50)

Thus, to calculate the time to reach equilibrium $(r_\theta - r)/r$ must be assigned an acceptably small number which is not zero but could be 0.001.

Overlap of Separated Isopycnic Zones

The total masses in any zone can be calculated by

$$m = \int_{r_0}^{r_b} c_r r^2 \, dr \tag{2.51}$$

and since

$$c_r = \frac{dc}{dr} \frac{RT}{M(1 - v\rho)r\omega^2} \, , \quad \text{eqn (2.41),}$$

then

$$m = \frac{RT}{M\omega^2} \int_{r_0}^{r_b} \frac{r_c}{(1 - \bar{v}_m \rho_s^{(r)})} \, dc \tag{2.52}$$

or when $\bar{v}_m$ is constant

$$m = \frac{RT}{\bar{v} M\omega^2} \int_{r_0}^{r_b} \frac{dc}{\left(\dfrac{d\rho}{dr}\right)_c} \tag{2.53}$$

Replacing $RT/(M\omega^2 \bar{v})$ by $\sigma^2 r_\theta (d\rho/dr)$ from eqn (2.46) gives

$$m = \sigma^2 r_\theta \int_{r_0}^{r_b} dc \tag{2.54}$$

which, if $c_{r_0} = c_{r_b} = 0$, the integral approximates to c_θ, to give

$$m = \sigma^2 r_\theta c_\theta \tag{2.55}$$

Hence, the width increases with the mass in the zone and inversely with the gradient.

Interactions Between Macromolecules and Solutes in the Gradient

Skerrett (1975) considered the reversible interaction between a protein and the small solute of the gradient. For a reaction between a dense solute (A) and a lighter macromolecule (P) of the type

$$P + iA = PA_i \tag{2.56}$$

the result is a single isopycnic zone if the reaction rates are fast relative to the sedimentation. Two separate peaks form if the reaction rates are slow. For example, for chemical equilibrium $K_i = c_i/c_p c_A^i$, the total concentration is

$$c_r = c_p \sum_i K_i c_A^i \tag{2.57}$$

Skerrett (1975) showed that the apparent molecular weight (M_{app}) can be defined by

$$M_{app} = M_Q + M_A c \left(\frac{2BRT}{M_A \omega^2 r_\theta^2 c_A} - 1 \right) \tag{2.58}$$

where M_Q is the local weight-average weight and $B = (c_A - a)/r$ the linear relationship between concentration and radius with $a =$ intercept.

$$c = \frac{\Sigma i K_i c_A^i}{\Sigma K_i c_A^i} - \frac{\Sigma i \beta_i K_i c_A^i}{\Sigma \beta_i K_i c_A^i}$$

where β is the weighted diffusion coefficient, $(a_i D_i)$, and a_i is the weight fraction of P in $PA_i, = M(M + iM_A)^{-1}$.

Similarly, the measured partial specific volume $\bar{v}_{app}$ will be less than the true value by

$$\bar{v}_{app} = \bar{v}_Q \left\{ \frac{1 - (M_A c/M_Q)(\bar{v}_A/\bar{v}_Q)}{1 - (M_A c/M_Q)[(2BRT/M_A \omega^2 r_\theta^2 c_A) - 1]} \right\} \tag{2.59}$$

Least-Squares Analysis for Curvilinear Polynomials

The general procedure for estimating coefficients of a curvilinear polynomial, eqn (2.60), is described by Davies (1961) and the formulae are summarized here:

$$y = \sum_{l=0}^{p} a_l x^l \tag{2.60}$$

For a record containing n pairs of co-ordinates (y_i, x_i) a problem arises in the choice of values of a_l which will minimize Q, eqn (2.61):

$$Q = \Sigma(y_i - \bar{y}_i)^2 \tag{2.61}$$

where $\bar{y}_i$ is the ordinate calculated from eqn (2.60). Thus, differentiating Q with respect to a_0 and equating to zero gives

$$\frac{\partial Q}{\partial a_0} = -2\Sigma(y - a_0 - a_1 x - a_2 x^2 - \ldots - a_p x^p) \tag{2.62}$$

or

$$a_0 = \bar{y} - a_1 \bar{x} - a_2 \bar{x^2} - \ldots - a_p \bar{x^p} \tag{2.63}$$

since multiplying eqn (2.62) throughout by n gives Σx_i^l and dividing by n gives the means x_i^l. Similarly, for $l \neq 0$, then the derivative of Q with respect to the parameter a_l is

$$\frac{\partial Q}{\partial a_l} = -2\Sigma x^l (y - a_0 - a_1 x - \ldots - a_p x^p) = 0 \tag{2.64}$$

Hence,

$$a_0 \Sigma x^l + a_1 \Sigma x x^l + a_2 \Sigma x^2 x^l + \ldots + a_p \Sigma x^p x^l = \Sigma y x^l \tag{2.65}$$

which when a_0 is eliminated using eqn (2.63) gives

$$a_1 \Sigma x^l (x - \bar{x}) + a_2 \Sigma x^l (x^2 - \bar{x}^2) + \ldots + a_p \Sigma x^l (x^p - \bar{x}^p) = \Sigma x^l (y - \bar{y}) \tag{2.66}$$

and x^l can be replaced by $(x^l - (\bar{x})^l)$ to give p simultaneous equations containing p coefficients. Matrix methods can be used to solve these equations in which eqn (2.66) can be written in the general form $\mathbf{AX} = \mathbf{Y}$, so $\mathbf{A} = \mathbf{X}^{-1}\mathbf{Y}$ and the column matrix $\mathbf{A}$ can be evaluated by inverting matrix $\mathbf{X}$ and multiplying by column matrix $\mathbf{Y}$. The errors of the coefficients are calculated from the diagonal elements of the inverse matrix $(\mathbf{A}^{-1})$ and denoted by $sl^1 I^{11}$, when it can be shown that $se(a_1) = (S^{11} sse)^{\frac{1}{2}}$, etc., where S^{11} is the first element of the diagonal of $\mathbf{A}^{-1}$; others would be S^{22}, etc.,

$$sse = \Sigma(y_i - a_0 - a_1 x_i - a_2 x_i^2 - \ldots - a_p x_i^p)^2 / (n - p - 1)$$

Earlier in the discussion it was suggested that an integral can be evaluated by fitting a record to a curvilinear polynomial and integrating the polynomial. This means the record fitted was for the differentials and the integral will be given by eqn (2.67), where G is the integration constant and is the value for y when $x = 0$:

$$y = G + \sum_{l=0}^{p} a_l x^{(l+1)} / (l+1) \tag{2.67}$$

The reverse procedure of differentiating eqn (2.60) is also possible but the result must be applied with caution, because precision decreases markedly with differentiation but increases or remains constant with integration.

REFERENCES

ANDERSON, N.G. (1955). *Expl Cell. Res.*, **9**, 446
ANDERSON, N.G. (1966a). *Natn. Cancer Inst. Monogr.*, No. 21, p. 9
ANDERSON, N.G. (1966b). *Science, N.Y.*, **154**, 103

BALDWIN, R.L. and SHOOTER, E.M. (1963). In *Ultracentrifugal Analysis*, p. 143. Ed. J.T. Williams. New York; Academic Press

BARBER, E.J. (1966). *Natn. Cancer Inst. Monogr.*, No. 2, p. 219

BERMAN, A.S. (1966). *Natn. Cancer Inst. Monogr.*, No. 21, p. 41

BIRNIE, G.D., RICKWOOD, D. and HELL, A. (1973). *Biochim. biophys. Acta*, **331**, 283

BISHOP, B. (1966). *Natn. Cancer Inst. Monogr.*, No. 21, p. 175

BRAKKE, M.K. (1951). *J. Am. chem. Soc.*, **73**, 1847

BRAKKE, M.K. (1964). *Archs Biochem. Biophys.*, **107**, 388

CASSASSA, E.F. and EISENBERG, H. (1964). *Adv. Protein Chem.*, **19**, 287

COHEN, R., GIRAUD, B. and MESSIAH, A. (1967). *Biopolymers*, **5**, 203

COX, D.J. (1966). *Science, N.Y.*, **152**, 359

COX, D.J. (1971). *Archs Biochem. Biophys.*, **142**, 514

DAVIES, O.L. (1961). *Statistical Methods in Research and Production*. Edinburgh; Oliver and Boyd

DISHON, M., WEISS, G.H. and YPHANTIS, D.A. (1966). *Biopolymers*, **4**, 449

FUJITA, H. (1961). *Mathematical Theory of Sedimentation*. New York; Academic Press

FUJITA, H. and MacCOSHAM, F.J.J. (1959). *J. phys. Chem.*, **30**, 291

GEHATIA, M. and KATCHALSKY, E. (1959). *J. chem. Phys.*, **30**, 1334

HALSALL, H.B. and SCHUMAKER, V.N. (1970). *Biochem. biophys. Res. Commun.*, **39**, 479

HOWLETT, G.J. and JEFFREY, P.D. (1973). *J. phys. Chem.*, **77**, 1250

McCALLUM, M.A. and SPRAGG, S.P. (1972). *Biochem. J.*, **128**, 389

MARTIN, R.G. and AMES, B.N.J. (1961). *J. biol. Chem.*, **236**, 1372

MESELSON, M., STAHL, F.W. and VINOGRAD, J. (1957). *Proc. natn. Acad. Sci. USA*, **43**, 581

NASON, P., SCHUMAKER, V.N., HALSALL, H.B. and SCHWEDES, J. (1969). *Biopolymers*, **7**, 241

NOLL, H. (1967). *Nature, Lond.*, **215**, 360

POLLACK, M.S. and PRICE, C.A. (1971). *Analyt. Biochem.*, **42**, 38

RICKWOOD, D., HELL, A., BIRNIE, G.D. and GILHUUS-MOE, C. Chr. (1974). *Biochim. biophys. Acta*, **342**, 367

SARTORY, W.K. (1969). *Biopolymers*, **7**, 251

SARTORY, W.K., HALSALL, H.B. and BREILLAT, J.P. (1976). *Biophys. Chem.*, **5**, 107

SCHACHMAN, H.K. (1959). *The Ultracentrifuge in Biochemistry*, p. 65. New York; Academic Press

SCHUMAKER, V.N. (1967). *Adv. biol. med. Phys.*, **11**, 245

SCHUMAKER, V.N. and HALSALL, H.B. (1969). *Analyt. Biochem.*, **30**, 368

SCHUMAKER, V.N. and HALSALL, H.B. (1971). *Biochem. biophys. Res. Commun.*, **43**, 601

SCHUMAKER, V.N. and ROSENBLOOM, J. (1965). *Biochemistry, N.Y.*, **4**, 1005

SKERRETT, R.J. (1975). *Biochim. biophys. Acta*, **385**, 28

SPRAGG, S.P. and RANKIN, C.T. (1967). *Biochim. biophys. Acta*, **141**, 164

SPRAGG, S.P., MORROD, R.S. and RANKIN, C.T. (1969). *Sep. Sci.*, **4**, 467

SVENSSON, H., HAGDAHL, L. and LERNER, K.D. (1957). *Sci. Tools*, **4**, 1

TRAUTMAN, R., SPRAGG, S.P. and HALSALL, H.B. (1969). *Analyt. Biochem.*, **28**, 396

TURNER, J.S. (1968). *J. Fluid Mech.*, **33**, 183

TURNER, J.S. and SHIRTCLIFFE, T.G.L. (1970). *J. Fluid Mech.,* **41,** 707
TURNER, J.S. and STERN, M.E. (1969). *Deep-Sea Res.,* **16,** 497
VAN HOLDE, K. (1971). *Physical Biochemistry.* Hemel Hempstead; Prentice-Hall
VINOGRAD, J. and BRUNER, R. (1966). *Biopolymers,* **4,** 131

3 Practical Aspects of Rate-Zonal Centrifugation

DAVID RIDGE

Department of Biochemistry, University College, London

Rate-zonal (zone velocity) centrifugation is used to separate two (or more) populations of particles from each other by virtue of a difference in their sedimentation rates. The particles may be macromolecules, or they may be as large as whole cells or cell nuclei. A solution or suspension of the particles to be separated is applied in a more or less narrow band to the top (centripetal end) of a density gradient, into which the particles sediment in a centrifugal field. The primary purpose of the gradient is to prevent convection, though it may have other functions.

The rate at which the particles sediment depends on their size, shape and density and on the density and viscosity at each point in the gradient. Sedimentation is opposed by viscous drag (frictional resistance), the magnitude of which is related to the surface area of the particle so that large or near-spherical particles (those with a small ratio of surface area to mass) will sediment faster than small or extended particles. In a family of particles the largest will sediment fastest. The sedimentation rate of a particle is also directly proportional to the difference in density between the particle and its surrounding medium. Lastly, the rate at which a particle sediments is proportional to the magnitude of the centrifugal field at each point in the gradient.

The maximum sample load for a rate-zonal separation is generally limited, so a differential centrifugation system (see pp. 2–4) should always be considered for a preparative application where both 100% purity and 100% yield are not required. Differential pelleting has been done also using a stabilizing gradient (Charlwood, 1963). Sedimentation is slower in density gradients, because local increases in density where particles have been concentrated against the wall are stabilized against convection. The effect will depend on the angle between the direction of sedimentation and the wall, and on the particle concentration. With low concentrations there may be no increase in the sedimentation rates of large or small particles due to wall effects (Anderson, 1968).

EXPERIMENTAL DESIGN

Rate-zonal separations can be done in tubes in swing-out rotors or in zonal rotors (Anderson, 1966a). The work of Charlwood (1963) and others suggested that fixed-angle rotors can also be used, and Vedel and D'aoust (1970) achieved good separations of RNA in a fixed-angle rotor. On the other hand, Castañeda, Sánchez and Santiago (1971) found serious wall effects when separating polyribosomes, and recommended the use of swing-out rotors. The conditions for centrifugation, including the choice of tube size or zonal rotor, will depend on some or all of the following factors.

Sedimentation Coefficients of Particles

The resolution (the completeness of separation of the zones from one another) depends on the width of the sample zone relative to the radial pathlength. This is particularly important if the particles to be separated have similar sedimentation coefficients. In this case one should use a narrow sample zone and a long pathlength; gradients in long tubes have the additional advantage that they are more stable during handling. The relative sedimentation rates of particles with different densities are also affected by the density of the surrounding medium.

Density Range of the Gradient

For best resolution, the sample suspension should be applied as a very narrow band. The gradient should have the minimum density and viscosity compatible with the separation required, so that the centrifugation is rapid and diffusion is kept to a minimum. However, since the sample particles must be more dense than all parts of the gradient, their presence increases the density of this region of the gradient. The narrower the band for a given load, the greater the increase and the steeper the gradient must be to support it; therefore, a compromise is usually necessary. Gradient shape is considered in more detail later (p. 40).

Volume

The volume of the gradient used (5 ml tube to 1600 ml zonal rotor) may depend partly on what apparatus is available. There may be constraints on the mass of sample particles or the volume in which they are contained. The volume of gradients may also be limited by the method of recovery of the particles and the rotors available for pelleting the fractions. The author knows of no successful rate-zonal separations in very small gradients, for example, in 10 μl capillary tubes. Anomalous sedimentation can occur in capillaries (Neuhoff, 1973) presumably because of the concentration-dependent wall effects.

Loading

The calculation of theoretical loads has occupied a number of workers, notably Svensson (1957), Berman (1966), Spragg and Rankin (1967) and Meuwissen (1973) and for simple mixtures, loads have been achieved which approach the theoretical maximum. No real rules of thumb can be given and, for complex mixtures, loads are best determined empirically. The minimum load is usually limited by the method of detection; continuous photometric monitoring of the gradient during unloading is difficult if the peaks of absorbance are less than about 0.1 O.D. (for RNA or DNA for example). In general, therefore, loads will normally be between 1 μg/ml and 1 mg/ml of gradient volume, but generally less than 0.1 mg/ml. Some materials can, of course, be detected in very low concentrations; for example, by radioactive labelling of the sample, monitoring enzymic activity, haemagglutination, etc. On the other hand, sample losses can become significant when separating microgram amounts of material. An example of this is the adsorption of DNA on to the surface of polyallomer tubes, although this can be reduced by presoaking the tubes in the gradient buffer.

Osmolarity of the Sample Medium

This can be critical in that cells, mitochondria and other membrane-enveloped particles are extremely sensitive to changes in the osmotic strength of the medium and they rapidly deteriorate in media of lower tonicity. Also, it may simply be convenient to load the sample in the medium in which it was prepared, rather than pellet and resuspend the particles. Such considerations may also influence the choice of solute used to make the density gradient.

Centrifugation time

The length of time taken by the centrifugation may be important. In order to minimize both the inactivation of labile particles and band broadening due to diffusion, the procedure should be as short as possible. However, in the case of mammalian cells rapid acceleration can damage the cells by enucleation. Moreover, the integrity of osmotically sensitive membranes of both cells and organelles can be damaged by the hydrostatic pressure generated by centrifugation at very high speeds (Collot, Wattiaux-De Coninck and Wattiaux, 1975). In the case of large particles, such as mammalian cell nuclei (Johnston *et al.*, 1968), a viscous gradient is needed to prevent excessive sedimentation even at very low speeds.

Temperature

With labile particles it is usually necessary to avoid high temperatures, although occasionally the use of high temperatures is a prerequisite condition for separation (see, for example, De Pomerai, Chesterton and Butterworth, 1974). Alternatively, higher temperatures are sometimes used to reduce significantly the viscosity of gradients (see Appendices I and II). However, not all ultracentrifuges can maintain temperatures above ambient.

GRADIENT MATERIALS

The properties of an 'ideal' solute for forming density gradients for rate-zonal and isopycnic centrifugation have been described (see, for example, Cline and Ryel, 1971; Hinton, Mullock and Gilhuus-Moe, 1974); they can be summarized as follows:

1. It should be stable in solution, and be soluble enough to give a solution of

   sufficient density (for isopycnic centrifugation, the density obtainable must be greater than the buoyant densities of the particles to be separated).
2. It should exert the minimum osmotic effect, and should cause the minimum change in viscosity, ionic strength and pH.
3. It should be totally inert towards biological materials; that is, it should not be surface-active in solution, or affect any biological particles by interacting with them, disaggregating or aggregating them, or otherwise altering their physical or chemical structures and compositions.
4. It should be capable of withstanding sterilization without any change in properties.
5. It should not be hydrated in aqueous solution (that is, the water activity of aqueous solutions should be unity).
6. Its solutions should not absorb light at wavelengths appropriate for photometric monitoring, or otherwise interfere with procedures for assaying biological materials, or their enzymatic activities.
7. It should not interact in any way with the materials from which gradient engines, rotors, seals, and centrifuge tubes and caps are constructed, or be at all toxic.
8. It should be readily and completely separable from fractionated particles.
9. It should be readily obtainable in a pure form, and be cheap, or easily recoverable (particularly when large amounts are required, for example, in zonal rotors).
10. For quantitative work in particular, its chemical, physical and thermodynamic properties should be known.

The following substances have frequently been used to form the density gradients for rate-zonal separations; although none of them has all the properties of the 'ideal' solute, some are much more suitable than others for particular applications: (*i*) simple sugars and analogous polyhydroxyl compounds (sucrose, sorbitol, glycerol, etc.); (*ii*) polysaccharides (Ficoll, dextran and glycogen); (*iii*) proteins (for example, bovine serum albumin); and (*iv*) deuterium oxide.
   In addition, a number of substances frequently used as solutes for isopycnic density-gradient centrifugation can be used for rate-zonal studies. These include inorganic salts (for example, CsCl and Cs_2SO_4), iodinated organic compounds (for example, metrizamide and Renografin) and colloidal silica (Ludox). A detailed description of the properties of these substances is given in Chapters 6 and 7. Finally, solvents other than water have been used; for example, 95% dimethyl sulphoxide (Bramwell, 1974) and 85% formamide (De Pomerai, Chesterton and Butterworth, 1974) as denaturing solvents for sucrose gradients in the study of RNA. Kennedy (1975) has used anhydrous gradients for the separation of nuclei prepared under anhydrous conditions.

Simple Sugars

Sucrose is the one compound which has been used universally for rate-zonal density-gradient separations. The density and viscosity of sucrose solutions are well documented by Swindells *et al.* (1942, 1958). Empirical relationships have been derived by Barber (1966) to calculate density and viscosity from the weight fraction of sucrose and the temperature, and tables have been prepared using these relationships (see Appendix I). *Figure 3.1* gives an indication of the relation-ship between viscosity and temperature, and provides a means of correcting viscosities for small temperature changes.

Solutions are prepared on a weight for weight (w/w) basis by weighing both water and sucrose. For concentrated solutions, sucrose should be added to water while stirring, and for concentrations over 60% the water should be heated to 100 °C (Austoker, Cox and Mathias, 1972). There is no accepted con-vention as to whether the water space of a sucrose gradient should be buffered

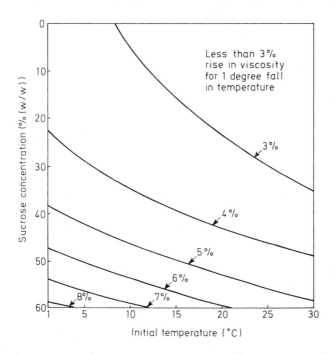

Figure 3.1 *Percentage viscosity change due to temperature. The diagram shows the approxi-mate increase in viscosity of sucrose solutions associated with a fall in temperature of 1 °C from the temperature shown. For example, 30% (w/w) sucrose at 8 °C will increase its viscosity by nearly 4% if the temperature falls to 7 °C. Uses: (1) as a vernier for viscosity tables – the viscosity of 55% (w/w) sucrose at 5 °C is 0.667 9 Poises (from tables). From the diagram, the increase in viscosity for a fall of 1 °C will be about 6.7% (by eye); therefore, the value at 4 °C $\simeq$ 0.667 9 × 1.067 $\simeq$ 0.712 6 Poises. (2) For correcting centri-fugation times. For example, if a separation is normally performed at 20 °C, but on this occasion the temperature is 22 °C, the run should be stopped early if the zone positions are to be comparable. For, say, a 15–35% (w/w) sucrose gradient, the run time will be about 3% less per degree of error, that is, about $(0.97)^2$ times the usual time. So a run of 4 h should be shortened by 15 min. (3) For correcting s-value calculations*

(that is, to make up the heavy and light solutions from similar buffer plus differ-ent amounts of sucrose) or whether a true buffer molarity should be maintained in the total volume. A graph of the data given in *Table 3.1* (from De Duve, Berthet and Beaufay, 1959) enables the molarity of a sucrose solution to be directly converted into its w/w concentration. The concentration of solutions can be measured using an Abbé refractometer with a sugar scale. RNAse-free sucrose is marketed by several companies specifically for density-gradient work, but is relatively expensive. In some instances analytical reagent grade sucrose,

Table 3.1 WEIGHT FRACTION AND MOLARITY OF SUCROSE AT 20°C

% (w/w)	Molarity at 20°	% (w/w)	Molarity at 20°	% (w/w)	Molarity at 20°
0.00	0.000	27.37	0.891	50.40	1.814
2.49	0.073	29.44	0.967	52.17	1.892
4.92	0.146	31.47	1.043	53.92	1.971
7.32	0.220	33.48	1.119	55.65	2.051
9.68	0.293	35.45	1.195	57.37	2.130
12.01	0.367	37.40	1.271	59.06	2.210
14.31	0.441	39.33	1.348	60.74	2.290
16.57	0.516	41.23	1.425	62.39	2.371
18.80	0.591	43.11	1.502	64.04	2.452
20.99	0.666	44.97	1.580	65.66	2.533
23.15	0.741	46.80	1.657	67.27	2.614
25.28	0.816	48.61	1.735	68.86	2.696

mineral water grade (from Tate & Lyle Ltd) and table sugar may all be of acceptable purity for gradient work, but they can contain significant amounts of heavy metals and ribonuclease activity (Reid, 1971). Ribonuclease activity can be eliminated by treating solutions with diethylpyrocarbonate (for example, 0.1% or 1 drop per 250 ml) although this treatment may alter the pH of solutions by the production of CO_2. This treatment also sterilizes sucrose solutions. It is important to destroy all traces of diethylpyrocarbonate, since it reacts strongly with proteins and the bases of nucleic acids (Leonard, McDonald and Reichmann, 1970), and this is usually done by heating the solutions to 60 °C for several hours prior to use. Heavy metal ions can be removed from sucrose solutions by passage through a chelating resin, for example Chelex-100. Contamination of the sucrose with ultraviolet absorbing materials can be reduced by treatment with activated charcoal (for example, Norit A, at 25 g/kg of dissolved sucrose). For autoclaving, sucrose should be adjusted to pH 5–6 to prevent caramelization.

When performing enzyme assays in the presence of high sucrose concentrations, it should be remembered that the reduction in water activity and (for diffusion-limited reactions) the increased viscosity of the solution may reduce enzyme activity and that some enzymes are more sensitive than others (Hartman *et al.*, 1974). The Lowry estimation is affected by sucrose (Schuel and Schuel, 1967). Other sugars, sugar alcohols, etc., may be found to suit specific particles or enzyme systems; gradients of sorbitol, for example, have been used for the separation of yeast mitochondria (Avers, Szabo and Price, 1969).

Glycerol

The density and viscosity of glycerol solutions at various temperatures are given in Appendix II. These properties are perhaps best described by comparison with sucrose. On a w/w basis, solutions of glycerol are less dense and less viscous than sucrose solutions. However, for solutions of the same density, the glycerol solutions are more viscous. As a rough guide, a sucrose solution and a glycerol solution which have the same viscosity at a given temperature will change in viscosity by similar amounts for a given change in temperature. Also, if diluted by equal amounts, they will give solutions which have similar viscosities.

Glycerol is readily available very pure (without, for example, ribonuclease activity). It stabilizes a number of enzymes against denaturation, although it can inhibit the activities of some. It has been used in place of sucrose, which inter-feres with DNA polymerase assays (Haines, Johnston and Mathias, 1970). It is also used when the samples to be fractionated contain enzymes which can modify or degrade sucrose. It can be autoclaved without degradation, and can be removed, for example from electron microscope grids, by evaporation under vacuum.

Polysaccharides

High-molecular-weight gradient solutes have the following advantages: (*i*) low osmotic pressure (this is particularly important when separating such osmoti-cally sensitive particles as whole cells and membrane-enveloped organelles); (*ii*) they are excluded from intact membrane-bound structures; and (*iii*) they diffuse slowly, thus minimizing the possibility of instability developing during loading of the sample (see p. 48).

A number of different polysaccharides have been used for rate-zonal separa-tions; these include a synthetic polyglucose polymer (Oroszlan *et al.*, 1964), dextrans (Mach and Lacko, 1968), glycogen (Beaufay *et al.*, 1964) and poly-sucrose (Ficoll). However, of these, only the last, Ficoll, has been widely used for separating biological materials. It is supplied by Pharmacia Fine Chemicals (Uppsala, Sweden) and is available with weight average molecular weights of 400 000 and 70 000 daltons. A solution of a polysaccharide is slightly less dense and much more viscous than a solution of sucrose of the same concentration. Some values for density and viscosity in Ficoll gradients are given by Pretlow *et al.* (1969) (see also *Figure 7.1*). They show that the density of Ficoll solutions vary with the w/v concentration in an almost linear fashion, and that the logarithm of the viscosity varies nearly linearly with con-centration. As supplied, Ficoll may contain up to 1% NaCl as impurity and also some other dialysable impurities which can interfere with cell metabolism (Boone, 1971).

Polymers in general can show charge interactions with the suspended sample particles, thus causing aggregation. Eliasson and Samelius-Broberg (1965), for example, found differences between species in the aggregation of red blood cells by various dextrans (polyglucose). Trial and error seems to be the only way to choose appropriately.

Proteins

The same general advantages and disadvantages apply as with other polymers, although an additional advantage is their protective effect against the denatura-tion of biological material. Sufficient density can be achieved using, for example, bovine serum albumin which forms stable gradients up to 1.106 g/cm^3 (Leif and Vinograd, 1964) but the viscosity is very high and ultraviolet monitoring of particles is impossible ($E_{280}^{1\%} \sim 7$). Relatively cheap grades of bovine serum albu-min can be utilized, although when used for cell separation some batches are found to cause significant aggregation. Other proteins have been investigated for the separation of blood cells (Mathias, Ridge and Trezona, 1969) and sperma-tozoa (Benedict, Schumaker and Davies, 1967).

Deuterium Oxide

In many ways this might seem the ideal gradient material, as it has very little effect on biological particles. Its density when pure is about 1.11 g/cm^3 and its viscosity is only slightly greater than that of water. It is not prohibitively expensive, at least for small gradients, but the content of tritium should be checked. Deuterium oxide gradients have been used, for example, by Levinthal and Davison (1961) for rate-zonal separations of RNA.

Mixed-solute Gradients

Generally these have not been used for rate-zonal separations (although most gradients contain buffers and specific ions, etc., to preserve the integrity of the particles being separated). The density gradient could in principle be formed from more than one solute, and in isopycnic centrifugation the use of two-solute gradients is not uncommon (see, for example, Mathias and Wynter, 1973). A gradient of constant osmotic pressure of 0.66 osm, with a decreasing NaCl concentration to balance the increasing sucrose concentration, has been used for fractionating avian erythrocytes (Trezona, 1969).

RESOLUTION AND GRADIENT SHAPE

Resolution

The degree of separation achieved by rate-zonal centrifugation is directly related to both the dexterity of operator and the design of the experiment. It is of the utmost importance that during the preparation, loading, running and analysis of gradients the degree of disturbance of the gradients is kept to an absolute minimum. In particular, it is important to load the sample in such a way as to prevent instability occurring (see p. 47). However, the main factors governing the resolution lie in the choice of centrifugation conditions and, in general, the following rules apply. Increasing the centrifugal force applied to the particles shortens the run and minimizes diffusion broadening of the bands. Increasing the slope of the gradient causes band sharpening, since the viscous drag increases enormously with increasing sucrose concentration. All of the early rotors, and some of the present-day long-bucket rotors, operate at intermediate speeds (25 000–40 000 rev/min) and their best separations are with either a shallow linear gradient (5–20% w/w sucrose) or an isokinetic gradient (see p. 43) which are capable of resolving particles whose sedimentation coefficients differ by at least 40%. Better separations can usually be obtained using steeper gradients (for example, a gradient of 10–50% (w/w) sucrose) in the newer rotors that operate at 65 000 rev/min (approx. 500 000g_{max}); these can resolve particles whose sedimentation coefficients differ by only 10%. However, the band compression that occurs at very high speeds may reduce the capacity of the zone (that is, the maximum mass of particles that can be carried in a stable band of a given width) to such an extent that density inversion occurs. For this reason, the best resolution is usually achieved by using relatively wide top zones and low sample concentrations rather than *vice versa*. Poor resolution is almost always the result of overloading. On the other hand, sharp bands can be obtained even with large loads, if the volume of the sample layer is expanded and the slope of the gradient increased (Noll, 1969a). Finally, when adapting the methods of other workers, which have been published several years previously, it is frequently worth considering whether changing the type of rotor or gradient might enhance the capacity or the resolution of the gradients to be used.

Design of Gradients

The 'shape' of a gradient normally refers to its concentration profile, that is, the variation of concentration along the tube. This is useful in terms of the method of generation, but may be misleading because concentration does not always have a linear relationship with either density or viscosity. *Figure 3.2* shows that a gradient with an exponential concentration profile can be convex, concave or sigmoid in terms of viscosity. Sedimentation rate varies inversely as the viscosity and so particles are slowed down by the increasing viscous drag as they move down the gradient; in practice this is usually more important than density in determining centrifugation time. Density is the factor which determines the stability of the gradient; the density gradient must be steep enough to prevent convection and to maintain the stability of sedimenting zones throughout centrifugation. A gradient which needs to be steep at the top to support the sample zone need not be so steep further down, where the zones have separated from each other and have spread by diffusion and because of particle heterogeneity. For this reason, so-called convex density gradients are often advocated, that is, gradients which are steep at the top and shallower towards the bottom. In zonal rotors, the paths of the particles diverge as they sediment on unobstructed radial paths; this 'radial dilution' further reduces the need for steepness towards the bottom of the gradient. Dilution of the zone can be counteracted by increasing viscosity towards the bottom, so that the front of the zone is slowed down more than the rear. However, a sharp increase in viscosity may concentrate a zone enough to cause instability.

There are advantages and disadvantages in the use of the various types of gradient shape, which will be dealt with briefly as step, constant slope, concave and convex, constant velocity, and complex. The mathematical derivations for defining the shape of gradients have been given elsewhere (Noll, 1969a).

Step Gradients

Gradients formed from two or more layers of differing densities are useful in isopycnic centrifugation and in many pelleting techniques, but are of limited use in rate-zonal centrifugation. Although they are called discontinuous gradients, the step rapidly diffuses away leaving alternating steep and shallow regions in the gradient. A one-step (or two-layer) gradient, for example, becomes an extreme kind of concave gradient in which the shallow portion has a very low sample capacity (see 'Complex Gradients', p. 43).

Constant Slope (Linear Gradients)

These gradients are easy to generate by diffusion techniques or by using simple apparatus. They are quite stable against diffusion. In a zonal rotor, a gradient which is linear with respect to volume is concave with respect to radius (see p. 74). A gradient which is linear with respect to radius in a zonal rotor can be produced with a programmable gradient generator (see p. 47); the concave viscosity profile of such gradients has been used to counteract zone broadening (Spragg, Morrod and Rankin, 1969).

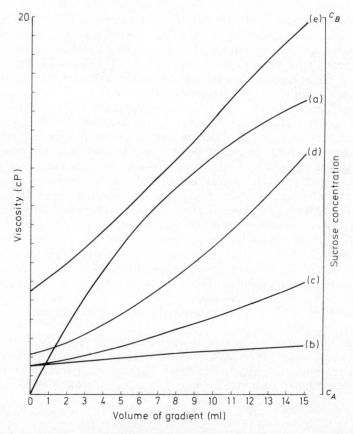

Figure 3.2 Gradient shape: concentration and viscosity profiles. Four sucrose gradients are shown, all with an exponential concentration profile (a). The starting concentration is c_A, and the concentration c_B is approached exponentially (see Figure 3.5a for details). The viscosity profile (b)–(e) adopts different forms depending on the concentrations of the solutions used, which are as follows (% w/w sucrose at 5 °C):

	c_A	c_B
(b)	0%	20%
(c)	0%	40%
(d)	10%	50%
(e)	30%	50%

The volume of the mixing vessel (v_A in Figure 3.5) is 10 ml in each case

Concave and Convex Gradients

These gradients can be prepared by similar types of apparatus as for linear
gradients (see p. 45). The gradients may either deviate only slightly from
linearity or may be of exponential shape. A concave sucrose concentration
gradient will have a relatively rapid increase in viscosity at the bottom; this can
be used to trap all particles over a given size, while the slower zones continue
to separate.

Constant Velocity

Martin and Ames (1961) suggested that a density and viscosity gradient could be
so constructed as to produce a uniform sedimentation velocity by compensating
for the changing centrifugal field along the tube. These gradients have been given
the name 'isokinetic gradients' (Noll, 1967). Various workers have computed
such gradients, including McCarty, Stafford and Brown (1968) and Spragg,
Morrod and Rankin (1969). Steensgaard has derived isokinetic gradients with a
non-iterative program which uses much less computer time (see Chapter 5). The
general shape of these gradients is convex and exponential because of the approximately logarithmic relationship between sucrose concentration and viscosity.
They are easy to generate (see p. 47) and have the additional advantage of the
high sample capacity of convex gradients. The main advantage of isokinetic
gradients is in the analysis of sedimentation behaviour. The distance travelled by
a given particle is proportional to the time of the run, the relative centrifugal
force applied and the sedimentation coefficient of the particle; thus, a single
point calibrates the whole gradient in terms of the *s*-values. However, to design
and calibrate an isokinetic gradient, the density of the sample particle must be
known and the molecular-weight markers must have the same density. For
tube gradients, a plot of equal fractions of the gradient against the *s*-values
of the particles found in each fraction will be a straight line. In a zonal rotor,
however, the fractions which correspond to equal increments of radius are progressively larger in volume towards the edge. In order to obtain a linear relationship between the collected volume and the *s*-value of the particles at each point
in a zonal rotor, 'equivolumetric' gradients have been devised (Pollack and
Price, 1971; Van der Zeijst and Bult, 1972; see also Chapter 4). A gradient can
also be designed to give a logarithmic distribution of sedimentation coefficients
(Brakke and Van Pelt, 1970).

Complex Gradients

Gradients having, for example, steeper or flatter regions to sharpen or separate
zones have been used successfully (Graham, 1973). Sharp angles in the concentration profile diffuse away quite quickly, particularly in zonal rotors in which
the radial distance is small (Birnie, 1973). A step in the gradient will distort the
shape of zones and may give misleading results (Spragg and Rankin, 1967).

Gradient Generation

Diffusion Techniques

The quickest way to make a density gradient is to pipette a dense and a light solution into a centrifuge tube and gently stir the interface (a little practice with coloured solutions establishes the amount of stirring necessary). This is not very reproducible, but may be adequate for some preliminary experiments and may be satisfactory for stabilizing sample layers for separations employing differential pelleting. A much more satisfactory method is to introduce several layers of solution of differing densities into a tube and allow them to diffuse into a continuous gradient, usually over 18–24 h. It is important to leave enough time for this diffusion, as steep regions in the gradient can produce spurious peaks in the distribution of sedimenting particles. The shape of the gradient can be checked by measuring the refractive index of each fraction. A variant of this method is to use only two layers, but to reduce the diffusion distance by turning the tube on its side (Stone, 1974). An apparatus has been devised to load the light layer automatically, while the tube is horizontal (A.B. Stone, personal communication). Baxter-Gabbard (1972) has reported that gradients can be made by repeated freezing and thawing of sucrose solutions, but he did not report the effect of this process on the distribution of buffering ions.

Simple Apparatus

There are several variations on the 'two-jars' theme, for large and small-scale gradients of various shapes, some of which need a pump. Centrifuge tubes can be loaded light end first, passing the gradient down narrow tubing to the bottom (see *Figure 3.3*), or dense end first, running the liquid down the side of the tube.

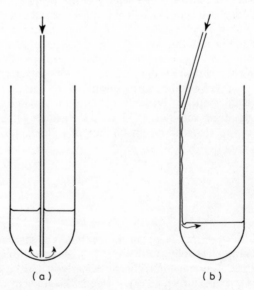

(a)   (b)

Figure 3.3 Loading tube gradients. (a) Light end first; (b) dense end first

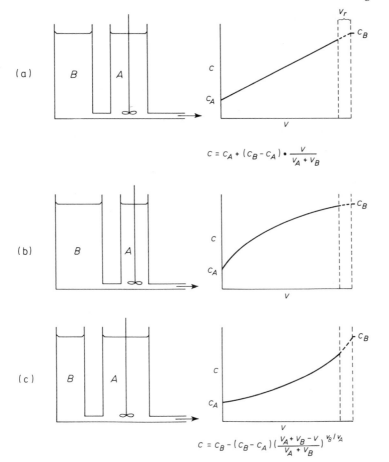

$$c = c_A + (c_B - c_A) \cdot \frac{v}{v_A + v_B}$$

$$c = c_B - (c_B - c_A)\left(\frac{v_A + v_B - v}{v_A + v_B}\right)^{v_B/v_A}$$

Figure 3.4 Two-jars gradient generators. (a) With equal cylinders; (b) with reservoir cylinder larger than mixing cylinder; (c) with reservoir cylinder smaller. B, dense liquid; A, light liquid; c_B, c_A, v_B, v_A, their concentrations and initial volumes, respectively; v, volume of gradient withdrawn; c, concentration of the gradient at volume v; v_r, residual volume. The lower equation applies to both separations (b) and (c)

Some tubes, for example, of polyallomer, are very hydrophobic and aqueous solutions do not form an even stream down the wall, although this can be overcome by pretreating the tubes with chromic acid (Wallace, 1969). An alternative procedure is to use a thin needle of glass (pulled out from a glass rod in a flame) clipped against the side of the tube, down which the gradient is run.

The simplest and most widely used gradient generator is shown in *Figure 3.4*. The gradient produced by such a device is linear if the vessels have equal cross-sections (*Figure 3.4a*). With vessels of different cross-sections, convex or concave gradients are produced (*Figures 3.4b,c*); the derivation of the equations is given elsewhere (Noll, 1969a). Construction and operation of simple apparatus of this type for small gradients has been described by Schumaker (1967) and Noll (1969b) and they are marketed by several companies, including the major centrifuge manufacturers; large-scale versions for use with zonal rotors are also available. Tubes can be filled by gravity or by using a peristaltic pump to control the flow. Suitable mixers can be made from a helix of stainless steel wire, driven by an overhead motor, or a magnetic stirrer can be used if it can be brought close enough to the bottom of the mixing vessel. Points to remember are:

1. There will always be a residual volume in the apparatus, so the density of the gradient will not reach the density of the denser solution (Noll, 1969a).
2. Very vigorous stirring increases the effective head of pressure in the mixing vessel, thus distorting the gradient.
3. The connecting channel needs to be quite wide to allow rapid equilibration of the liquid levels.
4. For the liquid levels to be at equilibrium initially, the heights (the volumes, if the vessels are identical) must be in inverse proportion to their densities.
5. With viscous solutions it is important to use a pump to ensure consistency in flow rate as the viscosity changes.

For smaller volumes (2 ml and less) Schumaker (1967) used a device with two syringes and a separate mixing chamber (*Figure 3.5a*). This is, in principle, the same as the two-jars system. Arcus (1967) used a double-syringe system, using

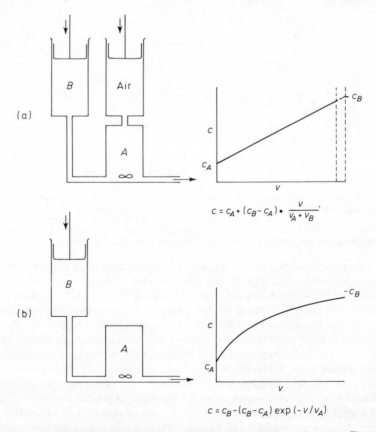

$$c = c_A + (c_B - c_A) \cdot \frac{v}{v_A + v_B}.$$

$$c = c_B - (c_B - c_A) \exp(-v/v_A)$$

Figure 3.5 Closed-vessel gradient generators. (a) With two syringes, analogous to Figure 3.4a; (b) with one syringe and closed mixing vessel, producing exponential gradient (after Bock and Ling, 1954). Key as for Figure 3.4

one syringe as a mixing vessel, whereas Ayad, Borsall and Hunt (1968) avoided gravity equilibration by using a three-channel pump. If the air syringe in *Figure 3.5a* is not used, and its entry to the mixing chamber is closed, this device produces a logarithmic or exponential gradient (*Figure 3.5b*). Several devices for producing exponential gradients have been described, all relying on the same principle (Bock and Ling, 1954). McCarty, Stafford and Brown (1968) describe an apparatus and its use in preparing tube gradients of this type with numerical examples. They also point out that with a multi-channel pump, two or more identical gradients can be produced at the same time. Birnie and Harvey (1968) and Paris (1968) describe large-scale systems for zonal rotors; the apparatus of Hinton and Dobrota (1969) is similar but more flexible in operation, and uses a two-channel piston pump. The Searle Isograd (Searle Instruments, Harlow, Essex) is provided with a hand calculator for designing isokinetic gradients and estimating sedimentation coefficients. The Isograd has the further refinement of a tube-holder with a drive unit to raise or lower the dispensing tube evenly, to avoid disturbing the gradient. De Duve, Berthet and Beaufay (1959) describe various gradient formers using syringes. Some of them are similar to those already described; others use cams to determine the shape of the gradient (see also Lakshmanan and Lieberman, 1953).

Complex Apparatus

Finally, there are the more sophisticated, programmable gradient generators, which rely on changing the relative pumping rates from two reservoirs. These are, in general, large-volume devices for zonal rotors, but the Dialagrad (Isco), for example, is also available in a small-volume version. The Ultrograd (LKB) is a gradient programmer which can be used for large or small volumes, depending on the pump. For zonal rotors, such a gradient generator can also be used to load the sample and overlay, and can be used as a pump for unloading the rotor. The program is determined by a cut-out template (Ultrograd, Beckman model 141 and MSE gradient pump) or by setting a number of dials (Dialagrad). Blattner and Abelson (1966) describe a small-volume gradient former using a program on punched paper tape, but this is not commercially available at the time of writing. Each type has its advantages: for example, the MSE model has peristaltic pumps which are easy to keep sterile; the Beckman model has a piston pump which is relatively unaffected by back-pressure; and the Dialagrad can be very quickly reprogrammed (it uses diaphragm pumps).

LOADING, RUNNING AND ANALYSING GRADIENTS

Sample Loading

The sample zone must be layered on top of the gradient gently enough to prevent mixing. Frequently, samples can be loaded on to gradients by using a pasteur pipette, although it is more difficult to obtain a stable sample zone, as is required to achieve maximum resolution. Qualitatively, there are three aspects to this stability. First, the concentration of the sample particles must not, of course, cause the density of the sample suspension to exceed the density at the top of the gradient. If it does, drops will stream down the gradient until they reach equilibrium. Secondly, the sample zone should not be bounded by sharp changes in osmotic pressure. Such steps encourage diffusion of the solvent, and can broaden the sample zone. Thirdly, the relative rates of diffusion of the sample particles, the solute and the solvent should not be such that the zone becomes

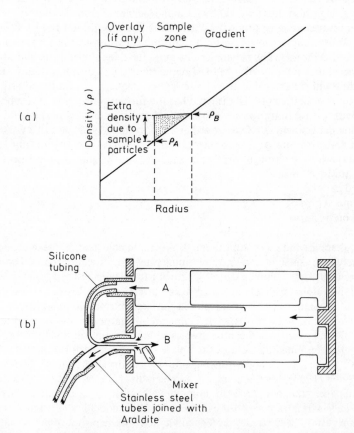

Figure 3.6 (a) Inverse sample gradient. Note that (i) the gradient slope is positive at all points, even with the additional density due to the sample particles; (ii) the density due to the gradient solute has no discontinuities; (iii) the sample zone is least concentrated at the bottom. ρ_A and ρ_B are the densities of the liquid at the top and the bottom of the sample zone. (b) Device for producing an inverse sample gradient. Syringe A contains liquid of density ρ_A plus sample; syringe B contains liquid of density ρ_B initially, without sample. If the jet of liquid into syringe B does not give enough mixing, a mixer made from glass rod can be used, and the whole apparatus shaken during loading

unstable. For example, a narrow zone of 1% bovine serum albumin (BSA) layered over 2% sucrose will be stable initially, but since sucrose diffuses much faster than BSA it will have diffused into the BSA zone before the BSA has diffused appreciably. The result, therefore, is a layer containing 2% sucrose and 1% BSA on top of 2% sucrose only, which is obviously unstable. This phenomenon is called, variously, streaming, droplet formation and turnover (see, for example, Schumaker, 1967). Such a situation in a gradient tends to be worse in the case of small particles because of their low sedimentation rates, and the extent of diffusion becomes significant. In addition, in low centrifugal fields large droplets can form, whereas in high centrifugal fields such droplets never achieve a sufficiently large size or a high enough density excess to affect seriously the resolution of the gradient. In some cases, however, the damage is done before the centrifuge is started. Instability is less likely to occur if the gradient is formed from high-molecular-weight substances, such as Ficoll, which diffuse only very slowly. An alternative procedure for loading tube gradients is to apply the sample while the rotor is spinning, using a band-forming cap (Beckman Instruments).

The way to increase stability is to load the sample as an inverse gradient, so that the leading edge of the sample zone grades into the top of the gradient (*Figure 3.6a*). For zonal rotors, where the sample zone may be from 10 to 200 ml, such a sample gradient can be made using a programmed gradient pump, or with a device such as the one shown in *Figure 3.6b*, with or without a pump. Loading slowly (5 ml/min for a zonal rotor sample) can improve resolution. Tube gradients can also benefit from an inverse sample gradient. For example, a sample of 0.01 to 0.1 ml of polysomes in buffer is taken into the tip of a pipette, and followed by an equal volume of 4% sucrose, while the pipette is kept at an angle. This forms a rough gradient in the pipette. The contents of the pipette are then layered on to the top of a 5–20% sucrose gradient (Williamson, 1971).

In many cases these techniques may not be necessary, but loading should always be done carefully, and large concentration steps should always be avoided. With zonal rotors, the overlay should be in the form of a gradient, particularly with slowly sedimenting samples. A zone which has broadened because of instability is usually asymmetrical, with a sloping leading edge.

Centrifugation Conditions

Temperature Control

The viscosity of sucrose and other aqueous solutions changes considerably with temperature (see *Figure 3.1* and Appendices I and II), so sedimentation time depends on the temperature. Centrifugation is seldom performed at room temperature because the particles under study are frequently labile. Density gradients can only change temperature slowly because they are stabilized against convection. We have found in experiments with zonal rotors that the temperature of the rotor contents during cooling can lag behind the indicated temperature by as much as 3 °C, and that temperature gradients can be set up thus affecting the viscosity profile of the gradient. Tube gradients are smaller, and are more remote from any heating through the drive shaft, but are usually in plastic tubes which tend to insulate them from the temperature controlling and measuring systems. In general, to minimize this problem the gradients should be at the temperature at which the sample is to be centrifuged. Tube gradients to be run in the cold are best made in a cold-room, or stored for 2–3 h in a refrigerator at the required temperature. For a uniform temperature run, the controls should be set to the temperature at which the gradient is believed to be at the start of the run.

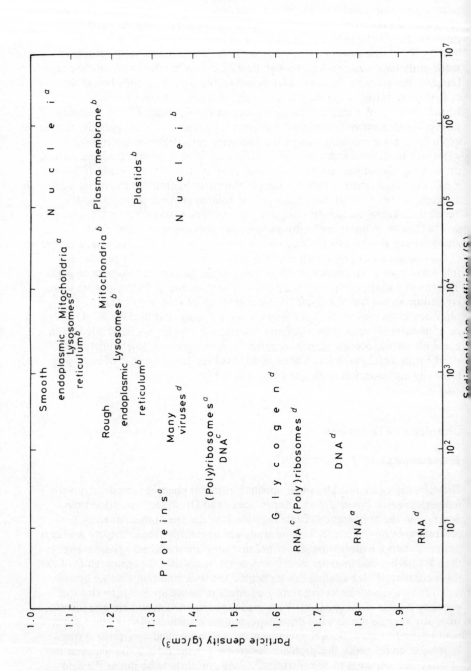

Chart 3.1 APPROXIMATE SEDIMENTATION COEFFICIENTS AND DENSITIES OF SOME BIOLOGICAL PARTICLES

The density of particles may change according to the medium, owing to hydration, permeability, osmotic or other interactions. Densities other than isopycnic (equilibrium) densities must be inferred from sedimentation or other behaviour.

a, inferred density in water or 0.25 M sucrose.
b, isopycnic density in sucrose solution.
c, isopycnic density in Cs_2SO_4.
d, isopycnic density in CsCl.

Adapted from Anderson (1966b). Additional sources include the following:

Nuclei, Johnston *et al.* (1968).
Plasma membrane (sheets), Evans (1970).
Mitochondria $\Big\}$ Cotman *et al.* (1970); De Duve (1963).
Lysosomes
(Nerve-end particles are approximately coincident with lysosomes; Cotman *et al.*, 1970.)
Endoplasmic reticulum (vesicles), Dallner, Bergstrand and Nilsson (1968).
Ribosomes and polyribosomes, McCarty, Stafford and Brown (1968); McConkey (1974).
DNA, Sober (1970).
RNA, Sober (1970); Boedtker (1968).

Speed and Time of Centrifugation

The particles separated by rate-zonal centrifugation are extremely diverse, ranging in size from single proteins of cells to whole cells themselves. Accordingly, the centrifugation conditions required for fractionations are equally diverse. This is especially true since the relative centrifugal force applied and time required for a particular fractionation are related not only to size of the particles but also to the viscosity and density range of the gradients, as well as to the shape of the gradients. Although it is not possible to specify exactly what are the best centrifugation conditions to achieve any particular separation, some guidance is given by *Charts 3.1* and *3.2*. *Chart 3.1* shows the approximate sedimentation coefficients and densities of a variety of biological particles, and *Chart 3.2* gives examples of conditions used to fractionate some of them.

One important point that should be considered is the rates of acceleration and deceleration of rotors at the beginning and end of runs. It has been found that sudden changes in the angular velocity, especially below 5000 rev/min, can cause a 'swirling' motion of the gradient solution. This effect is particularly pronounced when non-viscous gradients and/or wide-diameter tubes are employed.

Chart 3.2 CENTRIFUGATION CONDITIONS: EXAMPLES OF TIME, FORCE AND GRADIENT FOR VARIOUS RATE-ZONAL SEPARATIONS.

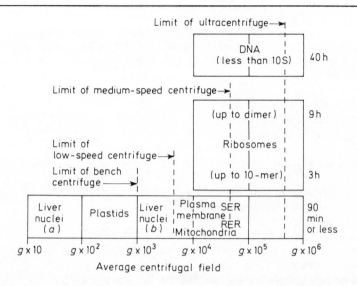

1. Liver nuclei (*a*) (aqueous). Rat and mouse nuclei of different ploidies separated from each other. 1 h; 600 rev/min; 20–50% (w/w) sucrose at 5 °C; 'A' zonal rotor (MSE Ltd). Johnston *et al.* (1968).
2. Plastids. Bean leaf chloroplasts at different stages of development separated according to size. 1500 rev/min; 13–55% (w/w) sucrose at 4 °C; Z15 zonal rotor (IEC). Price and Hirvonen (1967).
3. Liver nuclei (*b*) (non-aqueous). Rat liver nuclei subfractionated. 90 min; 2500 rev/min; concave gradient, from propane-1,3-diol to 3-chloro-1,2-propane diol at 5 °C; 6 × 15 ml rotor (MSE Ltd). Kennedy (1975).
4. Plasma membrane. Rat and mouse liver plasma membranes separated from mitochondria, etc. 40–50 min; 3900 rev/min; 24–54% sucrose (non-linear) at 4 °C; 'A' zonal rotor (MSE Ltd). Evans (1970).
5. Mitochondria. Yeast mitochondria subfractionated according to size. 15 min; 8000 rev/min; 13–55% (w/w) sucrose or sorbitol; Z15 zonal rotor (IEC). Avers, Szabo and Price (1969).
6. RER. Rough endoplasmic reticulum from rat liver subfractionated. 1 h; 58 500g; 0.59–1.02 M sucrose (19–31%, w/w) at 0–1.5 °C; SW25 rotor (Beckman Instruments). Dallner, Bergstrand and Nilsson (1968).
7. SER. Smooth endoplasmic reticulum from rat liver subfractionated. 75 min; 25 000 rev/min; 10–25% sucrose. Glaumann and Dallner (1968).
8. Polyribosomes. Polyribosomes and ribosomal RNA from rat liver, *E. coli* and sea urchin eggs subfractionated. For example, 3–9 h; various isokinetic sucrose gradients and temperatures; various rotors (swing-out and zonal). McCarty, Stafford and Brown (1968).
9. DNA. Fragmented DNA from mouse embryos subfractionated. For example, 41 h; 24 500 rev/min; 5–10% (w/w) sucrose at 20 °C; 6 × 15 ml rotor (MSE Ltd). Hell *et al.* (1972).

Hence, rotors should be accelerated as slowly as possible and for this many of the modern centrifuges possess rate controllers. At the end of the run the rotor should be allowed to come to rest without applying the brake. At speeds greater than 5000 rev/min the occurrence of swirling is minimal.

Unloading and Analysis

Close attention must be paid to the unloading technique, since resolution is easily lost at this stage. Tubes must be handled carefully, especially when they contain non-viscous or shallow density gradients (see p. 188). The advantages and disadvantages of a number of techniques for unloading gradients are discussed in Chapters 6 and 7. In general, however, the best approach is to unload the whole gradient into a number of fractions whenever this is possible, even if the particles to be recovered occupy only a single, narrow zone. Two versions of a technique which is especially suitable for unloading gradients in tubes, since it gives a continuous scan of the distribution of the fractionated particles in the gradient, are illustrated in *Figure 3.7.* Note particularly that the gradient is not inverted at any point (except just above the fraction tubes) and that the gradient is 'funnelled' out of the tube and into and out of the photometer cell. Unloading systems such as this have been investigated by Brakke (1963) and others, who found that this form of unloading gave better results than direct withdrawal of a zone through the side of the tube with a syringe needle. In addition, Moreton and Hirsch (1970) showed that resolution was lost by unloading the gradient from the bottom of the tube. Noll (1969b) has also pointed out that it is important that the flow is not turbulent and that the pumping action is pulse-free; he therefore recommends the use of syringe pumps, rather than peristaltic ones. If a peristaltic pump is used, it should never be between the tube and the flow cell. There are various complete unloading systems available commercially which more or less conform to these criteria.

The dense solution used to displace the gradient is passed through a narrow tube and so, if concentrated sucrose solutions are used, their high viscosity can result in high back-pressures; 55% (w/w) sucrose is about three times more viscous than 45% (w/w), and will therefore cause three times the pressure (see p. 57). It may be useful to displace very dense or concentrated sucrose gradients with a dense, immiscible liquid of lower viscosity, such as Fluorochemical FC43 (3M Chemical Company). If the gradient is to be displaced with a dense, aqueous solution, the solute must be the same as the gradient solute, otherwise significant diffusion at the interface rapidly occurs.

The most common method of analysing gradients is by direct spectrophotometry. Proteins and nucleic acids can generally be monitored at 280 nm and 260 nm, respectively. In addition, some proteins are associated with prosthetic groups (for example, NAD and FAD) that absorb at other wavelengths, which may be useful for determining the distribution of specific enzymes and organelles. The distribution of particulate material through gradients can be determined by light-scattering measurements at 500–600 nm. The best degree of resolution is obtained by passing the gradient effluent directly through a continuous-flow cell (*Figure 3.7*); the flow rate and design of the flow cell are very important for good resolution (Allington *et al.*, 1976). Dividing gradients into a number of fractions and reading the optical density of each markedly reduces the resolution of the gradient. Methods for dividing up only partially resolved peaks have also been described (see, for example, Allington, 1976).

The other analytical technique which has found widespread use is to use samples which have been labelled with radioisotopes. In some cases the distribution of radioactive material across the gradient can be determined by using a continuous-flow monitoring device, but more usually the gradient is divided up into

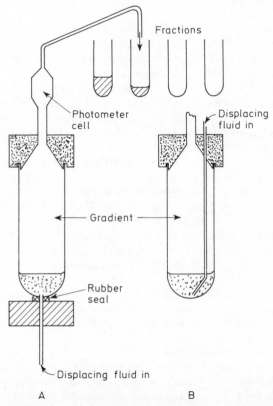

Figure 3.7 *Unloading of tube gradients by upwards displacement. Note particularly, (i) the gradient is not inverted at any point (except just above the fraction tubes); (ii) the gradient is 'funnelled' out of the tube, and into and out of the photometer cell; (iii) the flow rate must be slow, particularly at the beginning, to prevent turbulence; and (iv) the version shown in B allows the gradient to be displaced without the bottom of the tube being punctured. Another version of the upwards displacement method is illustrated in Chapter 6 (Figure 6.3)*

fractions and mixed with a water-miscible liquid scintillator. One problem with concentrated sucrose solutions is that in some scintillators the sucrose first precipitates out and then forms a two-phase mixture. Such a situation occurs when using scintillators based on Triton X-100/toluene recipes and it is accompanied by quenching of the sample (*Figure 3.8*). On the other hand, while Bray's scintillator has a lower overall efficiency it is able to tolerate much higher sucrose concentrations without quenching. Commercial scintillators that can tolerate high sucrose concentrations are also available. The degree of quenching is dependent not only on the physical state but also on the type of gradient solute; for

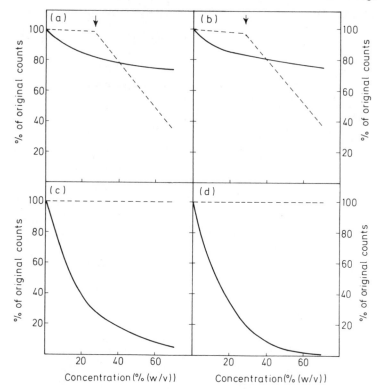

Figure 3.8 Quenching of ³H and ¹⁴C by sucrose and metrizamide solutions. The quenching of ³H-thymidine, (a) and (c), and ¹⁴C-thymidine, (b) and (d), by solutions of sucrose (————) and metrizamide (———) in (a) and (b) using Triton X-100 scintillator (Rickwood and Klemperer, 1971) and in (c) and (d) using Bray's scintillator (Bray, 1960). In each case 0.5 ml of sample was mixed with 4.5 ml of scintillator and the radioactivity of each was measured in a liquid scintillation counter. The arrows indicate the sucrose concentration at which precipitation first occurs

example, iodinated compounds quench radioactivity much more effectively in Bray's scintillator than in Triton/toluene scintillators (*Figure 3.8*).

APPENDIX I: DENSITY AND VISCOSITY OF SUCROSE SOLUTIONS

These data are calculated from equations and coefficients of Barber (1966)

% (w/w)	Density				Viscosity (cP)			
	5°	10°	15°	20°	5°	10°	15°	20°
0	1.000 0	0.999 7	0.999 3	0.998 2	1.520	1.310	1.140	1.002
1	1.004 3	1.004 0	1.003 4	1.002 6	1.558	1.343	1.170	1.030
2	1.008 2	1.007 9	1.007 3	1.006 4	1.603	1.380	1.202	1.057
3	1.012 1	1.011 8	1.011 1	1.010 2	1.650	1.420	1.235	1.086
4	1.016 1	1.015 7	1.015 0	1.014 0	1.700	1.462	1.271	1.116
5	1.020 1	1.019 6	1.018 9	1.017 9	1.753	1.506	1.308	1.148
6	1.024 1	1.023 6	1.022 9	1.021 9	1.809	1.553	1.348	1.182
7	1.028 2	1.027 7	1.026 9	1.025 8	1.869	1.603	1.390	1.217
8	1.032 3	1.031 7	1.030 9	1.029 8	1.933	1.655	1.434	1.255
9	1.036 5	1.035 8	1.035 0	1.033 9	2.001	1.711	1.481	1.295
10	1.040 6	1.040 0	1.039 1	1.038 0	2.073	1.771	1.531	1.337
11	1.044 8	1.044 1	1.043 2	1.042 1	2.150	1.835	1.584	1.382
12	1.049 1	1.048 4	1.047 4	1.046 2	2.231	1.902	1.641	1.430
13	1.053 4	1.052 6	1.051 6	1.050 4	2.319	1.974	1.701	1.481
14	1.057 7	1.056 9	1.055 8	1.054 6	2.413	2.051	1.765	1.535
15	1.062 0	1.061 2	1.060 1	1.058 8	2.513	2.134	1.833	1.592
16	1.066 4	1.065 5	1.064 4	1.063 1	2.621	2.222	1.906	1.654
17	1.070 8	1.069 9	1.068 8	1.067 5	2.736	2.316	1.985	1.720
18	1.075 3	1.074 3	1.073 2	1.071 8	2.859	2.417	2.068	1.790
19	1.079 8	1.078 8	1.077 6	1.076 2	2.992	2.525	2.158	1.865
20	1.084 3	1.083 3	1.082 0	1.080 6	3.135	2.642	2.255	1.946
21	1.088 9	1.087 8	1.086 5	1.085 1	3.290	2.768	2.358	2.033
22	1.093 5	1.092 3	1.091 1	1.089 6	3.456	2.903	2.470	2.125
23	1.098 1	1.096 9	1.095 6	1.094 1	3.636	3.049	2.590	2.226
24	1.102 8	1.101 6	1.100 2	1.098 7	3.831	3.206	2.719	2.333
25	1.107 5	1.106 2	1.104 8	1.103 3	4.043	3.377	2.859	2.449
26	1.112 2	1.110 9	1.109 5	1.107 9	4.272	3.562	3.010	2.575
27	1.116 9	1.115 7	1.114 2	1.112 6	4.523	3.763	3.174	2.710
28	1.121 7	1.120 4	1.118 9	1.117 3	4.796	3.982	3.352	2.857
29	1.126 6	1.125 2	1.123 7	1.122 1	5.094	4.221	3.546	3.016
30	1.131 5	1.130 1	1.128 5	1.126 8	5.422	4.481	3.757	3.189
31	1.136 4	1.134 9	1.133 4	1.131 6	5.781	4.767	3.987	3.378
32	1.141 3	1.139 8	1.138 2	1.136 5	6.177	5.080	4.239	3.583
33	1.146 3	1.144 7	1.143 1	1.141 4	6.614	5.425	4.515	3.808
34	1.151 3	1.149 8	1.148 1	1.146 3	7.099	5.805	4.818	4.053

Appendix 1 *continued*

%		Density				Viscosity (cP)		
(w/w)	5°	10°	15°	20°	5°	10°	15°	20°
35	1.156 3	1.154 8	1.153 1	1.151 3	7.637	6.225	5.153	4.323
36	1.161 4	1.159 8	1.158 1	1.156 3	8.236	6.692	5.522	4.621
37	1.166 5	1.164 9	1.163 1	1.161 3	8.905	7.211	5.932	4.949
38	1.171 7	1.170 0	1.168 2	1.166 3	9.656	7.790	6.386	5.311
39	1.176 8	1.175 2	1.173 4	1.171 4	10.50	8.438	6.894	5.714
40	1.182 1	1.180 3	1.178 5	1.176 6	11.45	9.167	7.461	6.163
41	1.187 3	1.185 6	1.183 7	1.181 7	12.53	9.988	8.097	6.664
42	1.192 6	1.190 8	1.188 9	1.187 0	13.76	10.92	8.813	7.226
43	1.197 9	1.196 1	1.194 2	1.192 2	15.16	11.97	9.621	7.858
44	1.203 3	1.201 4	1.199 5	1.197 5	16.76	13.17	10.54	8.570
45	1.208 7	1.206 8	1.204 8	1.202 8	18.60	14.54	11.58	9.376
46	1.214 1	1.212 2	1.210 2	1.208 1	20.73	16.11	12.77	10.29
47	1.219 5	1.217 6	1.215 6	1.213 5	23.18	17.92	14.13	11.33
48	1.225 0	1.223 1	1.221 1	1.218 9	26.03	20.01	15.69	12.52
49	1.230 6	1.228 6	1.226 5	1.224 4	29.28	22.37	17.45	13.86
50	1.236 1	1.234 1	1.232 0	1.229 9	33.16	25.17	19.52	15.42
51	1.241 7	1.239 7	1.237 6	1.235 4	37.73	28.45	21.93	17.23
52	1.247 4	1.245 3	1.243 2	1.240 9	43.16	32.32	24.75	19.33
53	1.253 0	1.250 9	1.248 8	1.246 5	49.62	36.89	28.06	21.79
54	1.258 7	1.256 6	1.254 4	1.252 2	57.39	42.34	31.98	24.67
55	1.264 5	1.262 3	1.260 1	1.257 8	66.79	48.87	36.64	28.07
56	1.270 2	1.268 1	1.265 8	1.263 5	78.24	56.75	42.22	32.12
57	1.276 0	1.273 9	1.271 6	1.269 3	92.30	66.35	48.95	36.96
58	1.281 9	1.279 7	1.277 4	1.275 0	109.7	78.11	57.12	42.79
59	1.287 7	1.285 5	1.283 2	1.280 8	131.6	92.65	67.13	49.85
60	1.293 4	1.291 4	1.289 1	1.286 7	159.0	110.8	79.48	58.50
61	1.299 6	1.297 3	1.295 0	1.292 6	194.0	133.6	94.85	69.15
62	1.305 6	1.303 3	1.300 9	1.298 5	239.1	162.7	114.2	82.39
63	1.311 6	1.309 3	1.306 9	1.304 4	297.9	200.0	138.7	99.01
64	1.317 6	1.315 3	1.312 9	1.310 4	375.6	248.6	170.2	120.1
65	1.323 7	1.321 4	1.318 9	1.316 4	479.4	312.5	211.0	147.0
66	1.329 8	1.327 5	1.325 0	1.322 5	620.4	397.7	264.6	182.0
67	1.336 0	1.333 6	1.331 1	1.329 6	814.4	512.9	336.0	227.8
68	1.342 2	1.339 8	1.337 3	1.334 7	1086.	711.2	432.2	288.5
69	1.348 4	1.346 0	1.343 4	1.340 8	1473.	891.7	563.8	370.2
70	1.354 6	1.352 1	1.349 7	1.347 0	2034.	1205.	746.7	481.8

APPENDIX II: DENSITY AND VISCOSITY OF GLYCEROL SOLUTIONS

These values are taken from Schmider (1973), calculated from coefficients of Van der Zeijst and Bult (1972).

% (w/w)	Density			Viscosity (cP)		
	5°	10°	20°	5°	10°	20°
0	0.999 4	0.999 1	0.997 7	1.505 0	1.307 0	1.002 0
1	1.002 0	1.001 6	1.000 1	1.555 1	1.349 3	1.031 9
2	1.004 5	1.004 1	1.002 5	1.606 5	1.392 6	1.062 5
3	1.007 1	1.006 6	1.005 0	1.659 3	1.436 9	1.093 8
4	1.009 7	1.009 1	1.007 4	1.713 5	1.482 4	1.125 8
5	1.012 2	1.011 7	1.009 9	1.769 1	1.529 1	1.158 6
6	1.014 8	1.014 2	1.012 3	1.826 4	1.577 9	1.192 2
7	1.017 4	1.016 7	1.014 8	1.885 3	1.626 3	1.226 6
8	1.020 0	1.019 3	1.017 2	1.945 9	1.676 9	1.261 9
9	1.022 5	1.021 8	1.019 7	2.008 4	1.729 0	1.298 2
10	1.025 1	1.024 4	1.022 2	2.072 9	1.782 6	1.335 4
11	1.027 7	1.026 9	1.024 7	2.139 4	1.837 8	1.373 8
12	1.030 3	1.029 5	1.027 2	2.208 1	1.894 8	1.413 2
13	1.033 0	1.032 1	1.029 7	2.279 1	1.953 6	1.453 9
14	1.035 6	1.034 6	1.032 2	2.352 7	2.014 4	1.495 8
15	1.038 2	1.037 2	1.034 7	2.429 0	2.077 3	1.539 1
16	1.040 8	1.039 8	1.037 2	2.508 2	2.142 5	1.584 0
17	1.043 4	1.042 4	1.039 7	2.590 5	2.210 2	1.630 4
18	1.046 1	1.045 0	1.042 2	2.676 2	2.280 5	1.678 5
19	1.048 7	1.047 6	1.044 7	2.765 5	2.353 7	1.728 5
20	1.051 4	1.050 2	1.047 3	2.858 8	2.430 0	1.780 5
21	1.054 0	1.052 8	1.049 8	2.956 3	2.509 7	1.834 7
22	1.056 7	1.055 4	1.052 4	3.058 5	2.593 0	1.892 3
23	1.059 3	1.058 0	1.054 9	3.165 8	2.680 3	1.950 4
24	1.062 0	1.060 6	1.057 5	3.278 6	2.771 9	2.012 4
25	1.064 6	1.063 3	1.060 0	3.397 4	2.868 3	2.077 4
26	1.067 3	1.065 9	1.062 6	3.522 9	2.969 9	2.145 8
27	1.070 0	1.068 5	1.065 2	3.655 7	3.077 2	2.217 9
28	1.072 7	1.071 2	1.067 7	3.796 4	3.190 8	2.294 0
29	1.075 4	1.073 8	1.070 3	3.946 0	3.311 3	2.374 5
30	1.078 1	1.076 5	1.072 9	4.105 3	3.439 4	2.459 9
31	1.080 8	1.079 1	1.075 5	4.275 4	3.575 9	2.550 6
32	1.083 5	1.081 8	1.078 1	4.457 5	3.728 1	2.647 4
33	1.086 2	1.084 5	1.080 7	4.652 8	3.878 0	2.750 7
34	1.088 9	1.087 2	1.083 3	4.863 0	4.045 8	2.861 3
35	1.091 6	1.089 8	1.086 0	5.089 7	4.226 4	2.980 0
36	1.094 3	1.092 5	1.088 6	5.334 9	4.421 5	3.107 9
37	1.097 1	1.095 2	1.091 2	5.600 9	4.632 7	3.245 8
38	1.099 8	1.097 9	1.093 8	5.890 4	4.862 1	3.395 2
39	1.102 5	1.100 6	1.096 5	6.206 4	5.111 9	3.557 3
40	1.105 3	1.103 3	1.099 1	6.552 4	5.385 0	3.733 8
41	1.108 0	1.106 0	1.101 8	6.932 5	5.684 4	3.926 7
42	1.110 8	1.108 8	1.104 5	7.351 6	6.013 8	4.137 9
43	1.113 5	1.111 5	1.107 1	7.815 3	6.377 4	4.370 2
44	1.116 3	1.114 2	1.109 8	8.330 3	6.780 3	4.626 6
45	1.119 1	1.117 0	1.112 5	8.904 5	7.228 6	4.910 4
46	1.121 9	1.119 7	1.115 1	9.547 4	7.729 2	5.225 9
47	1.124 6	1.122 4	1.117 8	10.270	8.290 7	5.578 1
48	1.127 4	1.125 2	1.120 5	11.087	8.923 3	5.972 8
49	1.130 2	1.127 9	1.123 2	12.014	9.639 3	6.417 1
50	1.133 0	1.130 7	1.125 9	13.071	10.454	6.919 5

REFERENCES

ALLINGTON, R.W. (1976). *Analyt. Biochem.*, **73**, 93

ALLINGTON, R.W., BRAKKE, M.K., NELSON, J.W., ARON, C.G. and LARKINS, B.A. (1976). *Analyt. Biochem.*, **73**, 93

ANDERSON, N.G. (1966a). *Natn. Cancer Inst. Monogr.*, No. 21

ANDERSON, N.G. (1966b). *Science, N.Y.*, **154**, 103

ANDERSON, N.G. (1968). *Analyt. Biochem.*, **23**, 72

ARCUS, A.C. (1967). *Analyt. Biochem.*, **18**, 383

AUSTOKER, J., COX, D. and MATHIAS, A.P. (1972). *Biochem. J.*, **129**, 1139

AVERS, C.J., SZABO, A. and PRICE, C.A. (1969). *J. Bact.*, **100**, 1044

AYAD, S.R., BORSALL, R.W. and HUNT, S. (1968). *Analyt. Biochem.*, **22**, 533

BARBER, E.J. (1966). *Natn. Cancer Inst. Monogr.*, No. 21, 219

BAXTER-GABBARD, K.L. (1972). *FEBS Lett.*, **20**, 117

BEAUFAY, H., BENDALL, D.S., BAUDHUIN, P. and DE DUVE, C. (1959). *Biochem. J.*, **73**, 628

BEAUFAY, H., JAQUES, P., BAUDHUIN, P., SELLINGER, O.Z., BERTHET, J. and DE DUVE, C. (1964). *Biochem. J.*, **92**, 184

BENEDICT, R.C., SCHUMAKER, V.N. and DAVIES, R.E. (1967). *J. Reprod. Fert.*, **13**, 237

BERMAN, A.S. (1966). *Natn. Cancer Inst. Monogr.*, No. 21, 41

BIRNIE, G.D. (1973). In *Methodological Developments in Biochemistry*, vol. 3, p. 17. Ed. E. Reid. London; Longmans

BIRNIE, G.D. and HARVEY, D.R. (1968). *Analyt. Biochem.*, **22**, 171

BLATTNER, F.R. and ABELSON, J.N. (1966). *Analyt. Chem.*, **38**, 1279

BOCK, R.M. and LING, N. (1954). *Analyt. Chem.*, **26**, 1543

BOEDTKER, H. (1968). In *Methods in Enzymology*, vol. 12B, p. 429. Ed. L. Grossman and K. Moldave. New York and London; Academic Press

BOONE, C.W. (1971). In *Separations with Zonal Rotors*, p. V1.5. Ed. E. Reid. Guildford, UK; Univ. of Surrey Press

BRAKKE, M.K. (1963). *Analyt. Biochem.*, **5**, 271

BRAKKE, M.K. (1964). *Archs Biochem. Biophys.*, **107**, 388

BRAKKE, M.K. and VAN PELT, N. (1970). *Analyt. Biochem.*, **38**, 56

BRAMWELL, M.E. (1974). *Biochem. J.*, **141**, 477

BRAY, G.A. (1960). *Analyt. Biochem.*, **1**, 279

BROWN, R.D. and HASELKORN, R. (1971). *Proc. natn. Acad. Sci. USA*, **68**, 2536

CASTAÑEDA, M., SÁNCHEZ, R. and SANTIAGO, R. (1971). *Analyt. Biochem.*, **44**, 381

CHARLWOOD, P.A. (1963). *Analyt. Biochem.*, **5**, 226

CLINE, G.B. and RYEL, R.B. (1971). In *Methods in Enzymology*, vol. 22, p. 168. Eds. L. Grossman and K. Moldave. New York and London; Academic Press

COLLOT, M., WATTIAUX-DE CONINCK, S. and WATTIAUX, R. (1975). *Eur. J. Biochem.*, **51**, 603

COTMAN, C., BROWN, D.H., HARRELL, B.W. and ANDERSON, N.G. (1970). *Archs Biochem. Biophys.*, **136**, 436

DALLNER, G., BERGSTRAND, A. and NILSSON, R. (1968). *J. cell. Biol.*, **38**, 257

DE DUVE, C. (1963). *Harvey Lect.*, **59**, 49

DE DUVE, C., BERTHET, J. and BEAUFAY, H. (1959). *Prog. Biophys. Biophys. Chem.*, **9**, 325

DE POMERAI, D.I., CHESTERTON, C.J. and BUTTERWORTH, P.H.W. (1974). *Eur. J. Biochem.*, **46**, 461

EIKENBERRY, E.F., BICKLE, T.A., TRAUT, R.R. and PRICE, C.A. (1970). *Eur. J. Biochem.*, **12**, 113

ELIASSON, R. and SAMELIUS-BROBERG, U. (1965). *Acta physiol. Scand.*, **64**, 245

EVANS, W.H. (1970). *Biochem. J.*, **116**, 833

FUNDING, L. and STEENSGAARD, J. (1971). In *Separations with Zonal Rotors*, p. M4.1. Ed. E. Reid. Guildford, UK; Univ. of Surrey Press

GLAUMANN, H. and DALLNER, G. (1968). *Abs. 5th FEBS Mtg (Prague)*, 819

GRAHAM, J.M. (1973). In *Methodological Developments in Biochemistry*, vol. 3, p. 205. Ed. E. Reid. London; Longmans

HAINES, M.E., JOHNSTON, I.R. and MATHIAS, A.P. (1970). *FEBS Lett.*, **10**, 113

HARTMAN, G.C., BLACK, N., SINCLAIR, R. and HINTON, R.H. (1974). In *Methodological Developments in Biochemistry*, vol. 4, p. 93. Ed. E. Reid. London; Longmans

HELL, A., BIRNIE, G.D., SLIMMING, T.K. and PAUL, J. (1972). *Analyt. Biochem*, **48**, 369

HINTON, R.H. and DOBROTA, M. (1969). *Analyt. Biochem.*, **30**, 99

HINTON, R.H., MULLOCK, B.M. and GILHUUS-MOE, C.C. (1974). In *Methodological Developments in Biochemistry*, vol. 4, p. 103. Ed. E. Reid. London; Longmans

HU, A.S.L., BOCK, R.M. and HALVORSON, H.O. (1962). *Analyt. Biochem.*, **4**, 489

JOHNSTON, I.R., MATHIAS, A.P., PENNINGTON, F. and RIDGE, D. (1968). *Biochem. J.*, **109**, 127

KENNEDY, D.F. (1975). *PhD Thesis*. Univ. of London

LAKSHMANAN, T.K. and LIEBERMAN, S. (1953). *Archs Biochem. Biophys.*, **45**, 235

LEIF, R.C. and VINOGRAD, J. (1964). *Proc. natn. Acad. Sci. USA*, **51**, 520

LEONARD, N.J., McDONALD, D.J. and REICHMANN, M. (1970). *Proc. natn. Acad. Sci. USA*, **67**, 93

LEVINTHAL, C. and DAVISON, P.F. (1961). *J. molec. Biol.*, **3**, 674

McCARTY, K.S., STAFFORD, D. and BROWN, O. (1968). *Analyt. Biochem.*, **24**, 314

McCONKEY, E.H. (1974). *Proc. natn. Acad. Sci. USA*, **71**, 1379

MACH, O. and LACKO, L. (1968). *Analyt. Biochem.*, **22**, 393

MARTIN, R.G. and AMES, B.N. (1961). *J. biol. Chem.*, **236**, 1372

MATHIAS, A.P. and WYNTER, C.V.A. (1973). *FEBS Lett.*, **33**, 18

MATHIAS, A.P., RIDGE, D. and TREZONA, N.St. G. (1969). *Biochem. J.*, **111**, 583

MEUWISSEN, J.A.T.P. (1973). In *Methodological Developments in Biochemistry*, vol. 3, p. 29. Ed. E. Reid. London; Longmans

MORETON, B.E. and HIRSCH, C.A. (1970). *Analyt. Biochem.*, **34**, 544

NEUHOFF, V. (1973). In *Molecular Biology, Biochemistry and Biophysics*, vol. 14, p. 205. Ed. V. Neuhoff. London; Chapman and Hall: Berlin, Heidelberg and New York; Springer-Verlag

NOLL, H. (1967). *Nature, Lond.*, **215**, 360

NOLL, H. (1969a). In *Techniques in Protein Biosynthesis*, vol. 2, p. 101. Ed. P.N. Campbell and J.R. Sargent. New York and London; Academic Press

NOLL, H. (1969b). *Analyt. Biochem.*, **27**, 130

OROZLAN, S.J., RIZUI, S., O'CONNOR, T.E. and MORA, P.T. (1964). *Nature, Lond.*, **202**, 780

PARIS, J.E. (1968). *Biochim. biophys. Acta.*, **165**, 286
POLLACK, M.S. and PRICE, C.A. (1971). *Analyt. Biochem.*, **42**, 38
PRETLOW, T.G. II, BOONE, C.W., SHRAGER, R.I. and WEISS, G.H. (1969). *Analyt. Biochem.*, **29**, 230
PRICE, C.A. and HIRVONEN, A.P. (1967). *Biochim. biophys. Acta.*, **148**, 531
PRICE, C.A. and KOVACS, A. (1969). *Analyt. Biochem.*, **28**, 460
REID, E. (1971). In *Separations with Zonal Rotors*, p. Z1.1. Ed. E. Reid. Guildford, UK; Univ. of Surrey Press
REID, E. (1972). *Sub-Cell. Biochem.*, **1**, 217
RICKWOOD, D. and KLEMPERER, H.G. (1971). *Biochem. J.*, **123**, 731
ROSÉN, C-G. and FEDORCSÁK, I. (1966). *Biochim. biophys. Acta.*, **130**, 401
SCHMIDER, P. (1973). In European Symposium of Zonal Centrifugation in Density Gradient, *Spectra 2000*, vol. 4, p. 299. Ed. M. Raynaud. Paris; Editions Cité Nouvelle
SCHUEL, H.S. and SCHUEL, R.S. (1967). *Analyt. Biochem.*, **20**, 86
SCHUMAKER, V.N. (1967). *Adv. biol. med. Phys.*, **11**, 245
SOBER, H.A. (1970). (Ed.) *Handbook of Biochemistry*, 2nd edn. Ohio; Chemical Rubber Co
SPRAGG, S.P. and RANKIN, C.T. Jr. (1967). *Biochim. biophys. Acta*, **141**, 164
SPRAGG, S.P., MORROD, R.S. and RANKIN, C.T. Jr. (1969). *Sepn Sci.*, **4**, 467
STEENSGAARD, J. (1970). *Eur. J. Biochem.*, **16**, 66
STONE, A.B. (1974). *Biochem. J.*, **137**, 117
SVENSSON, H. (1957). *Sci. Tools*, **4**, 1
SWINDELLS, J.F., JACKSON, R.F. and CRAGOE, C.S. (1942). In *Polarimetry, Saccharimetry and the Sugars*. Natn. Bur. Stand. Circ. No. 440. Washington; US Govt Ptg Off.
SWINDELLS, J.F., SNYDER, C.F., HARDY, R.C. and GOLDEN, P.E. (1958). Supplement to Natn. Bur. Stand. Circ. No. 440. Washington; US Govt Ptg Off.
SZYBALSKI, W. (1960). *Experientia*, **16**, 164
TREZONA, N.St. G. (1969). *PhD Thesis.* Univ. of London
VAN DER ZEIJST, B.A.M. and BULT, H. (1972). *Eur. J. Biochem.*, **28**, 463
VEDEL, F. and D'AOUST, M.J. (1970). *Analyt. Biochem.*, **35**, 54
WALLACE, H. (1969). *Analyt. Biochem.*, **32**, 334
WILLIAMSON, R. (1971). In *Separations with Zonal Rotors*, p. Z2.6. Ed. E. Reid. Guildford, UK; Univ. of Surrey Press

4 Fractionations in Zonal Rotors

JOHN M. GRAHAM

Department of Biochemistry, St George's Hospital Medical School, London

The term 'zonal centrifugation' is an extremely broad one and was originally used in contradistinction to 'differential centrifugation'. The essential difference is that in the former the particles sediment through a medium of increasing density, whereas in the latter the medium is of uniform density. Rate-zonal centrifugation describes the separation of particles into discrete zones on the basis of differences in sedimentation rate; isopycnic-zonal centrifugation describes the separation of particles on the basis of differences in buoyant or banding density (see Chapter 1). As such, these terms embrace all types of centrifugation, including those with fixed-angle and swing-out rotors. However, it has become common practice to associate the term 'zonal centrifugation' more particularly with zonal rotors, although the term is still often used in its original context.

Zonal rotors were developed to fulfil a number of requirements, first enumerated by Anderson (1966):

1. Ideal sedimentation and maximum resolution in sector-shaped compartments.
2. Rapid gradient formation in the rotor with minimal stirring or convection.
3. Sharp starting or sample zones.
4. High rotational speed.
5. Large capacity.
6. Rapid recovery of the gradient after centrifugation without loss of resolution.

A typical zonal rotor is in the form of a closed cylinder containing a central core and four or more radial vanes or septa which end close to the wall of the rotor, effectively dividing its interior into four or more sectors. The gradient fills the entire enclosed space; the capacity of the majority of zonal rotors varies from 300 to 1700 ml.

In a spinning rotor the centrifugal field is radial; thus, only that part of the centrifugal field passing through the centre of a tube in a swing-out rotor (*Figure 4.1a*) is parallel to the wall of the tube. Only the particles at the centre of the sample zone will sediment directly towards the bottom of the tube; all other particles will be deflected towards the wall of the tube to an extent proportional to their distance from the centre of the sample zone. This 'wall effect' leads to a loss of resolution which does not happen in the sector-shaped compartments of a zonal rotor (*Figure 4.1b*). To minimize disturbance of the gradient and to maintain sharp sample zones, these rotors were designed to be loaded and unloaded while rotating (that is, dynamically), so that the perturbation and reorientation of a gradient which occurs during acceleration and deceleration of swing-out and fixed-angle rotors are eliminated. However, subsequently reorienting ('Reograd') zonal rotors which can be loaded and unloaded while at rest (that is, statically) were designed, and others have been adapted for continuous-flow operation.

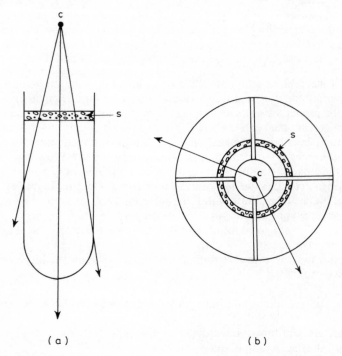

(a)   (b)

Figure 4.1   Radial lines of centrifugal force (→) from the centre of rotation (c) through a sample zone (s) in (a), a swing-out rotor and (b), a zonal rotor. (From Graham, 1975)

In this chapter, the basic forms of, and methods of operating, batch-type and continuous-flow zonal rotors will be considered, although a detailed description of the design and stability of zonal rotors is outside the scope of this review; for that, Barringer (1966) should be consulted. The application of zonal centrifugation techniques to a variety of fractionation problems is also discussed, with emphasis put on the suitability of particular rotors for certain tasks.

DESIGN AND OPERATION OF BATCH-TYPE ROTORS

Dynamic Loading and Unloading

These rotors are derived from the considerations outlined previously and are not designed for continuous-flow operation. There are a series of such rotors based on two models, the BIV and the BX, both of which are loaded and unloaded dynamically. Both types of rotor are capable of speeds in excess of 35 000 rev/min. Low-speed rotors (A-series) similar in shape to the BX rotor are also available; the principles of their design are very similar to those of the high-speed rotors. The types, designation, volume (capacity) and maximum speed of rotors used for zonal centrifugation are listed in the Appendix to Chapter 9.

The main consideration of dynamically-loaded rotors is access to the interior of the rotor (both to the edge and the core) while it is rotating. This is achieved by means of a fluid seal, within a bearing assembly, which comprises a stationary seal carrying a central fluid line (or passage) surrounded by an annulus. This seal is in contact, under pressure, with a rotating seal on top of the rotor's core. The rotating seal also contains a central channel and annulus and these are continuous with lines within the core which feed liquid to the surface of the core and to the edge of the rotor, respectively. The precise configuration of the fluid seal varies, and the reader is referred to Anderson *et al.* (1966a) and Barringer *et al.* (1966), and to the manufacturers' manuals for further details.

BIV-type Rotors

The upper lid of the rotor (*Figure 4.2*) is extended into a hollow shaft, the end of which is contained within the bearing and damper assembly. Since the height

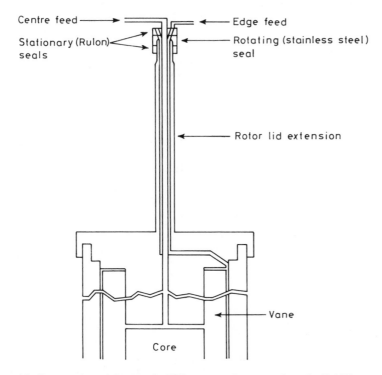

Figure 4.2 Cross-section of the top of a BIV-type zonal rotor to show the fluid lines. (From Beckman Instruction Manual, ZU-IM-2)

of a BIV-type rotor is more than twice that of its diameter, the bearing and damper assembly must remain in place during centrifugation to maintain stability (Anderson *et al.*, 1966a). The hollow shaft which ends in a stainless steel rotating seal also forms part of the double fluid-line into the rotor; a narrower tube attached to the centre of the rotating seal passes through the shaft and into the rotor (*Figure 4.2*). This central tube exits at the surface of the core, whereas the annulus between the inner tube and the shaft exits *via* a channel in the upper cap at the edge of the rotor. The static seal is made from a filled fluorocarbon (Rulon). A continuous supply of cooling water to the bearing and the fluid seal, and of oil for the bearing, are required for this particular rotor. Since the fluid seal remains in place during all phases of centrifugation, the two opposed faces experience considerable wear, and the rotor requires a special centrifuge for incorporating the bearing assembly. For these reasons a second series of zonal rotors, starting with the BX, were designed.

BX-type Rotors

These rotors have removable seals which are only in place during loading and unloading. During the actual centrifugation run the rotors are capped and can

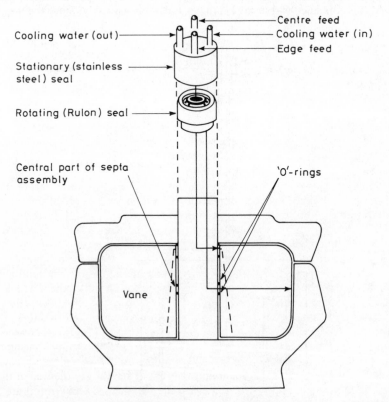

Figure 4.3 Cross-section of a BX-type zonal rotor and feed head to show the fluid lines. (From MSE Technical Publication No. 49)

thus be operated in normal centrifuges. These rotors have diameters greater than their height; they are therefore more stable than the BIV-type and do not require an upper bearing and damper assembly continuously in place. The core of these rotors carries the central channel and annulus, and terminates at a Rulon rotating seal (*Figure 4.3*). The static seal (stainless steel) carries four lines, two for circulating cooling water to the face of the seal and two leading to the central channel and annulus (*Figure 4.3*). The rotating seal is housed within a bearing assembly which is attached, along with the static seal, to the top of the core while the rotor is spinning at low speed (2000–3000 rev/min).

Unlike the BIV rotor, the core and septa are not part of the same assembly; instead, the septa are attached to a cylinder which fits over the core. The central line exits at the core of the rotor, just above and between the septa, whereas the annulus line continues downward to exit in a channel formed by a groove in the core and the inner surface of the septa assembly. Fluid in this channel reaches the edge of the rotor *via* holes drilled within the septa, 'O'-rings on either side of the groove in the core preventing fluid in this channel from egression at any point other than the rotor's edge. The modern forms of these dynamically-loaded rotors are the BXIV and BXV rotors.

The A-series zonal rotors are batch-type rotors whose relative dimensions are similar to the BX series. This series has culminated in the AXII whose fluid seal assembly is very similar to that of the BXIV, although because of the low maximum speed of this rotor (5000 rev/min) the seal, for convenience, remains in place throughout the run. One other difference is that the annulus line reaches the edge of the rotor through channels formed between the base of the rotor and each radial vane.

Operation of Dynamically-loaded Rotors

The routine procedures for operating these modern batch-type rotors are as follows. Variations in loading and unloading techniques will be discussed later with regard to specific fractionation problems.

The standard method of loading is to introduce the gradient, light end first, to the edge of the spinning rotor. As progressively denser medium reaches the edge, so the solution of lower density is displaced towards the core of the rotor, eventually emerging at the central line. Some workers prefer to fill the rotor, while it is stationary, with water or buffer prior to acceleration to loading speed and introduction of the gradient; the advantage of this approach is that it is possible to monitor the flow rate of the gradient from the volume of water displaced. To prevent accumulation of particles on the wall of the rotor during centrifugation of the sample it is normal practice to follow the gradient with a cushion of dense medium. When the rotor has been filled, the direction of flow is reversed; the sample is fed to the core of the rotor and displaced into the centrifugal field by an overlay of a low-density solution. The rotor is then accelerated to the required speed, after the fluid seal has been replaced by a vacuum cap in the case of BX-type rotors. With these rotors, unloading is always accomplished by feeding a dense solution (almost always of sucrose) to the wall of the rotor and collecting from the centre. Neither the BXIV nor the BIV rotors are suitable for unloading from the edge. Edge-unloading rotors are considered later in relation to specific fractionation requirements (see p. 84).

To detect material emerging from the rotor, the gradient effluent is passed through a flow cell and its absorption monitored continuously at a suitable wavelength (usually 260 or 280 nm).

Practical Details and Problems

A detailed description of the assembly and operation of these batch-type rotors can be obtained from the appropriate rotor manual. There are, however, a number of points which are worth emphasizing.

(1) In modern forms of the BXIV and BXV rotors, in which the Rulon seal forms part of a detachable feed-head assembly rather than being incorporated into the top of the rotor core (earlier models), it is necessary to grease lightly the 'O'-ring around the seal itself; this should be done very carefully to avoid contaminating the polished surface with vacuum grease.

(2) Immediately before assembly, the two surfaces of the fluid seal should be inspected for any fibres from the cleaning tissue; these must be removed.

(3) In modern rotors, the positioning of the feed head on the top of the spinning rotor requires the engagement of pins in the guard tray with grooves in the side of the feed head. The latter should be held below the screw cap and locking ring; if either of these threaded components are gripped tightly during placement of the feed head, the torque provided by the spinning rotor can cause them to loosen and consequently reduce the spring tension on the rotating seal. The screw cap and locking ring should be checked regularly to ensure that they have not become loosened; lack of tension on the fluid seal will result in cross-leakage, whereas excessive tension will cause the generation of undue heat by friction.

(4) Cross-leakage in the fluid seal is perhaps the major problem the operator will encounter. This can arise from (*a*) scratches across the face of the Rulon seal, and (*b*) high back-pressures generated usually from the flow of viscous media into the rotor and/or too high a flow rate. During loading of the gradient, cross-leakage from the annulus to the centre line results in liquid being expelled from the centre line by the air displaced from the rotor, and cross-leakage centrifugal to the annulus results in the appearance of liquid around the seal assembly. In completely enclosed seals, this centrifugal leakage is not apparent until liquid emerges from two drain holes on the upper side of the assembly. A small amount of cross-leakage is tolerable, but anything in excess of 0.3% of the total rotor capacity should be considered undesirable. Unless excessive, cross-leakage from the annulus outwards is more of a nuisance than a real problem, particularly in the case of an enclosed seal assembly in which gradient medium then finds its way into the bearing. During unloading, cross-leakage from the annulus to the centre line is a real problem. The incomplete mixing of the high-density displacing medium and the low-density effluent results in a streaming effect within the effluent and an erratic absorption pattern from the continuous-flow cell. Another possibility is that, in spite of the effluent flow, the cross-leaking displacing medium may also sink through the low-density medium and back into the rotor. Once a pathway for a cross-leak has been established across the face of the seal, reducing the flow rate may not alleviate the problem. However, with the older types of rotor in which the Rulon seal is incorporated on top of the core, reducing the flow rate is the only course of action. In the more recent models, the Rulon seal forms part of the removable feed-head assembly; consequently,

the feed head may be removed and the fluid seals cleaned and, if necessary, repolished. In the author's experience, cross-leakage from the centre line to the annulus during loading of the sample has never been a problem and indeed it is unlikely to occur because the recommended flow rates for the sample are usually much lower than for the gradient (see *Table 4.1*). If such a cross-leakage does occur it is immediately apparent by the appearance of turbidity in the high-density effluent.

Table 4.1 RECOMMENDED FLOW RATES FOR LOADING ZONAL ROTORS AT 4 °C*

Fluid	Flow rate ml/min
Low-density gradient (0–20% w/w sucrose)	25–35
Medium-density gradient (20–40% w/w sucrose)	20–25
High-density gradient (40–60% w/w sucrose)	10–15
Sample and overlay	5–10

*These flow rates maintain a sufficiently low pressure (<0.5 atm) at the fluid seal, using tubing of about 3 mm internal diameter. Halving the diameter will increase the pressure at the fluid seal 16-fold for the same flow rate.

(5) Obviously, the inadvertent introduction of a large air bubble into the edge line could cause a serious disturbance in the gradient if it is allowed to enter the rotor; the incorporation of a bubble trap in the edge line will overcome this problem. Alternatively, a simple plastic tube-connector inserted into this line enables the operator to clamp off the tubing on the feed-head side of the connector, uncouple the tubing and remove the air bubble. Once a continuous line of fluid is re-established to the connector, the tubes can be carefully rejoined. Clearly, this simple remedy is effective only if the fault is noted before the bubble reaches the connector. If, however, a large column of air is about to enter the feed head, the pump must be stopped, the lines clamped off and air pumped to the centre until a continuous line of fluid is established from the feed head to the edge line connector.

(6) After the sample and overlay have been introduced, and the pump has been switched off, a few seconds should elapse before the fluid lines to the feed head are clamped off in order to permit pressure stabilization, and thus reduce extravasation of medium from rotor and feed head. However, some spillage of liquid is inevitable when the feed head is removed and this should be carefully cleaned from the collar of the guard tray.

(7) After removal of the feed head a metal cap is placed over the top of the core, and a large, deformable 'O'-ring on the core expands in the centrifugal field to seal the rotor effectively. Small air vent-holes in the cap prevent the formation of an air lock within the cap, thus allowing the latter to pass easily over the 'O'-ring. If the cap will not fit easily or completely over the top of the core (it should abut the top of the rotor) the air holes may be blocked; they can be cleared by insertion of a fine gauge syringe needle or fine wire. To facilitate positioning of the cap, its internal surface may be *very lightly* smeared with vacuum grease.

(8) It should be borne in mind that, although the gradient solutions may be maintained at 0–4 °C during loading, after they have passed through the gradient pump, tubing and fluid seal they will be 2–3 °C warmer when they enter the rotor. Steps can be taken to minimize this, namely by using ice-cold water to

cool the fluid seal, or inserting a cooling coil between pump and rotor, or by reducing the loading time by pumping at a high flow rate. However, only the first may be wholly beneficial; the others will certainly lead to a greater mixing of the gradient, and the last may also cause serious cross-leakage at the seal. If, on the other hand, the distance travelled by the inflowing gradient is made as short as possible, both temperature rise and gradient mixing will be minimized. If it is essential for the gradient to be at 4 °C, the loaded rotor can be centrifuged for 1–2 h at the required temperature prior to loading the sample.

If, during loading, the rotor's contents are at a higher temperature than at the moment of unloading, then contraction of the rotor's contents during centrifugation will cause small columns of air to form in the lines within the core. That in the central line is of little consequence since, even if introduced into the rotor chamber, it will remain at the core. If, however, high-density material is pumped immediately to the edge of the rotor, the air will enter the rotor and may disturb the gradient as it migrates centripetally. It is thus advantageous first to feed some low-density medium to the centre to expel any trapped air from the edge lines.

Another aspect of temperature control is that, in spite of improved sealing devices to divorce the air within the centrifuge chamber from the atmosphere, some condensation still occurs on the walls of the cold chamber. Also, it is often impossible to avoid the spillage of a few drops of gradient solution when removing the feed head after loading. It is essential in those centrifuges which use an infrared sensing device to record and control the rotor temperature that the vacuum is rapidly reduced below 10^{-2} torr. This cannot occur if liquid within the centrifuge chamber is frozen on the chamber wall. The poor vacuum results in friction at the surface of the infrared sensor which, consequently, records a temperature in excess of the correct rotor temperature. If refrigeration continues, the rotor's contents could freeze, with disastrous results. This can be overcome by setting the rotor temperature control about 10 °C above the temperature of the rotor when evacuation of the chamber begins. Residual fluid is swept from the chamber; after a few minutes, the temperature control can be turned down and sufficient vacuum is achieved as rapidly as with any other rotor.

(9) During the high-speed phase of the run, a transient fall in water pressure or loss of electrical power to the centrifuge which is of sufficient duration to cause it to fail will result in deceleration of the rotor. If there is no means of re-establishing power to the rotor drive, the rotor should be allowed to come to rest, with the brake off below 2000 rev/min. The run may be salvaged by controlled re-acceleration of the rotor by fine adjustment of the zonal loading speed. Since the gradient must reorient from a vertical to a horizontal position, the acceleration up to 1000 rev/min must be slow and steady (approximately 5 min to reach 1000 rev/min). Some centrifuges are designed to take reorienting zonal rotors (see p. 70) and with them a smooth, slow acceleration is readily achieved. With other centrifuges the zonal loading speed-control is usually too coarse and the initial 'kick' as the rotor starts to rotate is too strong to give satisfactory reorientation. The best way of overcoming this problem is to incorporate a variable rheostat ('Regovolt') into the rotor drive circuit so that the rate of acceleration may be finely controlled. Failing this, there are some very crude approaches which may work. For example, before increasing the zonal loading speed from zero, the rotor may be set gently in motion by hand; this at least minimizes the disturbing effect of the initial drive impulse. Thereafter, the

rate of acceleration may be reduced by adjusting the zonal speed control very slowly while applying a retarding force to the cap with the hand. Alternatively, a retarding force can be applied by replacing the feed head assembly prior to re-acceleration.

(10) If, at the end of the run, the rotor decelerates to rest rather than to the zonal loading speed, the rotor must be re-accelerated, as above, to unload. However, if the useful part of the gradient in a stationary rotor occupies the upper half of the rotor chamber only, the rotor can be unloaded statically. So long as the high-density cushion reaches, or is close to, the level of the septa exit ports, then passing dense unloading medium slowly into the static rotor will not affect appreciably that part of the gradient above the exit ports. These crude approaches will cause some disturbance, the severity of which is a matter of luck. They should only be employed as a last resort to recover valuable samples.

Cleaning and Maintenance of Rotor and Fluid Seal

It is essential that the rotor is removed from the centrifuge and disassembled upon completion of a run. The two parts of the rotor body, septa assembly, sealing ring and the fluid lines within the core and septa should be washed with copious volumes of cold tap water and then distilled water. All the fluid lines are then cleared of water by blowing air or nitrogen through them, and all parts dried. Failure to remove all traces of sucrose solution from the channels will result in sucrose crystallizing and, consequently, blocking the fluid lines. High back-pressure will thus be created at the fluid seal next time an attempt is made to feed gradient into the rotor. The enclosed fluid-seal units of the modern BXIV and BXV rotors should also be disassembled immediately and checked for the presence of sucrose in the space surrounding the static seal. More than a few drops of sucrose indicate a significant leakage in the seal and the Rulon face should be inspected for scratches. If there is a continuous film of liquid on top of the bearing housing the ball race should be exposed by removing the retaining clip and rubber washer, and washed several times in water, then 70% ethanol and finally benzene or petroleum spirit to remove the lubricant. If the bearing is washed directly in benzene or petroleum spirit before removing the excess sucrose solution, sucrose precipitates and is subsequently very difficult to remove. New bearing lubricant is then packed into the ball race. In the most recent models, the bearing is completely enclosed and cleaning is unnecessary.

The Rulon and static seals, and the fluid lines within them, should be washed sequentially with tap water, distilled water and 70% ethanol, and then dried in a stream of air or nitrogen. Failure to remove all traces of sucrose solution will cause blockage of the fluid lines with crystallized sucrose; moreover, any sucrose which crystallizes on the face of either seal will cause severe damage to the Rulon surface when it is used next. If the face of the Rulon seal has been scratched, it must be gently abraded using crocus paper (a very fine abrasive paper). It is essential that this is done on a flat, smooth surface (for example, a glass plate) and with an even pressure applied with the thumb on top of the seal. This is preferable to holding the seal between thumb and forefinger, since it is difficult to maintain an even pressure in that way. After the imperfections in the surface of the seal have been ground away, the face should be polished by rubbing it gently on glazed paper (the reverse side of graph paper is suitable), again using the downward pressure of the thumb. The feed head should be assembled and the fluid-seal faces relapped by circulating water through the spinning rotor for 3–4 h and then rechecked for cross-leakage by feeding high-density sucrose into the rotor. It is insufficient merely to pass water through the seal; even at high flow rates the low viscosity of the medium (compared to 60% sucrose) does not provide sufficient pressure within the flow lines to test the seal under experimental conditions.

The earlier models of the BXIV and BXV had titanium septa assemblies; the more recent ones are made from Noryl. One problem with this material is that it absorbs or reacts with iodine or iodide. If the cellular material being separated in the rotor is labelled with ^{125}I, then significant radioactivity can be detected on the Noryl septa and this activity is not easily washed off. Prolonged exposure to solutions of unlabelled iodide or sodium thiosulphate have relatively little effect. Also, Noryl septa cannot be sterilized by rinsing with aqueous diethyl carbonate (Baycovin) solutions; although the plastic does not appear to be affected by this reagent, septa so treated disintegrate when centrifuged. Ethanol solutions must never be used when cleaning HS or A-type rotors since they cause Perspex to crack and shatter.

Statically-loaded (Unloaded) Rotors

These rotors, also called reorienting (Reograd) rotors, are loaded with gradient while stationary and then very slowly accelerated to about 1000 rev/min. During this phase of the acceleration, the gradient reorients (*Figure 4.4*) from a vertical to a horizontal position (Hsu and Anderson, 1969). Elrod, Patrick and Anderson (1969) emphasized the smoothness rather than the actual rate of acceleration as being the important factor below 800 rev/min for the AXVI rotor. These rotors are also unloaded statically; the critical deceleration phase below 1000 rev/min (when reorientation occurs) must similarly be achieved very slowly. Some workers simply allow deceleration without the brake, others control deceleration by a steady reduction in the drive force.

In spite of these two reorientation phases, excellent linear gradients can be recovered from such rotors. Moreover, there are certain advantages of reorienting rotors over the dynamically-loaded and unloaded rotors: (*i*) no sophisticated rotating seal is required; (*ii*) because of this, it is easy to cope with very viscous gradients (for example, high concentrations of Ficoll and dextran) which are difficult to handle with a rotating seal; (*iii*) separation of large, rapidly sedimenting particles, such as intact cells or nuclei, is more convenient and reproducible since they are exposed only to unit gravitational field during loading and unloading; and (*iv*) the shearing forces across the fluid seal of a dynamically-loaded rotor can fragment DNA molecules (Anderson, 1967; Zadrazil, 1973) and such forces do not occur in a statically-loaded rotor. In reorienting rotors the centre line exits at the core of the rotor, whereas the other line exits at the bottom of the rotor rather than at the edge. Elrod, Patrick and Anderson (1969) compared the sedimentation-rate banding of rat liver nuclei in a dynamically-loaded BXV rotor, in which the core and edge lines were enlarged to cope with the necessary viscous gradient, with that in a statically-loaded AXVI rotor, and each gave equally good results.

Two versatile reorienting zonal rotors, the SZ20 and SZ14, are capable of maximum speeds of 15 000 and 20 000 rev/min respectively (Sheeler and Wells, 1969; Wells, Gross and Sheeler, 1970; Sheeler, Gross and Wells, 1971). The rotor chamber is divided by six septa, the minimum necessary to maintain the gradient during reorientation (Sheeler, Gross and Wells, 1971). The bottom of each septum is in the form of an asymmetric 'V', which is shallower on the centrifugal face than on the centripetal face, fitting in an identical groove in the base of the rotor. The septa lines exit at the point of the 'V'; this configuration is optimal for loading and reorientation. An additional sophistication of these

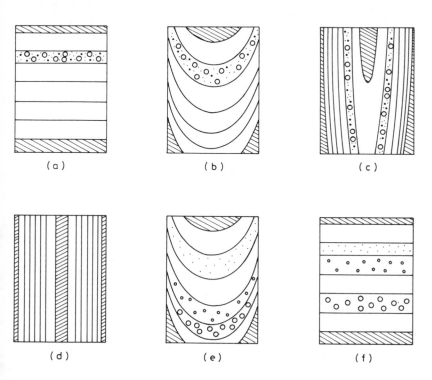

Figure 4.4 Schematic diagram of a fractionation in a reorienting gradient (Reograd) zonal rotor: (a) rotor at rest; (b) accelerating; (c) accelerating; (d) at speed; (e) decelerating; (f) at rest. (From Hsu and Anderson, 1969)

rotors is that the core lines exit immediately adjacent to each septum so that the density gradient can also be loaded dynamically *via* the centre line, dense end first, the septa directing the incoming gradient to the edge of the rotor. Sheeler, Gross and Wells (1971) found that the shape of the gradient was unaltered by reorientation. During reorientation the maximum shear occurs where the greatest changes in area occur (*Figure 4.4*); that is, those zones which are at the top and bottom of the rotor when at rest and which find themselves at the centre and edge, respectively, of the spinning rotor (Anderson *et al.*, 1964). Relatively little shearing occurs towards the centre of the gradient. It is common practice, therefore, to include significant amounts of cushion and overlay in such gradients.

Gradients and Sample Preparation

The effect of varying gradient shape and the generation of complex gradients will be described later in relation to the separation of specific particles. However, at the outset it should be pointed out that the most common forms of gradient are those which are linear with volume. Since within a zonal rotor equal volumes occupy a decreasing radial thickness towards the wall of the rotor, such gradients are not linear with radius (see Chapters 5 and 9). Noll (1967) has developed isokinetic gradients in which particles move at a constant speed through the gradient. Such a system possesses two advantages: first, it is easy to estimate sedimentation coefficients and, secondly, the width of the particle zone should remain constant. However, the application of such gradients to zonal rotors is difficult because a constant zone width is equivalent to an increasing volume towards the rotor edge (Pollack and Price, 1971). For this reason, these authors computed equivolumetric gradients, in which a particle zone of negligible radial thickness migrates through equal volumes of the gradient in equal time, that is, particles move through a constant volume per unit time at a rate proportional

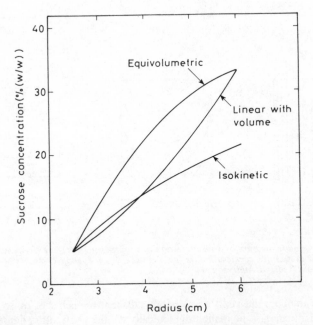

Figure 4.5 Equivolumetric, linear and isokinetic gradients for BXIV or BXXX zonal rotors. (From Pollack and Price, 1971)

to the sedimentation coefficient. Pollack and Price (1971) showed that the migration of ribosomal subunits from *Euglena* was a linear function of time and sedimentation coefficient in these gradients. *Figure 4.5* compares the shapes of equivolumetric, linear and isokinetic gradients. Van der Zeijst and Bult (1972) have computed equivolumetric gradients for the BXIV and BXV rotors; gradient profiles can be obtained from these authors.

Gradients linear with radius give better resolution at the top than at the bottom of the gradient. *Figure 4.6* shows the sedimentation of particles of increasing sedimentation coefficient in such a gradient; a convex gradient would give better resolution of some mixtures of particles (Birnie, 1973). However, it is impossible to make any generalizations because the optimum gradient depends upon the particles being separated; the linear gradient, for example, would be excellent for separating slower-moving components from a fast-moving one. Frequently, it is advantageous to develop empirically a discontinuous gradient to maximize particular separations (Graham, 1973a, 1973b; Prospero and Hinton, 1973). Diffusion of the solute will, however, tend to smooth out these discontinuities.

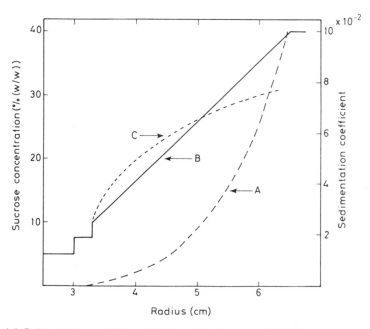

Figure 4.6 Sedimentation coefficients (A) of particles in a 10–40% (w/w) sucrose gradient (B) which is linear with radius in a BXIV rotor. The convex gradient (C) is more suitable for some purposes (see text). (From Birnie, 1973)

A shallow gradient will alter less than a steep gradient during centrifugation and, since the rate of diffusion is proportional to the surface area of the liquid, a 695 ml gradient occupying a radial distance of 10.3 cm in an HS rotor will diffuse less than a 640 ml gradient occupying a radial distance of 6.7 cm in a BXIV rotor (Prospero, 1973).

The most frequently adopted procedure in these zonal rotors is to introduce the sample at the core of the rotor, followed by an overlay. This is essential for rate-zonal or combined rate-zonal and isopycnic separations. For these it is also necessary to use as small a sample volume as possible to minimize band broadening. In isopycnic separations the sample volume is not important; indeed, the sample may be made part of the gradient so as to increase its capacity. Since a steep gradient has high capacity but low resolving power, and a shallow one low capacity and high resolving power, it may be advantageous to use a convex gradient so that the steepest part of the gradient is adjacent to the sample where the greatest concentration of particles occurs and high capacity is required (Birnie, 1973; Prospero, 1973). If the capacity of the gradient in the sample zone is exceeded, diffusion of the sample occurs; the causes and consequences of this are discussed in Chapter 5.

The sample and overlay (*Table 4.1*) must be loaded slowly (approximately 5 ml/min) to prevent mixing (Birnie, 1973); to prevent aggregation of the sample at the top of the gradient the density of the sample band should be close to (but obviously less than) that of the top of the gradient. Any instability at the interface between sample and gradient may be avoided by using an inverse sample gradient, that is, a sample in which the solute concentration increases while the particulate concentration decreases (Prospero, 1973). It is indeed advisable to check the density of the sample prior to loading it. This is particularly useful when the difference in density between the sample and top of the gradient is only 0.012–0.015 g/cm³ and when the sample is a total cell homogenate. In the latter case small increases in the concentration of particles and soluble proteins in the homogenate can quickly lead to overloading of the gradient and inversion of the density gradient at the interface between sample and the top of the gradient. Errors in the loading of the gradient should manifest themselves when the refractive index of the effluent gradient is checked.

Those rotors (for example BIV and A-type) whose fluid seals remain in place during the entire operation deform significantly during the acceleration phase due to the increasing centrifugal field. It is essential therefore to keep the line to the centre of the rotor open to a reservoir of buffer which can be taken up by the expanding rotor.

FRACTIONATIONS IN BATCH-TYPE ROTORS*

Separations of Different Cell Types

Since different cells have different sizes, and/or shapes, and/or densities, it is possible to separate them on the basis of either sedimentation rate or density. Since intact cells are sensitive to changes in the tonicity of the suspending medium it is common practice to use gradients which are more or less isotonic. This is frequently achieved by sedimentation through gradients of a high-molecular-weight solute dissolved in an isotonic solution. The high-molecular-weight solute can be Ficoll or dextran and the isotonic solution can be of sucrose or salt, or it could be a tissue-culture medium such as Eagle's. Relative to the isotonic solution the contribution of the polymer to the total osmolarity is negligible.

Figure 4.7 describes the sedimentation of three types of cell to their isopycnic densities: A and B cells have the same diameter but different density, B and C have the same density but differ in diameter, whereas A and C differ both in diameter and density. Boone, Harell and Bond (1968) calculated these parameters for rabbit thymocytes, HeLa cells, cultured human leukaemia cells and horse leukocytes, and devised gradients of Ficoll dissolved in minimal essential medium to separate various binary mixtures of these cells using an AIX rotor.

*In the following examples of fractionations, gradient solutions are described in terms of % (w/w) unless stated otherwise.

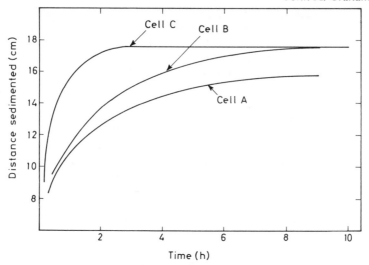

Figure 4.7 Sedimentation of three types of cell differing in diameter and/or density. Cell A: diameter 7.5 μm, density 1.060; cell B: diameter 7.5 μm, density 1.070; cell C: diameter 15.0 μm, density 1.070. (From Boone et al., 1968)

Table 4.2 gives the gradient and centrifugation conditions for the three mixtures. Because the loading and unloading speed of this rotor is 500 rev/min the unloading time has to be carefully standardized to obtain reproducible separations.

The HS rotor has been used to separate populations of cells from rat brain (Giorgi, 1971, 1973). Gradients consisted of either steps of 20, 40, 50 and 60% sucrose or a 40–60% linear sucrose gradient. The latter appeared to be more effective; after centrifugation at 4500 rev/min for 30 min a separation of glial cells from neurons was achieved and, indeed, resolution of astrocytes and oligodendrocytes also occurred. The use of sucrose for these separations is contrary to the rationale of using approximately isotonic media for cell fractionation and serves to illustrate the point that it is impossible to make any broad generalizations about techniques. Mathias, Ridge and Trezona (1969) also used a sucrose gradient (20–50%) in an AXII rotor (600 rev/min for 1 h) for fractionating nucleated erythrocytes.

It is quite possible that the separation of cells in sucrose gradients may partly reflect the differential sensitivity of different cells to shrinkage in a hypertonic

Table 4.2 CONDITIONS FOR SEPARATING MIXTURES OF CELLS
IN AN AIX ZONAL ROTOR*

Cells in mixture	Ficoll concentration in sample % (w/v)	Ficoll gradient % (w/v)	Centrifugation conditions
Thymocytes plus HeLa cells	7	10–20	1000 rev/min for 30 min
Thymocytes plus leukaemia cells	8	10–18	1000 rev/min for 60 min
Leukocytes plus HeLa cells	6	8–16	1000 rev/min for 22 min

*The rotor contained 90 ml of overlay, 40 ml of sample, 1000 ml of gradient and 160 ml of cushion

medium. This is emphasized in the binary gradient system used by Flangas and Bowman (1968) to separate rat-brain cells in the BXIV rotor. The two-phase gradient of 30% Ficoll and 58% sucrose, together with a cushion of potassium citrate, was centrifuged at 35 000 rev/min for 45 min. Neuronal perikarya sediment close to the citrate boundary, whereas glial cells, microcytes, astrocytes and oligodendrocytes were resolved either in the 30% Ficoll close to the position of the original boundary between the Ficoll and sucrose, or across the boundary region. The highly viscous Ficoll medium considerably retards the sedimentation of the cells, a prerequisite for cell separations in high-speed rotors. At the boundary region the cells meet three sharp gradients of decreasing viscosity, increasing density and increasing osmolarity. The relative importance of these parameters to the efficacy of the separation is impossible to estimate, but this work does serve to demonstrate the ability to improve separations by manipulating density, viscosity and osmolarity, all of which are important in determining the sedimentation characteristics of osmotically-active particles.

Separation of Cells According to Position in the Cell Cycle

During the cell cycle a cell increases its volume approximately twofold, so that after division two daughter cells are formed of the same size as the original cell. It is possible to separate these cells in suitable gradients on the basis of their sedimentation rates. Compared with the separation of different cell types, however, the differences in sedimentation characteristics as the cell advances through the cell cycle are small and the density of cells changes relatively little throughout the cell cycle. Since there is a continuous spectrum of sizes of cell, the cells will spread continuously down through a gradient so long as they do not reach their isopycnic density. For such cell fractionations the A-type or HS rotors not only provide the requisite low speed of centrifugation but also their Perspex construction allows continuous visual monitoring of the migration of the band of cells. The other type of rotor which has been extensively used for this purpose is the reorienting SZ14 rotor.

   Yeast cells have been very successfully separated on the basis of their sedimentation rate, so that the lightest fractions contain cells in the G_1 phase of the cell cycle and the heaviest fractions contain dividing cells. Sucrose gradients of 20–40% have been used for *Saccharomyces cerevisiae* Y185, which is diploid, and gradients of 10–25% for the haploid *S. lactis* Y123 (Sebastian, Carter and Halvorson, 1971). The heavier diploid cells obviously require a gradient of higher density and viscosity than the lighter haploid cells. Although at each point in the gradient there is a particular size distribution of cells, the median size of the cells increases through the gradient. The sharp increase in the amount of DNA per cell in the middle of the gradient indicates that these cells are in S-phase. Their position reflects the lack of nucleic acid synthesis by G_1- and G_2-cells which are smaller and larger than S-cells, respectively. Wells and James (1972) used the reorienting SZ14 rotor to fractionate *Schizosaccharomyces pombe*, making use of the ability of this rotor to be loaded dynamically and unloaded statically. A 1300 ml 15–30% sucrose gradient was loaded from the centre dynamically and the temperature of the rotor was allowed to stabilize before the sample was applied. The sedimentation of these cells at 1500 rev/min only requires 5.5 min and so the temperature of the gradient would not stabilize during such a short period. Sedimentation rate is dependent on viscosity which in turn is temperature-dependent, and the temperature effect becomes more significant in rate-zonal separations the smaller the difference between the sedimentation rates of the particles becomes. Because of this, good temperature control is essential for separating cells.

Cultured mammalian cells have also been separated on the basis of sedimentation rate for cell cycle studies. In many cases emphasis has been laid upon maintaining some degree of isotonicity within the gradient to prevent the cells shrinking, and gradient solutions containing some physiological media constituents have been used in order to maintain viability of the cells. Ehrlich ascites cells (Probst and Maisenbacher, 1973) were fractionated on the basis of cell size (position in the cell cycle) in gradients which were (*a*) linear with increasing sucrose concentration (2–31%); (*b*) linear with decreasing saline concentration (0.76–0%); and (*c*) constant with respect to Hanks' Balanced Salt Solution. The last provided a background of constant tonicity and the opposing gradients of salt and sucrose provided increasing density and viscosity without a significant change in osmolarity. The cells were applied to the gradient in an A-type rotor, in 1% sucrose with 0.86% NaCl in Hanks' Solution; consequently, the difference in density between the sample and the light end of the gradient was very small and great care had to be taken in loading of the sample. It is important, however, that the difference in density between sample and gradient be as small as possible in this type of experiment, since cells sediment rapidly and may aggregate at the interface of the sample and gradient if this density difference is not minimized. Cells taken from the gradient were cultured: the heaviest cells demonstrated a declining synthesis of DNA and were about to enter mitosis; the lightest started to synthesize DNA and entered mitosis after 11–12 h; cells recovered from the middle of the gradient showed a high level of DNA synthesis and entered mitosis after 5 h (Probst and Maisenbacher, 1973).

Pasternak and Warmsley (1970) also used a linear gradient of 2–10% (w/v) Ficoll dissolved in 50 mM tris buffer, pH 7.4 and 0.9% NaCl in an A-type zonal rotor to separate neoplastic mouse mast cells (P815Y). The cells were applied in 1% Ficoll (in tris-buffered saline) and centrifuged for 12–15 min at 500 rev/min. Ficoll was chosen as a suitable gradient solute since, at low concentrations, it has a very low osmolarity. *Figure 4.8* shows that cell volume increased through the gradient and that the cells towards the middle of the gradient were synthesizing DNA rapidly, whereas lighter (G_1) and heavier (G_2) cells were not; cells isolated from the gradient grew synchronously when recultured. A comparison with separations using a conventional swing-out rotor showed the superiority of the zonal separation both in terms of capacity and resolution (Warmsley and Pasternak, 1970). This separation system has also been used successfully for hamster sarcoma virus-transformed hamster-embryo fibroblasts (Pasternak and Graham, 1973; Graham *et al.*, 1973). Synchronizing tissue-culture cells by zonal centrifugation has many advantages over the standard chemical 'block and release' techniques. The latter certainly impose artificial constraints on the cell and they have been severely criticized by Petersen, Tobey and Anderson (1969). In addition, they cannot be applied readily to virus-transformed cells; many of the latter can continue growing in conditions used to arrest the growth of normal cells, for example, the removal of serum. Other methods such as mitotic selection (Petersen, Tobey and Anderson, 1969) are not suitable for cells which do not attach to glass. Against this must be set the requirement for a single-cell

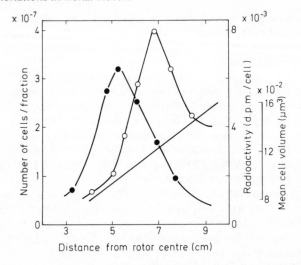

Figure 4.8 Separation of P815Y cells according to their position in the cell cycle in an A-type zonal rotor. Cell number, ●——●; mean cell volume, ——; incorporation of [¹⁴C] thymidine into DNA, ○——○; for centrifugation details see text. (From Warmsley and Pasternak, 1970)

suspension, making it unsuitable for monolayer cells which are difficult to disaggregate. Although aggregates will tend to sediment rapidly from the bulk of the single cells, large aggregates may sequester other cells and cause gradient disturbances.

Nuclei and Nuclear Membranes

Some of the first applications of the early AV zonal rotor were by Fisher and Cline (1963) for the isolation of calf-thymus nuclei. They used a linear gradient generated from 0.25 M sucrose and 30% dextran (in the presence of 6 mM Ca^{2+}). The sample was a total calf-thymus homogenate and after 20 min at 500 rev/min the nuclei were separated from all other cell components. The binary sucrose–dextran gradient minimized the damage caused by an all-sucrose gradient (Fisher and Cline, 1963). Nuclei, being rather dense and rapidly sedimenting, become exposed to the highest concentrations of the gradient solute. The damage caused by high concentrations of sucrose cannot be due entirely to the osmolarity effect because only 10% of the total nuclear volume is impermeable to sucrose (Johnston and Mathias, 1972). However, phase-contrast observation of nuclei transferred from isotonic to hypertonic sucrose solutions clearly show that these organelles shrink significantly in the latter.

These observations may only be relevant to calf-thymus nuclei; rat-liver nuclei have been successfully separated in sucrose gradients (Johnston *et al.*, 1968, 1969; Johnston and Mathias, 1972). These workers emphasized the need to purify partly the nuclei from the homogenate by centrifugation through 2.2 M sucrose before fractionating them. This concentration of sucrose (equivalent to about 75% w/v) would certainly remove any contamination by whole or partially-ruptured cells and it may be these components which interfere with the zonal separation of unpurified nuclei. Johnston *et al.* (1968) used a 20–50% sucrose gradient (1000 ml) buffered to pH 7.4 and containing 1 mM Mg^{2+}; the sample volume was 10–20 ml (15% sucrose) and the overlay about 100 ml. After 1 h at 600 rev/min two well-separated major zones containing diploid (slower moving) and tetraploid (faster moving) nuclei were recovered. The tetraploid nuclei reached approximately 31% sucrose; thus, in spite of using a less viscous gradient and higher centrifugal fields than Fisher and Cline (1963), Johnston *et al.* (1968) found that the majority of the nuclei did not sediment into high concentrations of sucrose. Whether this reflects the different tissue source or the lower concentration of divalent cations used by Johnston *et al.* is uncertain. Hexaploid and octaploid nuclei can be resolved in the more dense regions of the gradient. Johnston and Mathias (1972) also reported the ability of this type of gradient at pH 6.0 (or with 3 mM Mg^{2+}) to separate diploid stromal and diploid parenchymal nuclei, and also to separate partially nuclei from avian reticulocytes and erythrocytes. Although reticulocyte nuclei are larger than those of the erythrocytes, they are considerably less dense.

For studies of certain enzymes in nuclei (such as DNA polymerase) it may be advantageous to modify the zonal centrifugation conditions. By using an HS rotor and a 26–66% (w/w) glycerol gradient, the time can be reduced to 15 min if the speed is increased to 1200 rev/min (Johnston and Mathias, 1972). Interestingly, nuclei from diploid cells in S-phase sediment between the diploid and tetraploid zones and as the cells move from S to G_2 the nuclei trail increasingly into the tetraploid peak. Obviously, the increasing DNA content of the nuclei is raising their sedimentation rate and density. In this connection it might be pertinent to mention that metaphase chromosomes from human diploid lymphocytes have been fractionated in an AXII rotor (Schneider and Salzman, 1970). The gradient of 15–40% buffered sucrose was able to resolve single chromosomes of different sizes, large chromosomal aggregates and intact nuclei from a 20 ml sample by centrifugation at 2000 rev/min for 20 min. These authors provided a clear indication that the mean chromosomal length increased through the gradient.

Price, Harris and Baldwin (1972) used a discontinuous sucrose gradient in an AXII rotor to isolate nuclear membranes. The sucrose solutions were buffered to pH 7.6 with 1 mM $NaHCO_3$, and the sample (a crude nuclear pellet from rat liver or rat hepatoma, previously stored in 1 mM $NaHCO_3$ overnight) was applied in 50–100 ml, together with 150 ml of overlay. After centrifugation at 3000 rev/min for 40 min the 'nuclear ghosts' formed a broad band (28–38% sucrose) between the membrane vesicles plus mitochondria and the intact nuclei plus cells. To prevent aggregation due to pelleting, the total 'ghost' fraction (approximately 400 ml) was transferred to a BXIV rotor. After centrifugation in a very shallow (40–44%) sucrose gradient at 30 000 rev/min for 16 h the nuclear ghosts had formed a band at 1.21 g/cm^3. Contaminating plasma membranes banded above this at 1.18 g/cm^3. The higher density probably reflects residual chromatin material bound to the nuclear membranes.

Mitochondria, Lysosomes and Peroxisomes

In Core-unloading Rotors

Since the more complex task of isolating plasma membrane and endoplasmic reticulum from cell homogenates is, by implication, also concerned with the separation of mitochondria and lysosomes, it is somewhat artificial to consider the resolution of these organelles separately. However, there have been developed a few rather specialized techniques for this purpose. Under normal conditions, that is, in linear sucrose gradients, the separation of native lysosomes, mitochondria and peroxisomes is exceedingly difficult because the isopycnic densities and sedimentation coefficients of these organelles overlap considerably. The use of tube gradients is thus very restricted and the greater resolving power of zonal rotors seems to offer greater scope for improving separations. For example, a crude mitochondrial fraction from 12.5 ml of a total rat-liver homogenate (25% w/v) was fractionated on a 17–55% sucrose gradient (total volume 1200 ml) in a BII rotor (Schuel, Tipton and Anderson, 1964). Various centrifugation conditions (10 000–30 000 rev/min) were used to effect a partial separation of mitochondria from other membranes on a rate-zonal basis; the mitochondria were then further purified by isopycnic banding.

The major contaminant of mitochondria (other than lysosomes and peroxisomes) in rate-zonal separations is co-sedimenting membrane fragments. If this contamination is avoided by the use of nitrogen cavitation as a means of homogenization (particularly useful for tissue culture cells), a crude mitochondrial fraction can be separated simply from all other membranes. The gradient that has been used (Graham, 1972) consists of a 190 ml linear sucrose gradient (10–60%) containing a plateau of 150 ml of 30% sucrose in a BXIV rotor. The sample (a postnuclear supernatant fraction) is introduced in 0.25 M sucrose (40 ml) and moved into the centrifugal field by a 50 ml overlay. Centrifugation at 30 000 rev/min for 90 min results in a complete separation of microsomes and mitochondria across the plateau (Graham, 1972).

The use of high-speed rotors to subfractionate the mitochondria is limited because these organelles will approach their isopycnic density under the conditions employed, so that separation from lysosomes is difficult. However, the BXIV rotor, centrifuged at 25 000 rev/min for 1 h, has been used (Withers, Davies and Wynn, 1968) to separate a crude rat-liver mitochondrial fraction into two major peaks on a 0.5–2.0 M linear sucrose gradient. Under these conditions, a peak of acid phosphatase activity was obtained at about 1.2 g/cm^3. The only other peak (around 1.12 g/cm^3) also contained some acid phosphatase, although the major constituent was NADPH–cytochrome c reductase. Withers, Davies and Wynn (1968) used this as a mitochondrial marker, although it would be more reasonable to assume that this light material was microsomal. Schuel, Schuel and Unakar (1968) preferred the AXII rotor for lysosome preparation. A total liver homogenate was centrifuged at 3500 rev/min for 285 min through a discontinuous gradient of 12% and 32% sucrose followed by a 35–40% gradient. While the mitochondria banded around 37.1% sucrose, the lysosomes were largely arrested in the sharp gradient formed between the 12% and 32% sucrose. A peak of acid phosphatase banding broadly within the 12% sucrose was considered to be microsomal. This supports the hypothesis that the light material observed by Withers, Davies and Wynn (1968) was indeed due to microsomes and not mitochondria.

Almost without exception, other workers also observe that in sucrose gradients (approximately 15–55%) the lysosomes band centripetally to the mitochondria (Schuel and Anderson, 1964; Schuel, Tipton and Anderson, 1964; Corbett, 1967). The incorporation of a sharp density discontinuity within the gradient to

Table 4.3 SEPARATION OF LYSOSOMES FROM MITOCHONDRIA

Source	Sucrose gradient %	Volume ml	Centrifugation rev/min	time (h)	Rotor	Ref.
Rat liver*	15–33‡ and 41–60‡	1300	3700	1	AXII	Hartman and Hinton (1971)
Rat hepatocarcinoma*	15–33‡ and 41–60‡	1300	4300	1.25	AXII	Hartman and Hinton (1971)
Rat liver†	10–33‡ and 33–42§	750	9000	0.75–1	HS	Burge and Hinton (1971)

* Fraction pelleted by centrifuging a homogenate at 10 000g for 10 min.
† Crude mitochondrial fraction.
‡ Linear gradient.
§ Exponential gradient.

arrest the movement of the lysosomes without appreciably affecting further sedimentation of the mitochondria has been widely adopted. Although the precise configuration of the gradient and the centrifugation conditions vary (see *Table 4.3*), the aim is to band the lysosomes around 30–32% sucrose within a steep gradient. Hartman and Hinton (1971) found that, because of the more slowly sedimenting properties of hepatocarcinoma mitochondria (compared with those of liver), it is necessary to increase the speed or time of centrifugation. Burge and Hinton (1971) were, moreover, able partly to separate their lysosomal fraction into lysosomes and peroxisomes in an isopycnic sucrose gradient.

Hinton *et al.* (1970) used a similar gradient and centrifugation system to separate lysosomes, peroxisomes and mitochondria from rats treated with Triton X-100. As in the experiments of Burge and Hinton (1971), after centrifugation the microsomes had barely left the sample zone and the mitochondria had reached the sucrose cushion. The Triton-loaded lysosomes banded at a lower density than the mitochondria, whereas the peroxisomes overlapped the lighter lysosomes. Thomson, Nance and Tollaksen (1974) stressed that damage to peroxisomes may arise from exposure to high concentrations of sucrose during isopycnic centrifugation and they devised a more elaborate density–perturbation system to overcome this problem. Mice were injected with Triton WR-1339, which is taken up by lysosomes and makes them less dense, and prednisolone, which causes an increase in the sedimentation rate of mitochondria. The gradient consisted of a 0.29–0.58 M linear sucrose gradient, together with a 1.46 M sucrose cushion. The sample was a pellet, prepared by centrifuging a liver homogenate at 42 000g for 20 min, suspended in 10 ml of 0.25 M sucrose–0.02 M ethanol. After centrifugation until $\omega^2 t$ equalled 1.4×10^9 rad^2/s in a Ti14 rotor, most of the mitochondria had reached the cushion, lysosomes and microsomes remained near the top of the gradient, whereas the peroxisomes were broadly distributed in the middle of the gradient. Since the lighter components overlapped the peroxisomes to some extent, the peroxisome fraction was harvested, suspended in 10 ml of 0.44 M sucrose, applied to a 0.58–1.17 M sucrose gradient and centrifuged until $\omega^2 t$ equalled 4×10^9 rad^2/s to reduce the contamination.

Clearly, the difference in sedimentation rate between these subcellular organelles is so small that, even with the increased capacity and resolution of centre-unloading zonal rotors, under normal conditions it is difficult to separate these components adequately by a single centrifugation without resorting to density perturbation.

In Edge-unloading Rotors

The design of centre-unloading rotors makes them unsuitable for unloading *via* the edge and considerable loss of resolution occurs if this is attempted (Anderson *et al.*, 1968). The BXXIII was developed (Anderson *et al.*, 1968) to overcome this problem; it has a tapered wall (*Figure 4.9*) such that the rotor volume centripetal to the taper volume (350 ml) is 1100 ml. Anderson *et al.* (1968) showed that the bands were only slightly broader when this rotor was unloaded *via* the edge. Two such rotors have been produced commercially, the BXXIX and the BXXX, which are equivalent to the core-unloading BXV and BXIV, respectively. Beckman manufacture BXXIX cores and septa assemblies which can convert either a BXIV or BXV rotor to an edge-unloading rotor, but in doing so the total available volume is reduced by 100 and 300 ml, respectively.

Although these edge-unloading rotors have never attained the popularity of the core-unloading rotors, they do possess certain advantages which make them useful for certain separations. First, they can be unloaded by feeding water to the core of the rotor, which is certainly more convenient than using the very dense, viscous sucrose solutions necessary for core unloading. Second, rapidly sedimenting particles can be removed from the edge and replaced by fresh gradient, thus extending considerably the effective length of the gradient, which is thus able to cope with particles possessing a wide range of sedimentation coefficients. Third, they can be used for flotation separations, in which the sample is introduced at the edge of the rotor, the tapering of the rotor wall providing the necessary sharply defined starting zones. Fourth, the rotor can be loaded either in the conventional way or by feeding overlay, sample and gradient sequentially to the edge of the rotor.

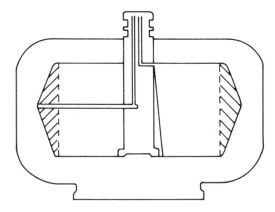

Figure 4.9 Schematic diagram of an edge-unloading BXXIII-type zonal rotor; the volume of the taper at the wall (shaded) is 350 ml (From Anderson et al., 1969a)

A report by Brown *et al.* (1973) demonstrates the usefulness of the BXXIX rotor for separating particles on the basis of rate-zonal sedimentation by the process they call 'sequential product recovery'. A linear 20–30% sucrose gradient (1 litre) and a 45% sucrose cushion were loaded dynamically; the sample, a postnuclear supernatant fraction of a rat-liver homogenate in 8.5% sucrose, was introduced at the core in a total volume of 100 ml and moved radially 3.3 cm from the centre by a 6% sucrose overlay. The rotor was accelerated from 2500 rev/min to its running speed of 10 000 rev/min and then decelerated to 2500 rev/min. The cushion (240 ml) plus 40 ml of the dense end of the gradient were displaced by feeding low-density solution to the core; fresh cushion was fed to the edge and the rotor re-accelerated to the running speed. These operations of partial unloading, gradient replacement and sedimentation were repeated five times, with the final sedimentation phase done at 25 000 rev/min. During such a procedure it is essential to use a digital integrator to compute the centrifugal force in terms of $\omega^2 t$ (rad^2/s). Six fractions of displaced material were thus obtained, the mitochondria occupying the third and fourth fractions, peroxisomes the fifth and sixth fractions, whereas lysosomes occurred in all fractions (Brown

et al., 1973). These workers considered that improvements in resolution might
be obtained by using a smaller sample volume and media of a lower osmolarity.

Contractor (1973) also used the BXXIX rotor to purify rat-liver lysosomes,
using a linear 25–40% sucrose gradient and a cushion of 54% sucrose. The sample (20 ml) was introduced in 23% Ficoll and overlaid with 100 ml of 20%
sucrose. After centrifugation until $\omega^2 t$ equalled 3.1×10^{10} rad^2/s ($476 \times 10^3 g$
$\times$ min), the mitochondria were banded around the position of the original 40–
54% interface; they were displaced from the edge of the rotor by passing 720 ml
of water to the centre, and 400 ml of 54% sucrose was replaced at the edge of
the rotor. The lysosomes remaining in the gradient were then banded isopycnically by further centrifugation.

Plasma Membrane and Endoplasmic Reticulum

The separation of plasma membrane from cell or tissue homogenates by zonal
centrifugation cannot be discussed without reference to many other problems,
such as the method of homogenization, the composition of the isolation medium
and the gradient, and the identification of the products. Lack of space precludes
a comprehensive assessment of these factors, for which reviews by Hinton (1972),
De Pierre and Karnovsky (1973) and Graham (1975) should be consulted. There
are, however, three fundamental problems which can be stated briefly as follows:
(*a*) the separation of plasma membranes from other membranes, primarily the
smooth endoplasmic reticulum, which possess a very similar density; (*b*) the
maintenance of all the osmotically-sensitive organelles such as mitochondria,
lysosomes and nuclei in an intact state; and (*c*) the recovery of a significant proportion of the total plasma membrane. The relative success of a particular fractionation technique depends not only on the centrifugation schedule but on all
the other parameters mentioned previously. There is a profusion of techniques
in the literature, yet there remain just two basic approaches to the task of isolating the surface membrane: (*i*) maintaining the surface membrane as large
sheets or fragments which are more rapidly sedimenting than the vesicles derived
from the endoplasmic reticulum; and (*ii*) changing all the plasma membrane and
endoplasmic reticulum into vesicles, and relying on sophisticated density gradients to capitalize on the slightly higher density and sedimentation rate of the
vesicles from endoplasmic reticulum. A more recent approach is that of affinity-
density–perturbation (Wallach *et al.*, 1972), in which the density of the plasma
membrane is artificially increased by coupling to a large high-molecular-weight
ligand, but such techniques have not, as yet, been used in association with zonal
centrifugation.

Centrifugation in zonal rotors has been widely used to isolate and fractionate
membranes from a variety of tissues and cells. The fractionation methods used
are almost as varied as the sources of the membranes, because each tissue presents
its own problems, many of which have been solved in different ways. It is not
possible to discuss all of these properly here; consequently, an indication only
of the approaches taken is given by the following descriptions of some fractionations of material from rat liver and hepatoma, and tissue-culture cells. Details of
the fractionation procedures for these and several other tissues also are given in
the references cited in *Table 4.4.*

Table 4.4 PREPARATIONS OF PLASMA MEMBRANE AND ENDOPLASMIC RETICULUM

Starting material	Products	Gradient	Centrifugation	Rotor	Ref.
Rat liver and hepatoma*:					
Post-mitochondrial fraction	Rough and smooth ER	10–55% (w/w) sucrose in 15 mM CsCl	25 000 rev/min for 2 h	BXXIX	Lee et al. (1969)
Homogenate	PM, ER, Mit.	Complex 2-stage sucrose gradient	(1) 3000 rev/min for 25 min (2) 20 000 rev/min for 1 h	BXV	Weaver and Boyle (1969)
Homogenate	Ribosomes – Mit.	Disc. 8.9–51% (w/w) sucrose	20 000 rev/min for 1 h	JCF-Z	Griffith and Wright (1972)
Sarcoplasmic reticulum:					
Microsomes (canine cardiac)	Mic. and Mit.	Two-step sucrose density 1.12–1.14 and 1.14–1.22 g/cm^3 in 0.5 M LiBr	5000 rev/min for 2 h	BXIV	Katz et al. (1970)
Homogenate (rabbit skeletal muscle)	Mic. and Mit.	0.35–1.4 M sucrose	18 000 rev/min for 1 h	BXIV	Headon and Duggan (1971)
Brain†					
Crude mitochondrial fraction (rat)	Membrane fractions	15–50% (w/w) sucrose	27 000 rev/min for 2 h	BXV	Cotman et al. (1968)
Homogenate	Myelin fragments	10–50% (w/w) sucrose	40 000 rev/min for 2 h	BIV	Shapira et al. (1970)
Crude mitochondrial fraction	Synaptic complexes	Self-forming CsCl in 0.14 M sucrose	40 000 rev/min for 66 h	BXIV	Korngurth et al. (1971)
Intestinal brush borders‡					
Homogenate	Brush borders and membranes	30–40% (w/w) sucrose	1500 rev/min for 1 h	HS	Connock et al. (1971)
Platelet membranes:					
Mitochondrial fraction	PM and ER	18–36% (w/w) sucrose	47 000 rev/min for 18 h	BXIV	Taylor and Crawford (1974)

Abbreviations: ER, endoplasmic reticulum; PM, plasma membrane; Mit., mitochondria; Mic., microsomes; disc., discontinuous.
Other fractionations: Tissue-culture cells (see text for details), Warren et al. (1966) and Graham (1972, 1973a, 1973b); kidney cortex nuclear fraction, Robinson et al. (1973); Escherichia coli lysate, Quigley and Cohen (1969); Saccharomyces carlsbergensis homogenate, Cartledge et al. (1969).
* See text for descriptions of fractionations by El Aaser et al. (1966a, 1966b), Anderson et al. (1966b), Evans (1970, 1971), Hinton et al. (1967, 1971), Nelson et al. (1971), and Prospero and Hinton (1973); see also Gavard et al. (1974).
† See also Mahaley et al. (1968), Rodnight et al. (1969), Spanner and Ansell (1970), Day et al. (1971) and Ansell and Spanner (1972).
‡ See also Spenney et al. (1973).

Rat Liver and Hepatoma

One of the most successful techniques for rat liver is derived from the methods of Neville (1960) and Emmelot *et al*. (1964). Rat liver gently disrupted by Dounce homogenization in 1 mM NaHCO₃ produces large sheets of surface membrane which are rapidly sedimenting; the pellet obtained by centrifuging at 1000g for 10 min contains most of these membranes, together with nuclei, heavy mitochondria and any trapped lighter components. This material is then resuspended and washed several times, which is not only time-consuming but also results in a gradual loss of material.

Because of the relative ease of homogenizing and fractionating rat liver and because of the central importance this organ has gained in biochemistry, numerous zonal techniques have been designed to prepare plasma membrane from this tissue; some of these are summarized in *Table 4.4*. Both El Aaser *et al*. (1966a) and Evans (1970, 1971) fractionated the total nuclear fraction on the basis of sedimentation rate in sucrose gradients in a low-speed A-type rotor. El Aaser *et al*. (1966a) applied the nuclear pellet material to a non-linear 8–68% (w/v) sucrose gradient which contained steep and shallow regions, and centrifuged at 4000 rev/min for 1 h; Evans (1970) used a complex sucrose gradient generated from 300 ml 6% (w/v), 100 ml 24% (w/v), 100 ml of a 24–36% (w/v) gradient, 300 ml of a 36–54% (w/v) gradient and 100 ml 54% (w/v), and centrifuged at

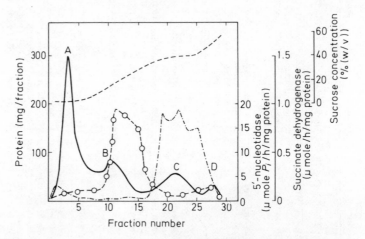

Figure 4.10 Resolution of a nuclear fraction from mouse liver in an A-type zonal rotor. Density gradient, - - -; protein, ———; 5'-nucleotidase, –.–.–.; succinate dehydrogenase, ○- - -○; for centrifugation details see text. (From Evans, 1970)

3900 rev/min for 40–50 min. The methods are very similar and both produced four major bands. Since, however, the densest part of the gradient used by Evans (1970) was 54% (w/v) and that by El Aaser *et al*. (1966a) was 68% (w/v), the nuclei reached the wall of the rotor in the former, whereas they formed a band in the latter. *Figure 4.10* is taken from the results of Evans (1970); band A consisted of small endoplasmic reticulum particles which had obviously been trapped within the original nuclear pellet, B consisted of mitochondria, C contained plasma membrane sheets and D contained nuclei and cell debris. The yield of surface membranes was significantly better with this method than with non-zonal methods.

Rather than using a nuclear fraction, Anderson *et al.* (1966b) used a total liver homogenate. This obviates the problem of resuspending the nuclear pellet which could lead to further fragmentation of the large sheets of surface membrane. Hinton, Norris and Reid (1971) compared the recovery of plasma-membrane sheets from a whole liver homogenate with that from a nuclear fraction, using a discontinuous 8–68% (w/v) sucrose gradient at 3700 rev/min for 1 h. With the whole homogenate the capacity of the gradient was equivalent to approximately 4 g of liver, whereas with a nuclear pellet up to 20 g of liver could be successfully fractionated. These workers also emphasized the advantage of removing red blood cells by perfusion of the liver.

In this type of fractionation it is essential that the mitochondria are not allowed to move towards their isopycnic density, otherwise they will move through and overlap the plasma-membrane band. It is clear, however, that these observations only apply to liver. Dounce homogenization of hepatomas tends to produce smaller plasma-membrane fragments and, in a non-linear 8–68% (w/v) gradient in the AXII rotor, these fragments overlapped the mitochondria (Prospero and Hinton, 1973); that is, they were too small to be separated on the basis of sedimentation rate in a low-speed rotor. In this situation, banding the plasma membrane and mitochondria isopycnically may be the answer.

To harvest the smaller plasma-membrane fragments from rat hepatoma, Prospero and Hinton (1973) separated the material which pelleted on centrifugation at 10 000g for 10 min (with or without a preliminary low-speed centrifugation to remove nuclei) on gradients of density 1.045–1.24 g/cm^3 or 1.039–1.24 g/cm^3, plus a cushion of density 1.265 g/cm^3 contained within an HS rotor. After centrifugation at 9000 rev/min for 90 min, the mitochondria, nuclei and cell debris formed a major peak at the 1.24/1.265 g/cm^3 interface, whereas a peak of 5′-nucleotidase activity (plasma membrane) formed centripetally to this at a density of about 1.16 g/cm^3. The high density of the mitochondria is due to the presence of 2 mM Ca^{2+} in the gradient, which causes them to shrink. Under these conditions the small fragments of plasma membrane approach their isopycnic density, whereas the microsomal fraction (vesicles of mainly endoplasmic reticulum) have only just migrated into the gradient. Prospero and Hinton (1973) were able to recover more 5′-nucleotidase in the plasma-membrane band from material from which nuclei and debris had been removed by a preliminary centrifugation (400g for 3 min). This may reflect the propensity of rapidly-sedimenting particles such as nuclei and partially-disrupted cells to aggregate and trap other material. The major contaminant of the plasma-membrane fraction is lysosomes, which can be removed by taking this fraction and recentrifuging to isopycnic density in a BXIV rotor.

Isopycnic centrifugation is far more adaptable than rate-zonal centrifugation as regards sample preparation. Prospero and Hinton (1973) pelleted the material from the HS rotor and suspended it in 50 ml of sucrose of density 1.19 g/cm^3. This was sandwiched between two gradients, the first of density 1.10–1.185 g/cm^3 (400 ml), the second 1.195–1.23 g/cm^3 (150 ml), all the gradient solutions and sample being fed to the edge of the rotor. Centrifugation at 47 000 rev/min for 2 h separated the 5′-nucleotidase (plasma membrane: A) and acid phosphatase (lysosomes: B) (*Figure 4.11*). Placing the sample in the middle of the gradient probably maximizes the separation between plasma membrane and lysosomes, since these two particles move in opposite directions towards their isopycnic densities. In fact, it is not necessary to sediment the material from the HS rotor at all; the fraction may simply be made part of the sucrose gradient after appropriate adjustment of its density, and it may thus occupy almost the entire volume of the rotor.

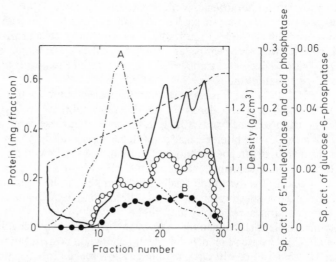

Figure 4.11 Resolution of hepatoma plasma-membrane fragments from a mitochondrial (and lysosomal) fraction in a BXIV zonal rotor. Density gradient, ---; protein, ——; acid phosphatase, ●——●; glucose-6-phosphatase, ○——○; 5'-nucleotidase, —·—·—·; for centrifugation details see text. (From Prospero and Hinton, 1973)

One of the first groups of workers to report the fractionation of rat-liver microsomes in high-speed zonal rotors was El Aaser *et al.* (1966b). They used a BIV rotor, a sucrose gradient of density 1.08–1.2 g/cm³ and the sample was a 20 000g supernatant fraction from a liver homogenate. Centrifugation for 6 h at 40 000 rev/min resolved a number of fractions containing ATPase, 5'-nucleotidase and glucose-6-phosphatase activities. Microsomes from liver are derived predominantly from the endoplasmic reticulum and the plasma membrane from the sinusoidal and bile canalicular surfaces of the parenchymal hepatocytes. The overlap of 5'-nucleotidase and glucose-6-phosphatase activity is not markedly improved by making the gradient more shallow; however, the latter activity was found to be moved centrifugally by addition of 4×10^{-3} M Mg^{2+} (El Aaser *et al.*, 1966b; Hinton *et al.*, 1967). Hinton *et al.* (1967) achieved similar separations in the BXIV and BXV rotors; the use of sigmoid-shaped gradients did not eliminate overlap of plasma membrane and endoplasmic reticulum vesicles. Low-density, 5'-nucleotidase-containing vesicles from rat liver have also been isolated from a post-nuclear supernatant fraction by Evans (1971), using a 20–50% sucrose gradient in a BXV rotor spun at 25 000 rev/min for 3.5 h.

Nelson, Masters and Peterson (1971) used the Ti15 rotor for bulk preparation of rat-liver microsomes; these workers filled the entire rotor, at rest, with a previously purified microsome suspension and centrifuged the particles to the wall of the rotor. A simple, more effective method would be almost to fill the rotor statically with the microsomal suspension, then introduce a small amount of 60% sucrose cushion dynamically so as to trap the microsomes at the interface and obviate the inconvenience of scraping the particles off the wall of the rotor. This would be particularly convenient using an edge-unloading BXXIX-type rotor.

Tissue-culture Cells

One of the major problems associated with the fractionation of material from tissue-culture cells is that Dounce homogenization produces a broad range of sizes of surface membrane fragments. Unless the surface of the cell is stabilized by the addition of hardening agents such as Zn^{2+} or fluorescein-mercuric acetate (FMA) (Warren, Glick and Nass, 1966) it is difficult to devise suitable fractionation techniques to harvest all of the surface membrane. An alternative means of homogenization, nitrogen cavitation, converts all the plasma membrane to vesicles, and Graham (1972, 1973a, 1973b) has devised a zonal gradient system for separating the plasma-membrane vesicles from those of the endoplasmic reticulum. It should be pointed out, however, that nitrogen cavitation is only suitable for those cells whose nuclear volume is small relative to that of the whole cell. Tissue-culture cells whose nuclei occupy a large part of the total cellular volume are notoriously difficult to rupture by this method without simultaneous breakage of the nuclei themselves. The other major problem regarding nitrogen cavitation is that the plasma membrane and smooth endoplasmic reticulum vesicles which are produced have very similar densities and only slightly different sedimentation rates.

Linear sucrose gradients are quite inadequate to achieve any degree of resolution. Discontinuous gradients were designed empirically to capitalize on the slightly greater sedimentation rate of the endoplasmic reticulum vesicles. A gradient was formed in a BXIV rotor from 100 ml of 5% sucrose, 40 ml of a post-nuclear supernatant fraction from a homogenate of baby hamster kidney (BHK) cells in 0.25 M (8.5%) sucrose, 90 ml of 15% sucrose, 40 ml of 19% sucrose, 100 ml of 22.5% sucrose, 200 ml of 35% sucrose, the remainder being 60% sucrose (all solutions were buffered with 5 mM tris-HCl, pH 7.4 and contained 1 mM EDTA).

Centrifugation at 40 000 rev/min for 90 min produced a slight separation of the Na^+/K^+-ATPase and NADH diaphorase (oxidase) activities around 22% sucrose. Substituting 50 ml of 22.5% dextran (mol. wt 40 000) for the 100 ml of 22.5% sucrose successfully resolved the plasma membrane and endoplasmic reticulum vesicles (*Figure 4.12*) while having very little effect on the mitochondria, which sediment through the 35% sucrose plateau (Graham, 1973a). The dextran provides a region of high viscosity and low osmolarity (relative to sucrose); this appears to retard the movement of the plasma membrane vesicles while having little effect on those of the smooth endoplasmic reticulum. The low osmolarity might also be important in modifying the sedimentation rate of osmotically-active vesicles.

This gradient system was designed for BHK cells and it must not be applied blindly to any cell grown in tissue culture. Like all fractionation techniques, it requires modification before it can be applied successfully to material other than that for which the technique was designed. For example, for a line of hamster-embryo fibroblasts (Nil 8) it is necessary to increase the speed of centrifugation to 45 000 rev/min to obtain an efficient separation of the microsomes. Although other types of tissue-culture cell have not been thoroughly studied by this method, it is apparent that some virus-transformed derivatives of BHK cells require adjustment of the concentration of the dextran region of the gradient.

This technique illustrates the advantages of gradient solutes other than sucrose and, in particular, the judicious combination of different gradient solutes to achieve the desired resolution. Other compounds, such as metrizamide (Rickwood and Birnie, 1975), give solutions of low viscosity and osmolarity. Since

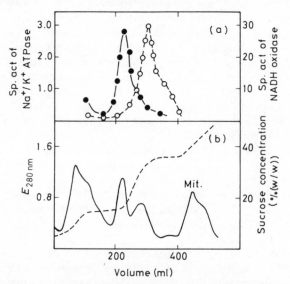

Figure 4.12 Resolution of a post-nuclear supernatant fraction from BHK cells in a BXIV zonal rotor. (a) Na^+/K^+ ATPase, ●——●; NADH oxidase, ○----○; (b) density gradient, - - - - ; E_{280nm}, ——; Mit., mitochondria; for centrifugation details see text. (From Graham, 1973a)

solutions of sucrose combine high osmolarity with high viscosity, those of dextran (or Ficoll) combine low osmolarity (at concentrations less than 25%) with high viscosity and those of metrizamide combine low osmolarity with low viscosity, gradients composed of mixtures of these solutes can supply a whole range of sedimentation conditions for the separation of particles based on sedimentation rate, or isopycnic density, or both.

Ribosomes and Polysomes

Birnie, Fox and Harvey (1972) have fractionated post-mitochondrial supernatant fractions of mouse-embryo extracts in a variety of rotors. *Figure 4.13* shows the resolution of soluble protein (A), ribosomes (B) and polysomes of 2–5 ribosomes (C) in a 10–50% linear sucrose gradient in a BIV rotor, after centrifugation at 40 000 rev/min for 90 min. All solutions contained 5 mM Mg^{2+}, 10 mM K^+ and 10 mM tris-HCl, pH 7.5. So long as the size of the sample volume relative to that of the gradient and the position of the sample relative to the centrifugal field remained constant, similar resolution was obtained in BIV, BXIV, BXV and BXXIX rotors (Birnie, Fox and Harvey, 1972). Membraneous material is largely eliminated in the pre-zonal centrifugation if a post-mitochondrial fraction is used; any residual material sediments beyond the polysomes (band D in *Figure 4.13*). The material for the separation shown in *Figure 4.13* had been stored at −20 °C. Polysomes appear to be stable under these conditions but the membranes probably aggregate and may therefore be separated more readily than are unaggregated membranes from unfrozen material.

To fractionate polysomes (on the basis of the number of ribosomes) and sub-ribosomal particles (on the basis of the difference in their size), rate-zonal centrifugation in shallow sucrose gradients is often used. The centrifugation times required for these separations are usually 2–4 h at high speed (at least 30 000 rev/min). Norman (1971) and Birnie, Fox and Harvey (1972) noted that the resolution in a zonal rotor was much better than in a swing-out rotor. It is also

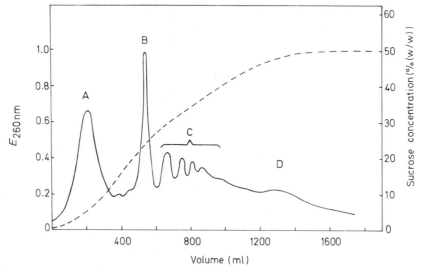

Figure 4.13 Fractionation of a post-mitochondrial extract of mouse embryos in a BIV zonal rotor. Density gradient, - - -; E_{260nm}, ——; for centrifugation details see text. (From Birnie et al., 1972)

common practice to use a small sample volume to optimize resolution and a large overlay (say, 200 ml) to expose the sample to the high centrifugal field required (Birnie, Fox and Harvey, 1972). *Table 4.5* gives the details of some zonal systems designed to separate polysomes, ribosomes and ribosomal subunits from various sources; such gradients normally contain K^+ and Mg^{2+} at low concentrations to stabilize the particles.

Eikenberry *et al.* (1970) devised a hyperbolic sucrose gradient for the BXV rotor, extending from 7.4% at 808 ml (6.35 cm from the centre) to 38% at 1600 ml (8.66 cm from the centre); a large overlay (708 ml) was used for reasons given previously. An inverse sample gradient was generated using 50 ml of the ribosome suspension and 50 ml of 7.4% sucrose in a gradient-forming machine. Thus, a continuous gradient of sucrose was formed from the overlay to the cushion with the sample in the first 100 ml. Centrifugation at 31 000 rev/min for 10 h completely separated the 30 S and 50 S subunits. A convex 15–30–40% sucrose gradient in a BXIV rotor centrifuged at 40 000 rev/min for 16 h has been used by Williamson (1973) to separate ribosomal subunits, ribonucleoprotein particles and tRNA from chick-lens polysomes.

Table 4.5 SEPARATION OF RIBOSOMES, POLYSOMES AND RIBOSOMAL SUBUNITS IN SHALLOW SUCROSE GRADIENTS

Source	Sucrose gradient %	Centrifugation rev/min	time (h)	Rotor	Products	Ref.
Rat liver	15–35	40 000 ($\omega^2 t = 6 \times 10^{10}$ rad²/s)	1	BIV	Ribosomal subunits and post-ribosomal particles	Kedes et al. (1966); see also Parish et al. (1966)
Rat liver	15–35	30 000	3	BXIV	Ribosomal subunits	Norman (1971)
Rat liver and hepatoma	10–25	{47 000, 21 000}	{2.5, 16}	{BXIV, BXV}	Ribosomal subunits	Hinton and Mullock (1971)
Tetrahymena	10–30	40 000	2.5	BXV	Ribosomes and polysomes	Cameron et al. (1966)
Mouse embryo	10–20	30 000	4	BXIV	Ribosomal subunits	Birnie et al. (1972)
Reticulocytes	15–30	40 000	4	BIV	Ribosomal subunits	Birnie et al. (1972); see also Klucis and Gould (1966), Kedes et al. (1966)
Escherichia coli	5–22	40 000	4	BXIV	Ribosomal subunits	Jonák and Mach (1971)

Nucleic Acids

After centrifugation of a sample containing DNA, which had been loaded on to
a sucrose gradient in a *spinning* BXIV rotor, the peak of DNA showed an obvious
asymmetry towards the low-molecular-weight side (Zadrazil, 1973). The skew of
the peak was entirely eliminated when the BXIV rotor was used in a reorienting mode, with the sample loaded while the rotor was at rest. Zadrazil (1973)
concluded that the DNA had been sheared at the interface between the static
and rotating seals when the sample was introduced into the spinning rotor.
Klucis and Lett (1970) also used the BXXV–Ti rotor in a reorienting mode,
loading the rotor at rest with 1430 ml of a 10–30% alkaline sucrose gradient
plus a 30% sucrose cushion. An interesting sophistication was added; on top of
the gradient were layered in sequence: 40 ml of lysing medium (an alkaline
saline–EDTA medium), 20 ml of saline, 10 ml of cell suspension and 80 ml of
overlay. During a 'storage period' prior to acceleration, the cells sedimented
slowly into the lysing medium. The rotor was then accelerated to 450 rev/min
over a period of 30 min, longer than is normally taken for reorientation because
of the susceptibility of DNA to hydrodynamic shear. This system has been used
to fractionate DNA from phage T4 and from mammalian cells.

   In instances in which the DNA has already been sheared by sonication or treatment with deoxyribonuclease (Birnie *et al.*, 1973a) to produce small, denatured
fragments, the problem of shearing at the rotating seal surface is not important.
Similarly, Williamson (1973) encountered no problems when isolating various
types of RNA from sea-urchin eggs using a 10–30% sucrose gradient in a BXIV
rotor centrifuged at 42 000 rev/min for 16.5 h. A similar system with a 10–20%
sucrose gradient containing 50 mM K^+ and 1.5 mM Mg^{2+} has been used to purify
small quantities of RNA; although small amounts of RNA can be purified in a
swing-out rotor, Williamson (1973) considered that a zonal system gave improved
resolution of 7 S, 9 S and 12 S RNA from mouse reticulocyte RNA.

Cell Surface Coat (Glycocalyx)

The cell surface coat is composed chiefly of carbohydrate and protein with little
or no lipid and, consequently, it is denser than any other cellular component
except the nucleus. Its isolation from tissue-culture cells depends upon the use
of nitrogen cavitation to rupture the cell membrane. Although a relatively minor
component of the cell, it can be isolated in high yield by isopycnic centrifugation
in a zonal rotor (Graham and Hynes, 1975). The density of a post-nuclear supernatant fraction from a homogenate of hamster-embryo fibroblasts was made 1 mM
and 5 mM with respect to EDTA and tris–HCl (pH 7.4), respectively, and
adjusted to 47.5% sucrose and a total volume of about 450 ml. A BXIV rotor
was loaded sequentially with 25 ml of 20% sucrose, 80 ml of 40% sucrose, the
sample and enough 60% sucrose to fill the rotor. Centrifugation at 30 000 rev/
min for at least 4 h resulted in the banding of the glycocalyx within the sharp
gradient formed across the interface between the 47.5% and 60% sucrose, whereas
the membraneous material moved centripetally. The large volume is required to
minimize any aggregation of the glycocalyx with other cellular material.

Plasma Lipoproteins

The classical approach to the isolation of plasma lipoproteins is by flotation from serum whose density is increased by the addition of a mixture of NaCl and KBr. Wilcox and Heimberg (1970) employed a NaCl–KBr gradient ranging in density from 1.0 to 1.4 g/cm^3 (together with a water overlay) in BXIV and BXV zonal rotors. The serum sample (50 ml for the BXIV, 150 ml for the BXV), density 1.4 g/cm^3, was introduced at the edge of the rotor which was then spun at 37 000 rev/min for 1 h (BXIV) or 27 000 rev/min for 2 h (BXV). Good separations of the very low-density lipoprotein (VLDL) and low-density lipoprotein (LDL) from the high-density lipoprotein (HDL) were obtained, but HDL usually overlapped the plasma proteins. Even if the VLDL and LDL were removed, fresh gradient introduced at the centre in the BXV rotor, and the material centrifuged for a further 48 h, the HDL and plasma proteins still overlapped. Wilcox and Heimberg (1970) suggested that the higher speeds obtainable with the Ti14 and Ti15 rotors (172 000 and 121 000g_{max}, respectively) might resolve the HDL from residual proteins. Moreover, they considered that a sucrose–KBr gradient would be better because of the reduced diffusion in the gradient. Using sucrose–KBr gradients (density 1.0–1.20 or 1.28 g/cm^3), Wilcox and Heimberg (1968) obtained good separations of VLDL, LDL and HDL from 25 ml of serum introduced at the edge of the rotor and centrifuged at 27 000 rev/min for 24 h. There was still overlap of the HDL and plasma proteins, and some dilution did occur as compared with the normal methods of flotation in centrifuge tubes, but the authors considered that the benefit of a one-step separation of the three classes of lipoprotein outweighed these problems.

Sucrose–NaBr gradients have been used by Mallinson and Hinton (1973); the two-part gradient consisted of (*a*) a linear sucrose gradient of density 1.0–1.15 g/cm^3 (300 ml), and (*b*) a linear NaBr gradient of density 1.2–1.4 g/cm^3. The first was designed to introduce a viscosity element to delay the flotation of the chylomicrons and VLDL and thus separate them. A BXIV rotor was used at 47 000 rev/min for 45 min.

Although BXIV- and BXV-type rotors are designed for loading of the sample at the core, they do appear to be adequate for edge loading in these cases. It would be interesting to compare the BXXIX rotor with the BXIV for these separations; the generally poor resolution of HDL from the plasma proteins may be due, at least partly, to the non-ideality of the BXIV and BXV rotors for loading samples *via* the edge.

Viruses

Most viruses become inactivated to some extent when they are sedimented into a pellet, although whether this is due to aggregation of the material or to the drastic shearing techniques which are frequently required to disaggregate the virus pellet, or to both, is not clear. Whatever the cause, in view of the ease with which batch-type zonal rotors can be used for harvesting virus it is surprising that this technique is not practised more widely.

Harvesting virus has two objectives, namely, concentration and purification. If the starting material is complete culture fluid, often two centrifugation steps are required; the first is a concentration and partial purification and the second is predominantly a purification. After removal of large contaminants such as cells, nuclei and cellular debris by an initial low-speed centrifugation, sucrose is mixed with the virus-containing fluid. This suspension can then be introduced into the rotor as part of a discontinuous gradient, or used as one component in the generation of a linear gradient, or simply introduced between an overlay and a cushion.

If a region of clean (not virus-containing) sucrose solution can be included between the leading edge of the sample zone and the position at which the virus bands in the gradient, this will result in both concentration and purification. When the size of sample precludes this, and the virus bands at, or close to, the interface between sample and cushion, then purification is sacrificed. To complete the purification, the band containing the virus particles is harvested and recentrifuged in a second zonal gradient. This may be done conventionally, that is, by loading at the centre of the rotor on top of a linear or discontinuous sucrose gradient or, for larger volumes, the virus may again be made part of the gradient. However, to achieve a satisfactory purification the sample must be restricted to the lower density part of the gradient so that the virus can band well away from soluble protein remaining in this region. Thus, without resorting to pelleting, the virus can be purified and concentrated. Frequently, this purification step is used after harvesting of virus from large (>10 litres) quantities of virus-containing fluid by continuous-flow zonal centrifugation (see p. 104). *Table 4.6* summarizes the zonal centrifugation schedules for the isolation of a number of viruses, some of which will now be considered in detail.

Fox *et al.* (1968) were among the first to indicate the usefulness of using batch-type rotors for harvesting viruses. Batches of culture fluid containing either polyoma or Semliki Forest virus (SFV) were used as the solvent for sucrose (20–50% in 5% steps). The BXIV rotor was loaded with 20 ml of buffer, 120 ml of culture fluid, 60 ml of each of the virus-containing sucrose solutions and 100 ml of 67% sucrose. After centrifugation at 29 000 rev/min for 18 h, 90% of the virus had banded in 70 ml (SFV) or 100 ml (polyoma virus) (*Figure 4.14a*). For the second purification and concentration step of polyoma virus, sucrose was added to the virus-containing fractions to make 50 ml each of 20, 25, 30, 35 and 40%. The rotor was reloaded with 200 ml of buffer, the five virus-containing solutions and 200 ml of 66% sucrose; after centrifugation the virus was now contained in 25–30 ml (*Figure 4.14b*). To increase the purification of SFV, Fox *et al.* (1968) included a short virus-free linear sucrose gradient (38–50%) between the discontinuous virus-containing gradient and the position at which the virus banded (see *Table 4.6*). With the larger BXV rotor it is possible to process up to 1.3 l of tissue-culture fluid in this way. Although Fox *et al.* (1968) used five or six different sucrose steps in these experiments, fewer steps do not affect the resolution in any way. Furthermore, since the sedimentation coefficients of many viruses are in the range 10^2–10^3 S, the long centrifugation times used by Fox *et al.* (1968) are not necessary.

To demonstrate the capacity of this technique, a BXIV rotor was loaded with the following: 100 ml of buffer, 100 ml of tissue-culture fluid containing Sendai virus, 200 ml of 20% sucrose (containing virus), 100 ml of 35% sucrose (containing virus) and about 150 ml of 55% sucrose (Kohn, 1975). After centrifugation at 45 000 rev/min for 3 h, the virus banded sharply in the steep gradient formed across the interface between the 35% and 55% sucrose steps. Sendai virus from this, and a second identical run, was harvested and diluted to give

Table 4.6. HARVESTING OF VIRUS IN BATCH-TYPE ROTORS

Virus	Original volume of virus suspension ml	First centrifugation			Second centrifugation			Rotor	Ref.
		sucrose gradient % w/w	rev/min ($\times 10^{-3}$)	time h	sucrose gradient % w/w	rev/min ($\times 10^{-3}$)	time h		
Polyoma	440	0^V + disc.* $20-50^V$ + 67	29	18	Disc.* $20-40^V$ + 66	29	18	BXIV	Fox et al. (1968)
Semliki Forest	500 (× 2)	0^V + disc.* $20-50^V$ + 67	29	17	Disc.* $20-32^V$ + linear 38–50 + 55	29	16	BXIV	Fox et al. (1968)
Sendai	350	0^V + 20^V + 35^V + 55	45	3	20^V + 30^V + 55	40	15	BXIV	Kohn (1975)
Influenza	1000	0^V + 66	25	2	–	–	–	BXXIX	Anderson et al. (1969b)
Rous sarcoma	130	0^V + 10^V + 20 + 60	45	0.75	–	–	–	BXIV	J.M. Graham (unpublished data; see text)

*Discontinuous, i.e. stepped, gradient (see text).

^{V} Virus-containing fluid or sucrose solutions prepared from virus-containing medium.

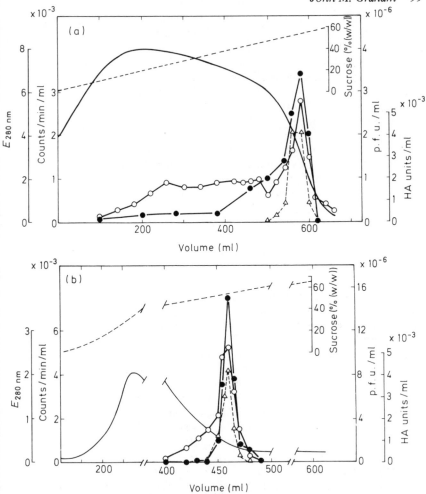

Figure 4.14 Isolation and purification of [³H] thymidine-labelled polyoma virus from 440 ml of tissue-culture fluid: (a) isolation in first gradient; (b) purification in second gradient. Density gradient, - - -; E_{280nm}, —; haemagglutination titre, △---△; infectivity titre, ●—●; radioactivity, ○—○; for centrifugation details see text. (From Fox et al., 1968)

200 ml of virus in 20% sucrose and 250 ml of virus in 30% sucrose. These solutions were introduced into a BXIV rotor together with 50 ml of buffer and approximately 150 ml of 55% sucrose, and centrifuged at 40 000 rev/min for 15 h; the virus banded at 42% sucrose, away from any contaminating material. The virus, originally in 700 ml of tissue-culture fluid, was concentrated to the extent that more than 90% was contained within 50 ml of gradient (Kohn, 1975). Since the gradient diffuses with time, it would be advantageous to centrifuge for shorter periods so as to maintain a sharp gradient in the region at which the virus bands, thus minimizing the volume occupied by the virus. However, if the virus-containing fluid is contaminated with microsomal membranes the centrifugation schedule must allow these membranes to band isopycnically separately from the virus. Reimer, Baker and Newlin (1966a) observed that the microsomal membranes from a crude sample of influenza virus still contaminated the

virus band after centrifugation at 25 000 rev/min for 30 min in a 12–34% sucrose gradient; the contaminant was only separated from the virus by centrifugation at 40 000 rev/min for 4 h in a 34–50% sucrose gradient.

There have been reports (Anderson, Nunley and Rankin, 1969a) of using edge-unloading BXXIX and BXXX rotors for harvesting virus. These rotors would seem to be well suited for concentrating virus particles in short, steep gradients near the wall of the rotor, since there is no need to use large volumes of dense sucrose to unload the rotor. Batch-type rotors can also be used conventionally to purify and concentrate virus in one simple step from a relatively small sample placed on top of a gradient. If a BXIV rotor is loaded with 100 ml of saline, 55 ml of Rous sarcoma virus (RSV) in saline, 78 ml of RSV in 10% sucrose, 148 ml of 20% sucrose and a 60% sucrose cushion, and spun at 45 000 rev/min for 45 min, the virus moves through the 20% sucrose layer to band sharply at about 38% sucrose in 10 ml (J.M. Graham, unpublished data).

It is possible to fractionate virus particles in zonal gradients. Murakami *et al.* (1968) separated polyoma virus into infective virions and empty capsids in a 15–30% sucrose gradient in a BIV rotor spun at 40 000 rev/min for 1 h; 86% of the total infectivity was recovered in the virion fraction. Herpes virus particles from Lucké Frog Kidney Tumour can be separated in a linear 10–60% sucrose gradient (in a BXV rotor spun at 23 000 rev/min for 1 h) into non-enveloped nucleated particles (48–55% sucrose), empty capsids (33–40% sucrose) and enveloped nucleated virus (39–43% sucrose); any microsomal membranes in the fraction overlap the empty capsids (Toplin *et al.*, 1971).

To purify Herpes virus, Pertoft (1970) devised an unusual technique which obviates the possibility of inactivating the virus by exposing it to solutions of sucrose or carbohydrate polymers. A stock gradient solution of 9.4% (w/v) colloidal silica gel (Ludox HS), pH 7.5, containing 5% (w/v) polyethylene glycol, tris–HCl, pH 7.5, 0.1% calf serum and Hanks' Salt Solution (1:20 dilution) was prepared; 800 ml of virus-containing fluid mixed with 600 ml of the stock gradient solution was placed in the rotor while it was at rest, then 350 ml of the stock gradient solution was fed to the edge of the rotor after it had been accelerated to 3000 rev/min. After centrifugation at 30 000 rev/min for 30 min, the virus had moved out of the sample zone to band sharply within the colloidal silica medium. The capacity of the rotor was used to good effect and, when the virus from two such separations was pooled and recentrifuged through a third colloidal silica barrier, a 66-fold purification over the starting material was obtained (Pertoft, 1970).

CONTINUOUS-FLOW ZONAL CENTRIFUGATION

Design and Operation of Rotors

The basic design requirements for effective continuous-flow operation of zonal rotors are best illustrated by reference first to the early continuous-flow rotors developed at Oak Ridge by Anderson and his colleagues; the various modern equivalents which incorporate important modifications will then be discussed.

The primary consideration is that the configuration of the rotor must allow sample to pass continuously over the core while the rotor is spinning at high speed. Thus, (*a*) the fluid-lines into a continuous-flow rotor differ from those into a batch-type rotor, and (*b*) the rotating fluid-seal must remain in place during the entire operation. Because of the latter requirement, the natural ante-cedent for continuous-flow rotors was the BIV rotor, which already possessed a high-speed rotating seal. Adaptation of the BIV to continuous-flow operation produced the BVIII, BIX and BXVI rotors. *Figure 4.15* illustrates diagrammati-cally a BIX-type rotor. The obvious difference (cf. *Figure 4.2*) is that there are

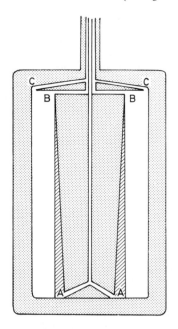

Figure 4.15 Schematic diagram of a BIX-type continuous-flow zonal rotor; the shaded area represents the taper volume. (From Birnie, 1969)

two separate lines to the core of the rotor; one exits at the bottom of the tapered core at A while the other, which exits at the top of the core at B, ultim-ately forms a common path with the line which leads to the edge of the rotor at C.

Operation of Continuous-flow Rotors

Gradient is introduced into the rotor in the conventional manner by pumping it (low-density end first) through the annular line to C, eventually to emerge at the centre through A, while the rotor is spinning at about 3000 rev/min. The rotor is then accelerated to the running speed during which time water or buffer is allowed to flow through the core line to A; the water or buffer exits at B after passing upwards over the surface of the core. During this process the lightest portion of the gradient contained within the core-taper volume (shaded area in *Figure 4.15*) is removed by the flow across the surface of the core. At the running speed, virus-containing fluid is pumped into the rotor at A and, as it passes over the surface of the core, the virus particles sediment out of it into the adjacent gradient. After all the virus-containing solution has been pumped through the rotor, buffer is passed over the core surface to remove any residual virus from the lines and the rotor is allowed to spin for a sufficient length of time to achieve either pelleting or banding of all the virus. The rotor is then decelerated to the unloading speed; dense sucrose is passed to the wall (C) of the rotor and the effluent emerges from the centre of the rotor *via* A.

To expose the stream of virus particles to a sufficiently high gravitational force to sediment them out of the stream across the surface of the core requires the core to have a large diameter and thus occupy a significant proportion of the total volume of the rotor. Consequently, the radial path length for the sedimenting material may be of the order of only 1–2 cm. Conditions must be designed such that the maximum possible removal ('clean-out') of virus particles from the input tissue-culture fluid is obtained. For a particular virus there are two basic parameters which affect the clean-out factor: (*i*) the gravitational field experienced by the sample flowing over the core, which is proportional to the speed of the rotor and the radius of the core; and (*ii*) the time spent by the virus-containing fluid within the rotor, which is proportional to the length of the flow path across the core and the reciprocal of the sample flow rate. Continuous-flow rotors of the BIX type, which are shaped like the BIV, have longer flow paths than do those of the squat BXV type. The latter therefore require a significantly larger diameter of core to achieve tolerable clean-out factors.

Operational Problems

All the problems associated with the running and maintenance of batch-type zonal rotors (see pp. 68–72) apply equally to continuous-flow rotors. There are, however, a few operational problems specific to these rotors. It is absolutely essential to ensure that no air bubbles whatsoever enter the rotor during the passage of virus-containing fluid over the core of the rotor. With either the BVIII or BIX rotor, such air bubbles can block the flow of fluid into the rotor (D.R. Harvey, personal communication). In modern forms of this type of rotor (SZ14 and JCF-Z) it is important to maintain a high enough centrifugal field to sediment all the particles in the sample to prevent removal of smaller particles from the rotor by the fluid flow. This process is called elutriation and is discussed further on pp. 107–109. Like all rotors whose fluid seals remain in place during the entire operation, the line to the centre of a continuous-flow rotor must be open to a reservoir of liquid which can be taken up by the rotor during acceleration. It is also advantageous to fill a continuous-flow rotor with buffer while it is stationary and before loading it with gradient at 3000 rev/min. This prevents entrapment of air at the bottom of the core within the volume of the taper (see *Figure 4.15*).

Harvey (1970) observed that small amounts of the displacing solution leaked through the fluid line leading to the top of the core at B during unloading of the BVIII and BIX rotors. This is potentially quite serious, since the gradient will be disturbed by the high-density sucrose solution passing through it. Chervenka and Cherry (1968) used a technique of introducing a small column of air just ahead of the unloading sucrose solution to overcome this problem. The bubble of air will tend to remain at the core of the rotor, effectively sealing line B; Beckman indeed recommend this procedure for unloading the modern CF-32 and JCF-Z rotors. In practice, however, Harvey (1970) found this to be a difficult technique and some rotors, for example, the BXVI (Cline, Nunley and Anderson, 1966) and the BXX (Birnie *et al.*, 1973b) have some sort of centrifugal valve incorporated (either an 'O'-ring which expands at speeds in excess of the loading speed or a spring-controlled valve, whose tension is overcome by high centrifugal forces) such that the line to the wall is sealed at high rotational speeds and the line to the top of the core is sealed at low speeds. An alternative approach with continuous-flow rotors which do not incorporate such systems is to use them in a reorienting mode (Harvey, 1970). The rotor (for example, the BVIII) can be loaded with overlay, gradient and cushion statically, accelerated for use as a normal continuous-flow rotor, then decelerated and unloaded statically.

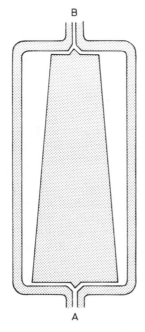

Figure 4.16 Schematic diagram of a KH-type continuous-flow zonal rotor. (From Anderson et al., 1969b)

Clearly, problems of cross-leakage in the fluid seals of continuous-flow rotors such as the BVIII, BIX, BXVI and BXX are more serious than in the batch-type rotors. To overcome these problems for large-scale virus harvesting, the JI, KII and RKII rotors were developed (Cline, Anderson and Fennell, 1971; Cline and Dagg, 1973; Anderson *et al.*, 1969b). These rotors are driven by an air turbine, and no sophisticated fluid-seal assembly is required. They can be loaded and unloaded either statically or dynamically, although Anderson *et al.* (1969b) considered the former to produce the better resolution. The rotor is filled at rest from A (*Figure 4.16*) and accelerated to the running speed, buffer being introduced at B after reorientation has occurred. As with all continuous-flow rotors, the diameter of the core increases in the direction in which the sample flows over the surface of the core, and the light part of the gradient within the tapered volume is displaced by this flow. The sample is then pumped into the rotor, the virus is allowed to band (or pellet) and the rotor is decelerated at a controlled rate and unloaded statically. The total volume of fluid which can be processed is dictated by the slow continuous wash-out of the gradient (Anderson *et al.*, 1969b). The volume of liquid in the JI rotor is 780 ml, that in the RKII, 1800 ml and that in the KII, 3600 ml.

The tall continuous-flow rotors of the BXVI and KII variety require specially constructed centrifuges to accommodate them. On the other hand, the Beckman JCF-Z and CF-32 rotors and the MSE BXX rotor, which are all squat rotors of the BXV type, can be used in standard centrifuges modified to incorporate a fluid-seal and bearing mounted on the lid. All these rotors have large cores so that the distances between the surface of the core and the wall of the rotor is approximately 1 cm. Generally, therefore, these rotors cannot be used to effect purification of viruses from contaminants other than soluble protein, and they are designed for sample volumes of 10–20 l, as compared to the 50–60 l of the BXVI and the 100 l of the JI and KII rotors.

Fractionation with B-type Rotors*

Some of the earliest continuous-flow separations were done with the BV rotor. This rotor (not available commercially) was used for continuous flow without banding, that is, there was no provision for removal of the contents from the centre by the normal unloading process. Separations made with this rotor are briefly discussed, because the modern continuous-flow-with-banding rotors can be used in the same mode. Reimer *et al.* (1966a, 1966b) used this rotor to pellet polio virus; at 40 000 rev/min they obtained a clean-out of 100% at a flow rate of 1 l/h and 98% at 2–3 l/h. Reimer *et al.* (1966b) finally resuspended the pelleted material in the medium remaining within the rotor by repeated acceleration and deceleration of the rotor. Rhino viruses have also been purified by sedimenting in a BV rotor using flow rates of 4 l/h; at 40 000 rev/min a clean-out of 95% was recorded (Gerin *et al.*, 1968).

Using either a two-step gradient (density 1.1 and 1.4 g/cm^3) or a linear gradient (density 1.1–1.5 g/cm^3) of potassium citrate, Gerin *et al.* (1968) obtained a clean-out of rhino virus of about 95% with a flow rate of 1.3 l/h, but this decreased rapidly above 4 l/h. The material was harvested after a 1–2 h banding spin and the rhino virus further purified in a BIV rotor using a 10–55% sucrose gradient. Respiratory syncytial virus was harvested by Cline *et al.* (1967) using a 10–55% sucrose gradient in a BIX rotor; flow rates of 4.15 l/h at 40 000 rev/min produced a clean-out of 99.9%. The effluent contained some small non-infectious particulate material which had not sedimented out of the sample under the conditions used. This demonstrates the importance of choosing the maximum flow rate which will permit adequate recovery of virus, since slower flow rates can lead to increased retention within the rotor of contaminants smaller than the virus particles.

Fox *et al.* (1967) used the BXVI rotor to isolate polyoma virus and Semliki Forest virus (SFV) from large volumes of tissue-culture fluid. *Figure 4.17* shows the separation of SFV from 4.6 l of tissue-culture fluid, pumped through the rotor at 1.4 l/h. More than 90% of the harvested virus was contained in 100 ml

*In the following examples, gradient solutions are described in terms of % (w/w) unless stated otherwise.

of gradient after a banding spin of 2 h at 40 000 rev/min. It is clear that considerable purification from soluble protein and light contaminants (incomplete virus particles) has been achieved and the harvested material has been concentrated about 50-fold. With viruses which are more rapidly sedimenting than either polyoma or SFV (for example, influenza virus), the flow rate can be increased to about 4 l/h without sacrificing clean-out of the virus (G.D.Birnie and D.R. Harvey, personal communication). Reimer *et al.* (1967) increased this rate to 8 litres/h and processed 105 litres of fluid containing influenza virus. After the continuous-flow run, the material was further purified in a BIV rotor to achieve an overall concentration of 618-fold. Reimer *et al.* (1967) emphasized

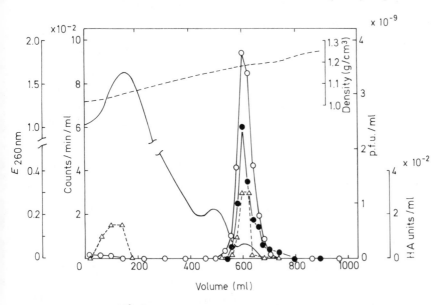

Figure 4.17 Isolation of [³H] *thymidine-labelled Semliki Forest virus from 4.6 l of tissue-culture fluid by centrifugation in a BXVI continuous-flow zonal rotor. Density gradient,* ---; E_{260nm}, —; *haemagglutination titre,* △·-·△; *infectivity titre,* ●—●; *radioactivity,* ○—○; *for centrifugation details see text. (From Fox* et al., *1967)*

that it is essential to remove the larger contaminating cell debris prior to any zonal centrifugation when using such large amounts of starting material. If this is not done, the debris overloads the gradient and blocks the narrow channels within the rotor. Apart from this, the volume which can be processed is largely dictated by the time available. However, it should also be borne in mind that, if a contaminant of the virus has a similar density but is more slowly sedimenting, longer periods of centrifugation will result in a progressive loss of purity of the virus. By using a combination of continuous-flow rotor harvesting and subsequent purification and concentration in a batch-type rotor, Fox *et al.* (1967) were able to concentrate Sindbis virus from a starting volume of 6.5 l down to 35 ml.

Other viruses which have been harvested in the BXVI rotor include infectious canine hepatitis (ICH) and canine distemper virus (CDV); Elliot and Ryan (1970) used a 5–50% sucrose gradient and, instead of allowing buffer to flow over the surface of the core during acceleration from the gradient-loading speed to the separation speed, they pumped through the virus solution at 1–2 ml/min. At

40 000 rev/min, Elliot and Ryan (1970) obtained a clean-out of about 99% for both viruses at a flow rate of 3.2 l/h; at 4.8 l/h this decreased to 95% for ICH and 78% for CDV. A back-pressure of 2 lb/in^2 maintained against the rotating seal while the sample was pumped through the rotor minimized any cross-leakage from influent to effluent lines. CDV was found to band very broadly from 32% to 48% sucrose after a banding spin of 1 h (Elliot and Ryan, 1970). This may reflect the presence of particles with a wide range of sedimentation coefficients which would account for the relatively rapid drop in clean-out of this virus as the flow rate was increased.

The BXX, JCF-Z and CF-32 zonal rotors have only recently become available commercially and data on fractionations with these rotors are, as yet, rather sparse. During evaluation of the BXX rotor, Birnie *et al.* (1973b) studied its efficacy in the recovery of influenza virus and SFV from relatively small volumes of tissue-culture fluid. Using a two-step gradient (150 ml of 30% and 100 ml of 60% sucrose) and 75 ml of buffer, 1 l of influenza virus suspension was pumped through the rotor (which was spun at 25 000 rev/min) at 1500 ml/h; 90% of the virus was recovered in 80 ml of gradient, and the clean-out was also 90%. For the more slowly sedimenting SFV, the sucrose solutions were of lower density (10% and 55%) and clean-out values of 80–90% were obtained only at flow rates of 1 l/h or less at 25 000 rev/min (Birnie *et al.*, 1973b).

Fractionation with J- and K-series Rotors*

Cline (1971) pointed out that, with large rotors whose sedimentation paths were short, particularly those such as the JI and KII rotors whose trapped gradients are exposed to large volumes (100 litres) of a flowing liquid, the large surface area of the core and the short sedimentation path means that the gradient is rapidly smoothed out. Cline (1971) emphasized the importance of using several steep diffusion interfaces within the rotor, with the centre-point density isopycnic with that of the particle to be separated. Although stepped shallow gradients in a KII rotor have allowed separation and fine resolution of insect viruses, the main object of using a rotor of this type is surely concentration, not fractionation. Cline (1971) also suggested that, to minimize back-diffusion of the gradient into the flowing sample fluid, it might be advantageous to increase the density of the sample fluid to near that of the light end of the gradient. However, as shown later, this can have a drastic effect on the clean-out rate.

Reimer *et al.* (1967) investigated the feasibility of using the KII rotor to harvest influenza virus, and worked out some important practical details concerning the use of the rotor. The rotor was loaded at rest with sucrose solution of a density only slightly greater than that of the sample; the rotor was then accelerated to 2000 rev/min and the direction of flow through the rotor reversed several times in order to remove trapped air. The rotor was brought to rest again and a heavy sucrose step (60%) was introduced; the rotor was then accelerated slowly to 2000 rev/min to permit reorientation. Fluid flow over the core was first established with buffer during the acceleration to 20 000 rev/min; flow of the virus-containing fluid was begun during the final acceleration to 27 000 rev/min.

*In the following examples, gradient solutions are described in terms of % (w/w) unless stated otherwise.

Reimer *et al.* (1967) recommended a flow rate during this phase which kept the input line back-pressure below 15 lb/in². At 27 000 rev/min, they found acceptable clean-out at 4–5 l/h, although some 17–29% of the influenza virus was lost in the effluent under these conditions and some 6–16% lost in the gradient outside the main virus band. Gerin and Anderson (1969) also used the KII for purifying influenza virus in a discontinuous 0–55% sucrose gradient. At 27 000 rev/min and a flow rate of 12 l/h, a clean-out of 90% was recorded, dropping to 80% at 30 l/h. A reasonable compromise was 16 l/h and, under these conditions, a 125-fold concentration was achieved. The greatly improved clean-out rates reported by Gerin and Anderson (1969) compared to those of Reimer *et al.* (1967) are probably at least partly due to the lower density of their virus-containing medium which enabled the particles to sediment more rapidly from the sample stream. However, when the influenza virus was suspended in 0.9% NaCl and passed into an RKII rotor at 4.5 l/h over a gradient of 500 ml of 25% sucrose and 500 ml of 54% sucrose, the clean-out at 30 000 rev/min was only 70% (Cline and Dagg, 1973). In another report, the KII rotor was used at 35 000 rev/min for processing 100 l of chorioallantoic fluid containing influenza virus at a flow rate of at least 10 l/h (Anderson *et al.*, 1969b) and almost 100% clean-out was achieved. The higher centrifugal field used was clearly critical, and under these conditions the rotor is considered to be suitable for harvesting a broad range of particles from poliovirus to poxvirus and bacteria.

Cline and Dagg (1973) investigated the harvesting of poliovirus in a JI rotor using discontinuous CsCl gradients stabilized by 50% glycerol, and ranging in density from 1.2 to 1.45 g/cm³. At 50 000 rev/min and a flow rate of 4.5 l/h, the clean-out was found to be over 80%. These workers also reported on the isolation of *Escherichia coli*, *Chlamdia trachomitis*, insect viruses and chloroplasts in this rotor.

The K3–Ti rotor has a capacity of 3.5 l; two-step gradients of 1 l of 20% sucrose plus a volume of 55% sucrose dependent on the total sample volume have been used to concentrate and purify murine and feline leukaemia virus, murine, feline and Rous sarcoma virus and Rous-associated virus (Toplin and Sottong, 1972). Sample volumes of 20–30 l required 500–800 ml of 55% sucrose and volumes of 50–80 l required 1000–1400 ml of 55% sucrose in order to separate the virus from lighter contaminants. Centrifugation speed was 35 000 rev/min, the flow rate of the virus-containing medium was 9–11 l/h and a 1 h banding time was allowed after all the medium had been pumped through the rotor. Toplin and Sottong (1972) have also used the K10 rotor (capacity 8 l) for batch-type recovery of virus from 2–7 l of fluid. The rotor was loaded at rest with the sample, together with a two-step 20% and 55% sucrose gradient.

CENTRIFUGAL ELUTRIATION

This technique was first applied to the separation of large particles such as whole cells, by Lindahl (1962) and subsequently by McEwen, Stallard and Juhos (1968). Particles are separated according to their resultant sedimentation rates in a gravitational field opposed by liquid flow; a diagrammatic representation of the process is shown in *Figure 4.18*. Those particles whose rates of sedimentation under the imposed gravitational field are less than the rate at which the liquid is flowing in the opposite direction will move centripetally in the separation chamber. By increasing the flow rate, or decreasing the centrifugal field, cells of increasing sedimentation rate can be removed centripetally. Various modifications of this rotor have been investigated by Beckman. In the CR-1 rotor, the cells which are flushed out of the separation chamber are concentrated in a settling chamber. This chamber was eliminated in the CR-2 rotor but, although this allows sampling of a larger number of fractions and thus a finer resolution, the samples are very dilute. The method has been used to effect partial separation of different cell types from mixtures of cells. Its chief advantage over simple centrifugation is that less toxic and less viscous media can be used for separating

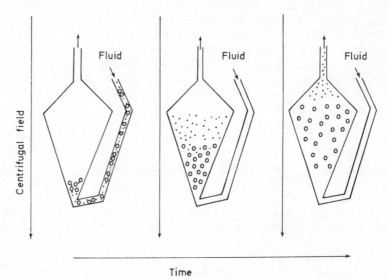

Time

Figure 4.18 Schematic diagram of fractionation of particles by centrifugal elutriation. (From Techniques of Preparative, Zonal and Continuous-flow Ultracentrifugation, Beckman Publication DS-468A)

rapidly sedimenting particles. The biggest problem is that of temperature control; to prevent convection currents and thus mixing in the separation chamber the temperature of the liquid in the chamber and that of the liquid flowing centripetally must be the same, and constant. The maximum speed of the currently available rotor is 7000 rev/min.

Centrifugal elutriation has been used successfully to separate various cell types from mammalian testis. Cells derived from mouse testis ($3 \times 10^8 - 2 \times 10^9$, in 10–17 ml of buffer) were injected into the separation chamber which contained 0.5% bovine serum albumin in phosphate-buffered saline (Grabske *et al.*, 1975a). During the run, which lasted for 2 h, the flow rate of buffer through the chamber was progressively increased from 11.2 to 20 ml/min and the rotor speed reduced from 5000 to 1500 rev/min. Some eleven fractions were obtained and populations of spermatids, spermatozoa and spermatocytes were resolved. Similar separations have been obtained for hamster-testis cells (Grabske *et al.*, 1975a) and rat-testis cells (Grimes *et al.*, 1975). Rat-brain cells have been fractionated in a comparable manner (Flangas, 1974). McEwen, Stallard and Juhos (1968) subjected whole blood (made 3% with respect to dextran) to elutriation, using flow rates of 23–26 ml/min and centrifugation at 3000 rev/min for 22 min; 92% of the cells remaining in the separation chamber were leukocytes, while the erythrocytes were removed by the buffer flow into the settling chamber. More recently, McEwen *et al.* (1971) have isolated leukocytes from malaria-infected monkey blood using 3000 rev/min and a flow rate of 16 ml/min. These workers found that optimal results were obtained with blood samples of less than 10^{10} cells and sample volumes of 1–4 ml. Red cells have also been separated into age groups by centrifugal elutriation (Sanderson, Palmer and Bird, 1975), and Glick *et al.* (1971) were able to isolate mast cells which were 79% pure from rat peritoneal washings (2.3×10^7 cells in 17 ml), using a rotor speed of 2610 rev/min for 25 min and a flow rate of 20 ml/min. The fraction was enriched eightfold over the starting material.

Centrifugal elutriation has been used to synchronize Chinese hamster-ovary cells (Grabske *et al.*, 1975b). These workers isolated 2×10^7 G_1-cells from 2×10^8 exponentially growing cells in 12 min. The enrichment of G_1-cells in the elutriate was at least as good as that obtained by mitotic selection.

ACKNOWLEDGEMENTS

I should like to thank Diane Jackson for her help in preparing the manuscript and Helen Withers Green for her patience and efficiency in its typing. I am grateful to the following for permission to republish diagrams:

Figure 4.1 (J. Wiley and Sons, London)
Figure 4.2 (Beckman)
Figure 4.3 (MSE Instruments Ltd)
Figure 4.4 (*Biophys. J.*)
Figure 4.5 (*Analyt. Biochem.*)
Figure 4.6 (Longmans, London)
Figure 4.7 (*J. Cell Biol.*)
Figure 4.8 (*Biochem. J.*)
Figure 4.9 (*Analyt. Biochem.*)

Figure 4.10 (*Biochem. J.*)
Figure 4.11 (Longmans, London)
Figure 4.12 (Longmans, London)
Figure 4.13 (Butterworths, London)
Figure 4.14 (*J. gen. Virol.*)
Figure 4.15 (*Lab. Equipment Digest*)
Figure 4.16 (*Analyt. Biochem.*)
Figure 4.17 (*J. gen. Virol.*)
Figure 4.18 (Beckman)

REFERENCES

ANDERSON, N.G. (1966). *Natn. Cancer Inst. Monogr.*, No. 21, 9
ANDERSON, N.G. (1967). In *Methods of Biochemical Analysis*, vol. 15, p. 271. Ed. D. Glick. New York; Interscience
ANDERSON, N.G., PRICE, C. A., FISHER, W.D., CANNING, R.E. and BURGER, C.L. (1964). *Analyt. Biochem.*, 7, 1
ANDERSON, N.G., BARRINGER, H.P., BABELAY, E.F., NUNLEY, C.E., BARTKUS, M.J., FISHER, W.D. and RANKIN, C.T. (1966a). *Natn. Cancer Inst. Monogr.*, No. 21, 137
ANDERSON, N.G., HARRIS, W.W., BARBER, A.A., RANKIN, C.T. and CANDLER, E.L. (1966b). *Natn. Cancer Inst. Monogr.*, No. 21, 253
ANDERSON, N.G., RANKIN, C.T., BROWN, D.H., NUNLEY, C.E. and HSU, H.W. (1968). *Analyt. Biochem.*, 26, 415

ANDERSON, N.G., NUNLEY, C.E. and RANKIN, C.T. (1969a). *Analyt. Biochem.*, **31**, 255

ANDERSON, N.G., WATERS, D.A., NUNLEY, C.E., GIBSON, R.F., SCHILLING, R.M., DENNY, E., CLINE, G.B., BABELAY, E.F. and PERARDI, T.E. (1969b). *Analyt. Biochem.*, **32**, 460

ANSELL, G.B. and SPANNER, S. (1972). *Progr. brain Res.*, **36**, 3

BARRINGER, H.P. (1966). *Natn. Cancer Inst. Monogr.*, No. 21, 77

BARRINGER, H.P., ANDERSON, N.G., NUNLEY, C.E., ZIEHLKE, K.T. and DRITT, W.S. (1966). *Natn. Cancer Inst. Monogr.*, No. 21, 165

BIRNIE, G.D. (1969). *Lab. Equip. Dig.*, 7, 59

BIRNIE, G.D. (1973). In *Methodological Developments in Biochemistry*, vol. 3, p.17. Ed. E. Reid. London; Longmans

BIRNIE, G.D., FOX, S.M. and HARVEY, D.R. (1972). In *Subcellular Components: Preparation and Fractionation*, 2nd edn, p. 235. Ed. G.D. Birnie. London; Butterworths

BIRNIE, G.D., HELL, A., SLIMMING, T.K. and PAUL, J. (1973a). In *Methodological Developments in Biochemistry*, vol. 3, p. 127. Ed. E. Reid. London; Longman

BIRNIE, G.D., HARVEY, D.R., GRAHAM, J.M., COOPER, A. and MOLLOY, J. (1973b). In *Methodological Developments in Biochemistry*, vol. 3, p. 51. Ed. E. Reid. London; Longmans

BOONE, C.W., HARELL, G.S. and BOND, H.E. (1968). *J. Cell Biol.*, **36**, 369

BROWN, D.H., CARLTON, E., BYRD, B., HARRELL, B. and HAYES, R.L. (1973). *Archs Biochem. Biophys.*, **155**, 9

BURGE, M.L.E. and HINTON, R.H. (1971). In *Separations with Zonal Rotors*, p. S5.1. Ed. E. Reid. Guildford, UK; Univ. of Surrey Press

CAMERON, I.L., CLINE, G.B., PADILLA, G.M., MILLER, O.L. and VAN DREAL, P.A. (1966). *Natn. Cancer Inst. Monogr.*, No. 21, 361

CARTLEDGE, T.G., BURNETT, J.K., BRIGHTWELL, R. and LLOYD, D. (1969). *Biochem. J.*, **115**, 56P

CHERVENKA, C.H. and CHERRY, K. (1968). *Analyt. Biochem.*, **22**, 65

CLINE, G.B. (1971). In *Separations with Zonal Rotors*, p. P5.3. Ed. E. Reid. Guildford, UK; Univ. of Surrey Press

CLINE, G.B. and DAGG, M.K. (1973). In *Methodological Developments in Biochemistry*, vol. 3, p. 61. Ed. E. Reid. London; Longmans

CLINE, G.B., NUNLEY, C.E. and ANDERSON, N.G. (1966). *Nature, Lond.*, **212**, 487

CLINE, G.B., COATES, H., ANDERSON, N.G., CHANOCK, R.M. and HARRIS, W.W. (1967). *J. Virol.*, **1**, 659

CLINE, G.B., ANDERSON, N.G. and FENNELL, R. (1971). In *Separations with Zonal Rotors*, p. A3.1. Ed. E. Reid. Guildford, UK; Univ. of Surrey Press

CONNOCK, M.J., ELKIN, A. and POVER, W.F.R. (1971). *Histochem. J.*, **3**, 11

CONTRACTOR, S.F. (1973). In *Methodological Developments in Biochemistry*, vol. 3, p. 187. Ed. E. Reid. London; Longmans

CORBETT, J.R. (1967). *Biochem. J.*, **102**, 43P

COTMAN, C., MAHLER, H.R. and ANDERSON, N.G. (1968). *Biochim. biophys. Acta*, **163**, 272

DAY, E.D., McMILLAN, P.N., MICKEY, D.D. and APPEL, S.H. (1971). *Analyt. Biochem.* **39**, 29

DE PIERRE, J.W. and KARNOVSKY, M.L. (1973). *J. Cell Biol.*, **56**, 275

EIKENBERRY, E.F., BICKLE, T.A., TRAUT, R.R. and PRICE, C.A. (1970). *Eur. J. Biochem.*, **12**, 113

EL-AASER, A.A., FITZSIMONS, J.T.R., HINTON, R.H., REID, E., KLUCIS, E. and
ALEXANDER, P. (1966a). *Biochim. biophys. Acta*, **127**, 553
EL-AASER, A.A., REID, E., KLUCIS, E., ALEXANDER, P., LETT, J.T. and SMITH, J.
(1966b). *Natn. Cancer Inst. Monogr.*, No. 21, 323
ELLIOT, J.A. and RYAN, W.L. (1970). *Appl. Microbiol.*, **20**, 667
ELROD, L.H., PATRICK, L.C. and ANDERSON, N.G. (1969). *Analyt. Biochem.*, **30**,
230
EMMELOT, P., BOS, C.J., BENEDETTI, E. and RÜMKE, Ph. (1964). *Biochim. biophys.
Acta*, **90**, 126
EVANS, W.H. (1970). *Biochem. J.*, **166**, 833
EVANS, W.H. (1971). In *Separations with Zonal Rotors*, p. S3.1. Ed. E. Reid.
Guildford, UK; Univ. of Surrey Press
FISHER, W.D. and CLINE, G.B. (1963). *Biochim. biophys. Acta*, **68**, 640
FLANGAS, A.L. (1974). *Prep. Biochem.*, **4**, 165
FLANGAS, A.L. and BOWMAN, R.E. (1968). *Science, N.Y.*, **148**, 1025
FOX, S.M., BIRNIE, G.D., MARTIN, E.M. and SONNABEND, A. (1967). *J. gen. Virol.*,
1, 577
FOX, S.M., BIRNIE, G.D., HARVEY, D.R., MARTIN, E.M. and SONNABEND, J.A.
(1968). *J. gen. Virol.*, **2**, 455
GAVARD, D., de LAMIRANDE, G. and KARASAKI, S. (1974). *Biochim. biophys.
Acta*, **332**, 145
GERIN, J.L. and ANDERSON, N.G. (1969). *Nature, Lond.*, **221**, 1255
GERIN, J.L., RICHTER, W.R., FENTERS, J.D. and HOLPER, J.C. (1968). *J. gen.
Virol.*, **2**, 937
GIORGI, P.P. (1971). *Expl Cell Res.*, **68**, 273
GIORGI, P.P. (1973). In *Methodological Developments in Biochemistry*, vol. 3,
p. 241. Ed. E. Reid. London; Longmans
GLICK, D., VON REDLICH, D., JUHOS, E.Th. and McEWEN, C.R. (1971). *Expl Cell
Res.*, **65**, 23
GRABSKE, R.J., LAKE, S., GLEDHILL, B.L. and MEISTRICH, M.L. (1975a). *J. cell
comp. Physiol.*, **86**, 177
GRABSKE, R.J., LINDL, P.A., THOMPSON, L.H. and GRAY, L.H. (1975b). *J. Cell
Biol.*, **67**, 142a
GRAHAM, J.M. (1972). *Biochem. J.*, **130**, 1113
GRAHAM, J.M. (1973a). In *Methodological Developments in Biochemistry*,
vol. 3, p. 205. Ed. E. Reid. London; Longmans
GRAHAM, J.M. (1973b). MSE Application Information, A7/2/73. Crawley,
Sussex, UK; MSE Instruments Ltd.
GRAHAM, J.M. (1975). In *New Techniques in Biophysics and Cell Biology*, p. 1.
Ed. R.H. Pain and B.J. Smith. London; Wiley
GRAHAM, J.M. and HYNES, R.O. (1975). *Trans. Biochem. Soc.*, **3**, 761
GRAHAM, J.M., SUMNER, M.C.B., CURTIS, D.H. and PASTERNAK, C.A. (1973).
Nature, Lond., **246**, 291
GRIFFITH, O.M. and WRIGHT, H. (1972). *Analyt. Biochem.*, **33**, 469
GRIMES, S.R., PLATZ, R.D., MEISTRICH, M.L. and HNILICA, L.S. (1975). *Biochem.
biophys. Res. Commun.*, **67**, 182
HARTMAN, G.C. and HINTON, R.H. (1971). In *Separations with Zonal Rotors*,
p. S4.1. Ed. E. Reid. Guildford, UK; Univ. of Surrey Press
HARVEY, D.R. (1970). *Analyt. Biochem.*, **33**, 469
HEADON, D.R. and DUGGAN, P.F. (1971). In *Separations with Zonal Rotors*,
p. V2.1. Ed. E. Reid. Guildford, UK; Univ. of Surrey Press

HINTON, R.H. (1972). In *Subcellular Components: Preparation and Fractionation,* 2nd edn, p. 119. Ed. G.D. Birnie. London; Butterworths

HINTON, R.H. and MULLOCK, B.M. (1971). In *Separations with Zonal Rotors,* p. P7.1. Ed. E. Reid. Guildford, UK; Univ. of Surrey Press

HINTON, R.H., KLUCIS, E., EL AASER, A.A., FITZSIMONS, J.T.R., ALEXANDER, P. and REID, E. (1967). *Biochem. J.,* **105,** 14P

HINTON, R.H., DOBROTA, M., FITZSIMONS, J.T.R. and REID, E. (1970). *Eur. J. Biochem.,* **12,** 349

HINTON, R.H., NORRIS, K.A. and REID, E. (1971). In *Separations with Zonal Rotors,* p. S2.1. Ed. E. Reid. Guildford, UK; Univ. of Surrey Press

HSU, H.W. and ANDERSON, N.G. (1969). *Biophys. J.,* **9,** 173

JOHNSTON, I.R. and MATHIAS, A.P. (1972). In *Subcellular Components: Preparatio and Fractionation,* 2nd edn, p. 53. Ed. G.D. Birnie. London; Butterworths

JOHNSTON, I.R., MATHIAS, A.P., PENNINGTON, F. and RIDGE, D. (1968). *Biochem. J.,* **109,** 127

JOHNSTON, I.R., MATHIAS, A.P., PENNINGTON, F. and RIDGE, D. (1969). *Biochim. biophys. Acta,* **195,** 563

JONÁK, J. and MACH, O. (1971). In *Separations with Zonal Rotors,* p. P8.1. Ed. E. Reid. Guildford, UK; Univ. of Surrey Press

KATZ, A.M., REPKE, D.I., UPSHAW, J.E. and POLASCIK, M.A. (1970). *Biochim. biophys. Acta,* **205,** 473

KEDES, L.H., KOEGEL, R.J. and KUFF, E.L. (1966). *J. molec. Biol.,* **22,** 359

KLUCIS, E.S. and GOULD, H.J. (1966). *Science, N.Y.,* **152,** 378

KLUCIS, E.S. and LETT, J.T. (1970). *Analyt. Biochem.,* **35,** 480

KOHN, A. (1975). *J. gen. Virol.,* **29,** 179

KORNGURTH, S.E., FLANGAS, A.L. and SIEGEL, F.L. (1971). *J. biol. Chem.,* **246,** 1177

LEE, T.C., SWARTZENDRUBER, D.C. and SNYDER, F. (1969). *Biochem. biophys. Res. Commun.,* **36,** 748

LIM, R.W., MOLDAY, R.S., HUANG, H.V. and YEN, S.P.S. (1975). *Biochim. biophys. Acta,* **394,** 377

LINDAHL, P.E. (1962). *Nature, Lond.,* **194,** 589

McEWEN, C.R., STALLARD, R.W. and JUHOS, E.Th. (1968). *Analyt. Biochem.,* **23,** 369

McEWEN, C.R., JUHOS, E.Th., STALLARD, R.W., SCHNELL, J.V., SIDDIQUI, W.A. and GEIMAN, Q.M. (1971). *J. Parasit.,* **57,** 887

MAHALEY, M.S., DAY, E.D., ANDERSON, N., WILFONG, R.F. and BRATER, C. (1968). *Cancer Res.,* **28,** 1783

MALLINSON, A. and HINTON, R.H. (1973). In *Methodological Developments in Biochemistry,* vol. 3, p. 113. Ed. E. Reid. London; Longmans

MATHIAS, A.P., RIDGE, D. and TREZONA, N.St.G. (1969). *Biochem. J.,* **111,** 583

MURAKAMI, W.T., FINE, R., HARRINGTON, M.R. and BEN SASSON, Z. (1968). *J. molec. Biol.,* **36,** 153

NELSON, E.B., MASTERS, B.S.S. and PETERSON, J.A. (1971). *Analyt. Biochem.,* **39,** 128

NEVILLE, D.M. (1960). *J. biophys. biochem. Cytol.,* **8,** 413

NOLL, H. (1967). *Nature, Lond.,* **215,** 360

NORMAN, M. (1971). In *Separations with Zonal Rotors,* p. P6.1. Ed. E. Reid. Guildford, UK; Univ. of Surrey Press

PARISH, J.H., KIRBY, K.S. and KLUCIS, E.S. (1966). *J. molec. Biol.,* **22,** 393

PASTERNAK, C.A. and GRAHAM, J.M. (1973). In *The Biology of the Fibroblast*, p. 461. 4th Internat. Sigrid Julius Foundation Symp. Finland, 1972. Ed. E. Kulonen and J. Pikkarainen. London; Academic Press

PASTERNAK, C.A. and WARMSLEY, A.M.H. (1973). In *Methodological Developments in Biochemistry*, vol. 3, p. 249. Ed. E. Reid. London; Longmans

PERTOFT, H. (1970). *Analyt. Biochem.*, **38**, 506

PETERSEN, D.F., TOBEY, R.A. and ANDERSON, E.C. (1969). *Fedn Proc. Fedn Am. Socs exp. Biol.*, **28**, 1771

POLLACK, M.S. and PRICE, C.A. (1971). *Analyt. Biochem.*, **42**, 38

PRICE, M.R., HARRIS, J.R. and BALDWIN, R.W. (1972). *J. Ultrastruct. Res.*, **40**, 178

PROBST, H. and MAISENBACHER, J. (1973). *Expl Cell Res.*, **78**, 335

PROSPERO, T.D. (1973). In *Methodological Developments in Biochemistry*, vol. 3, p. 1. Ed. E. Reid. London; Longmans

PROSPERO, T.D. and HINTON, R.H. (1973). In *Methodological Developments in Biochemistry*, vol. 3, p. 171. Ed. E. Reid. London; Longmans

QUIGLEY, J.W. and COHEN, S.S. (1969). *J. biol. Chem.*, **244**, 2450

REIMER, C.B., BAKER, R.S. and NEWLIN, T.E. (1966a). *Science, N.Y.*, **152**, 1379

REIMER, C.B., NEWLIN, T.E., HAVENS, M.L., BAKER, R.S., ANDERSON, N.G., CLINE, G.B., BARRINGER, H.P. and NUNLEY, C.E. (1966b). *Natn. Cancer Inst. Monogr.*, **21**, 375

REIMER, C.B., BAKER, R.S., VAN FRANK, R.M., NEWLIN, R.E., CLINE, G.B. and ANDERSON, N.G. (1967). *J. Virol.*, **1**, 1207

RICKWOOD, D., and BIRNIE, G.D. (1975). *FEBS Lett.*, **50**, 102

ROBINSON, D., PRICE, R.G. and TAYLOR, D.G. (1973). In *Methodological Developments in Biochemistry*, vol. 3, p. 199. Ed. E. Reid. London; Longmans

RODNIGHT, R., WELLER, M. and GOLDFARB, P.S.G. (1969). *J. Neurochem.*, **16**, 1591

SANDERSON, R.J., PALMER, N.F. and BIRD, K.E. (1975). *Biophys. J.*, **15**, 321a

SCHNEIDER, E.L. and SALZMAN, N.P. (1970). *Science, N.Y.*, **167**, 1141

SCHUEL, H. and ANDERSON, N.G. (1964). *J. Cell Biol.*, **21**, 309

SCHUEL, H., TIPTON, S.R. and ANDERSON, N.G. (1964). *J. Cell Biol.*, **22**, 317

SCHUEL, H., SCHUEL, R. and UNAKAR, N.J. (1968). *Analyt. Biochem.*, **25**, 146

SEBASTIAN, J., CARTER, B.L.A. and HALVORSON, H.O. (1971). *J. Bact.*, **108**, 1045

SHAPIRA, R., BINKLEY, F., KIBLER, R.F. and WUNDRAM, I.J. (1970). *Proc. Soc. exp. Biol. Med.*, **133**, 238

SHEELER, P. and WELLS, J.R. (1969). *Analyt. Biochem.*, **32**, 38

SHEELER, P., GROSS, D.M. and WELLS, J.R. (1971). *Biochim. biophys. Acta*, **237**, 28

SPANNER, S. and ANSELL, G.B. (1970). *Biochem. J.*, **119**, 45P

SPENNEY, J.G., STRYCH, A., PRICE, A.H., HELANDER, H.F. and SACHS, G. (1973). *Biochim. biophys. Acta*, **311**, 545

TAYLOR, D.G. and CRAWFORD, N. (1974). *FEBS Lett.*, **41**, 317

THOMSON, J.F., NANCE, S.L. and TOLLAKSEN, S.L. (1974). *Proc. Soc. exp. Biol. Med.*, **145**, 1174

TOPLIN, I. and SOTTONG, P. (1972). *Appl. Microbiol.*, **23**, 1010

TOPLIN, I., MIZELL, M., SOTTONG, P. and MONROE, J. (1971). *Appl. Microbiol.*, **21**, 132

VAN DER ZEIJST, A.M. and BULT, H. (1972). *Eur. J. Biochem.*, **28**, 463

WALLACH, D.F.H., KRANZ, B., FERBER, E. and FISCHER, H. (1972). *FEBS Lett.*, **21**, 29

WARMSLEY, A.M.H. and PASTERNAK, C.A. (1970). *Biochem. J.*, **119**, 493

WARREN, L., GLICK, M.C. and NASS, M.K. (1966). *J. Cell Physiol.*, **68**, 269

WEAVER, R.A. and BOYLE, W. (1969). *Biochim. biophys. Acta*, **173**, 377

WELLS, J.R. and JAMES, T.W. (1972). *Expl Cell Res.*, **75**, 465

WELLS, J.R., GROSS, D.M. and SHEELER, P. (1970). *Lab. Prac.*, **19**, 497

WILCOX, H.G. and HEIMBERG, M. (1968). *Biochim. biophys. Acta*, **152**, 424

WILCOX, H.G. and HEIMBERG, M. (1970). *J. Lipid Res.*, **11**, 7

WILLIAMSON, R. (1973). In *Methodological Developments in Biochemistry*, vol. 3, p. 135. Ed. E. Reid. London; Longmans

WITHERS, D., DAVIES, I. ab I. and WYNN, C.H. (1968). *Biochem. biophys. Res. Commun.*, **30**, 227

ZADRAZIL, S. (1973). In European Symposium of Zonal Centrifugation in Density Gradient, *Spectra 2000*, vol. 4, p. 192. Paris; Editions Cité Nouvelle

5 Rate-Zonal Centrifugation: Quantitative Aspects

JENS STEENSGAARD
NIELS PETER H. MØLLER
Institute of Medical Biochemistry, The University of Aarhus, Denmark
LARS FUNDING
Department of Clinical Chemistry, Vejle Hospital, Vejle, Denmark

The technique of rate-zonal centrifugation in density and viscosity gradients was primarily developed for preparative purposes (De Duve, 1965; Anderson, 1966). However, the data obtained from such centrifugations also allow the calculation of fairly accurate sedimentation coefficients of the components in the sample. This offers several advantages. First, it provides a method for estimation of sedimentation coefficients of such particles or substances whose physical properties make them unsuitable for an analytical centrifuge, such as subcellular organelles or substances which can be localized only by their biological activity. Secondly, overloading of the gradient, that is, the use of more sample material than can properly be supported by the gradient, lowers the power of resolution drastically in zonal centrifugation experiments. In the case of overloading the calculated sedimentation coefficients will appear too large. Hence, calculation of sedimentation coefficients may reveal whether the power of resolution can be increased by lowering the amount of sample material applied on a given gradient. Thirdly, such calculations may facilitate the choice of optimal experimental conditions. For instance, on the basis of the results of a pilot centrifugation it can be estimated whether a steeper gradient or possibly a longer run at another temperature would have led to a better resolution.

The main objectives of this chapter are to describe in detail the methods for manual and computerized calculations of sedimentation coefficients from zonal centrifugation data, and to discuss the factors which may influence the accuracy of results obtained in this way.

SOME IMPORTANT UNDERLYING ASSUMPTIONS

Svedberg originally defined the sedimentation constant (now preferably called the sedimentation coefficient) of a particle as its sedimentation velocity (dR/dt) in a unit centrifugal field (Svedberg and Pedersen, 1940). This definition implies that the sedimentation coefficient is a characteristic intrinsic parameter of a given particle in a given solvent at a given temperature.

The extension of Svedberg and Pedersen's original concepts to centrifugation in density and viscosity gradients creates some problems. In order to make sedimentation coefficients obtained in different gradients comparable it is necessary to convert them mathematically to a set of standard conditions, traditionally water at 20 °C ($s_{20,w}$). Two types of factor can be said to have a major influence on the sedimentation velocity in gradient centrifugation experiments. One type is that the steadily increasing density and viscosity of the gradient tend to slow down the movements of the particles. However, as discussed in Chapter 2 the influence of density and viscosity can easily be treated in physical terms, so that the sedimentation coefficient in gradient centrifugation experiments can be formally defined by

$$s_{20,w} \int_0^t \omega^2 \, dt = \frac{\rho_p - \rho_{20,w}}{\eta_{20,w}} \times \int_i^r \frac{\eta_{T,m}}{\rho_p - \rho_{T,m}} \times \frac{dR}{R} \qquad (5.1)$$

For definitions of symbols, see *Table 5.1.*

The other factors which can influence the sedimentation velocity in gradient centrifugation experiments are intrinsic to the particles in question. It is implied

Table 5.1 SYMBOLS AND ABBREVIATIONS USED IN THE TEXT AND APPENDICES

Symbol	FORTRAN name	Definition
A	A	Acceleration time, min
D	D	Deceleration time, min
i		Rotor radius corresponding to the original sample mass centre
P	P	Run time, min
Q	Q	Rotor speed, rev/min
r		Rotor radius corresponding to the final sample mass centre
R	R	Radius, cm
$s_{20,w}$	S, SEDCOF	Sedimentation coefficient in water at 20 °C
t		Time, s
V		Volume, ml
$\eta_{T,m}$	VISC	Viscosity of the medium at the temperature T, cP
$\eta_{20,w}$		Viscosity of water at 20 °C, cP
ρ_p	PDEN	Density of the particle, g/cm^3
$\rho_{T,m}$	DENS	Density of the medium at the temperature T, g/cm^3
$\rho_{20,w}$		Density of water at 20 °C, g/cm^3
ω		Angular velocity, rad/s

in Svedberg's definition of the sedimentation coefficient that the physical properties of the particles do not change while they are slowly sedimenting through a medium with an increasing concentration of gradient molecules and against a steadily increasing pressure. In other words it must be assumed that, first, the particle density is constant throughout a complete run; secondly, the particles do not polymerize or depolymerize while being diluted as the zone broadens; thirdly, the particles do not disintegrate or change shape with increasing pressure and, finally, that the degree of solvation of a macromolecule (or in case of a subcellular particle the water content) is independent of the osmotic pressure of the surrounding medium. No general rules for the fulfilment of these assumptions can be given, as they depend heavily on the system under investigation (see Wattiaux *et al.*, 1971, 1973; Steck, Straus and Wallach, 1970). Several proteins which are centrifuged in sucrose gradients behave as if their physical properties actually were independent of the presence of the gradient material (Steensgaard, Møller and Funding, 1975). On the other hand, certain nucleic acids show different particle densities when centrifuged in different gradients (see Chapters 6 and 7). In density-gradient centrifugation as well as in analytical ultracentrifugation it may be expected that the particle concentration itself has an influence on the sedimentation pattern because of its effects on the chemical potential of the sample particles.

GUIDE FOR THE COMPUTERIZED CALCULATION OF SEDIMENTATION COEFFICIENTS

Equation (5.1), which gives the formal definition of sedimentation coefficients, can be rewritten as

$$s_{20,w} \times C_1 = C_2 \int_i^r \frac{\eta_{T,m}}{\rho_p - \rho_{T,m}} \times \frac{dR}{R} \tag{5.2}$$

where $C_1 = \int_0^t \omega^2 dt$ and $C_2 = (\rho_p - \rho_{20,w})/\eta_{20,w}$ to emphasize that C_1 and C_2 can be regarded as constants for a given set of experimental conditions, and that they need to be calculated only once for each experiment.

Hence, the principal difficulties lie in the evaluation of the integral part of eqns (5.1) or (5.2). However, rate-zonal experiments are generally performed in a way that permits estimation of density and viscosity in a series of fractions which in turn can be taken as representing small radial steps. Therefore, the expression

$$Z_k = \frac{\eta_{T,m}}{\rho_p - \rho_{T,m}} \times \frac{dR}{R} \tag{5.3}$$

can be evaluated for each fraction, and the integral part of eqn (5.2) can be approximated as a summation over Z-values, starting with the fraction which contained the sample mass centre at the beginning of the run and giving one sedimentation coefficient corresponding to each fraction. Finally, the sedimentation coefficient of each of the separated components is obtained by matching the effluent profile with the calculated sedimentation coefficients of all fractions.

The principles for calculation of sedimentation coefficients, as outlined above, date back to Martin and Ames (1961). Since then, several slightly modified approaches have been published (Bishop, 1966; Leach, 1971; Hinton, 1971; Ridge, 1973; Funding, 1973; Steensgaard, 1974).

The flow chart in *Figure 5.1* illustrates how a computer program for calculation of sedimentation coefficients can be composed. Detailed instructions for each step will be given below. A complete program in FORTRAN for a BXIV rotor is provided in Appendix I (p. 135).

Step 1. Calculation of Force-time Integral (C$_1$)

Some centrifuges are equipped with an electronic force-time integrator on which this integral is displayed automatically. However, the integral can also be calculated from the knowledge of run duration (P, min), average rotor speed during

118

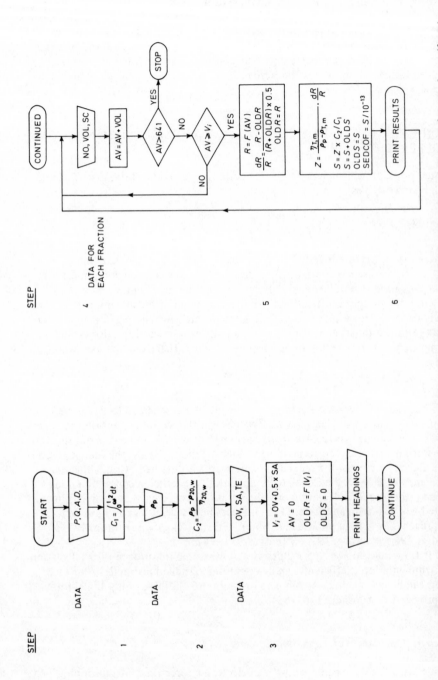

Figure 5.1 Flow chart for the FORTRAN program in Appendix I used for calculation of sedimentation coefficients after zonal centrifugation

the run period (Q, rev/min), acceleration time (A, min) and deceleration time (D, min) by use of the equation:

$$C_1 = \int_0^t \omega^2 \, dt = (\pi Q)^2 \times \left[P + \frac{(A+D)}{3} \right]\Big|15 \tag{5.4}$$

It should be noted that tachometers on commercially available centrifuges may not be accurate enough for this purpose. Hence, it is frequently better to calculate the average rotor speed from readings on the revolution counter at the beginning and the end of the run. Moreover, it should be noted that eqn (5.4) is based on the assumption that acceleration and deceleration are linear with respect to time. If they are not, the size of any error introduced depends on the time of centrifugation and will be important only for relatively short runs.

Step 2. Calculation of Standardization Term (C_2)

The density and viscosity of pure water at 20 °C is 0.998 2 g/cm^3 and 1.005 cP, respectively (Weast, 1966). *Table 5.2* shows some representative particle densities, and a further discussion of the importance of this parameter is given on p. 127.

Table 5.2 SELECTED PARTICLE DENSITY VALUES WITH SPECIFIC REGARD TO SUCROSE GRADIENTS

Particle	Density g/cm^3
Lipoproteins	1.00–1.21
Proteins	1.34–1.42
Mitochondria	1.13–1.19
Plasma membranes	1.14–1.17
Ribosomes/polysomes	1.4
Virus	1.2–1.4

Step 3. Initiation of Computations

To initiate the calculations it is necessary to know the radius which corresponds to the starting position of the mass centre of the sample. The radius in the rotor which corresponds to any fraction is related to the effluent volume in a simple way, as shown in *Figure 5.2* for a BXIV zonal rotor. Details of this relationship for the most commonly used rotors are given in Chapter 9.

The curve in *Figure 5.2* (for a BXIV rotor) can be approximated roughly (that is, the maximum error is about 1%) by the following expressions:

$$R = 0.594\,7 + (1.394 + 0.054\,95 \times V)^{\frac{1}{2}}$$

$$V = 18.20 \times R^2 - 21.64 \times R - 18.93$$

For a BXV rotor the expressions are:

$$R = 0.512\,6 + (1.480 + 0.040\,4 \times V)^{\frac{1}{2}}$$

$$V = 24.75 \times R^2 - 25.37 \times R - 30.12$$

These expressions have been derived by parabolic curve fit, using data from Steensgaard (1970) and Cline and Ryel (1971), respectively. Data from different laboratories on the radius–volume relationship differ somewhat, and slightly different expressions have been calculated by Norman (1971b). These are given in Appendix II. Alternatively, the radius–volume relationship can be obtained with sufficient accuracy by linear interpolation in a table containing paired radius and volume values for each rotor (Funding, 1973).

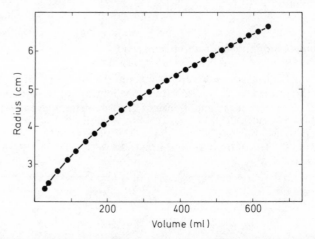

Figure 5.2 The relation between radius and volume in a BXIV rotor. The individual points are determined experimentally. (From Steensgaard, 1970)

The original sample mass centre is defined by the volume of the overlay plus half the sample volume in the case of a so-called block zone (that is, a zone introduced between the gradient and the overlay solution). Hence, the starting point of the integration procedure is given by the radius (i) corresponding to V_i = overlay volume (OV) + 0.5 × sample volume (SA).

The rotor temperature during the run is conveniently inserted at this point as the last of the data which are needed to describe the experimental conditions.

Step 4. Data Input for Each Fraction

In the following calculations it is necessary to know or to calculate the density and the viscosity of the solvent in each fraction. In the case of sucrose gradients, density as well as viscosity can be calculated when the sucrose concentration and the temperature are known (see step 5). It might be helpful also to insert the fraction number for identification purposes. Moreover, the volume of each fraction is needed. The calculations are ended when the accumulated fraction volume exceeds the total rotor volume.

Step 5. Integration

The integral part of the calculations starts with the appearance of data from the original sample mass centre; this fraction is identified as described in step 3. The radial position of each fraction is taken as the radius which corresponds to the last drop of the fraction. The accumulated fraction volume is conveniently summed up after inserting the data of each fraction; dR/R is calculated as the difference between the radius of a given fraction and the radius of the previous fraction, divided by the average of these two radius values. The fraction in which the original sample mass centre is located presents a special case, because this sample zone sediments through only a part of the fraction volume.

Barber (1966) has developed a series of equations for calculating the density and viscosity of sucrose solutions as functions of their composition and temperature. These expressions are included in the FORTRAN program in Appendix I.

If the data are treated as described, the program can be conveniently arranged to finish the calculations for each fraction immediately after insertion of the data for the fraction, and the results printed out. Hence, it is not necessary to store data concerning sucrose concentration and fraction volume.

GUIDE FOR THE MANUAL CALCULATION OF SEDIMENTATION COEFFICIENTS

The complexity of eqn (5.1) could, at first sight, suggest that the need for a computer for these calculations was imperative, but this is not so. Different attempts to simplify the necessary calculations have been made. McEwen (1967) has published tables which facilitate calculation of sedimentation coefficients from data obtained with linear sucrose gradients in swing-out rotors. Halsall and Schumaker (1969) have devised a semi-graphical method to be used when the values of density and viscosity of each fraction can be measured or calculated. Young (1972) has calculated a revised version of McEwen's tables for use with shallow sucrose gradients. Fritsch (1973) has described a simple method for calculating sedimentation coefficients from data obtained using a number of Beckman rotors, but with only one gradient (a linear sucrose gradient, 5–20% w/w) and only at 5 °C and 20 °C.

The following approach is restricted to sucrose gradients also, but the main advantages are that it can be used for any sucrose gradient (up to 44.5% sucrose), that it can be used directly for zonal rotors, and that it can easily be adapted to fixed-angle and swing-out rotors.

The foundation of this approach is the definition of a new function:

$$\text{sedim}(y_k) \;=\; \frac{\rho_p - \rho_{20,w}}{\eta_{20,w}} \times \frac{\eta_{T,m,k}}{\rho_p - \rho_{T,m,k}} \tag{5.5}$$

where y is the sucrose concentration in fraction k. The value of sedim depends only on the particle density, the sucrose concentration and the temperature (in the rotor during a run). Sedim values for several different particle densities and over a wide range of temperatures have been calculated and tabulated (Funding and Steensgaard, 1973). Appendix III contains such sedim tables for particle densities from 1.1 to 1.4 g/cm^3, for temperatures between 4 °C and 25 °C and for sucrose concentrations up to 44.5% (w/w) (pp. 139–162).

By use of the sedim function, eqn (5.1) can be approximated by

$$s_{20,w} = 1/C_1 \times [\text{sedim}\ (y_j) \times (\ln R_j - \ln R_i)$$

$$+ \sum_{k=j}^{r-1} \text{sedim}\ (y_{k+1}) \times (\ln R_{k+1} - \ln R_k)] \tag{5.6}$$

noting that special attention should be given to the fraction which contained the starting position of the sample mass centre (fraction *j*). Furthermore, these calculations can be simplified by the use of tables which give the natural logarithm of the radius corresponding to a given accumulated fraction volume (see Appendix IV).

The following example will demonstrate how sedimentation coefficients can be calculated by this method. The data used are given in *Table 5.3*, and a similar example is illustrated in *Figure 5.3*. An entirely similar approach can be used for gradients in swing-out rotors, provided that the correct radius-volume data are used at steps (5) and (6) in the following procedure.

1. The volume of each fraction is written in column B. From these values the corresponding accumulated fraction volume for each fraction is worked out and written in column C.
2. The measured biological activities of each of the fractions are recorded in column D in order to identify the final position in the rotor of the activity.
3. The sucrose concentration in each fraction is noted in column E. Only the values which are necessary for the calculations are given in this example.
4. On the basis of knowledge of the particle density and the rotor temperature, the proper sedim values are found in the tables using sucrose concentrations in column E as entrance values. Sedim values are written in column F.
5. The volume (V_i) corresponding to the sample mass centre at the beginning of the run is taken as the sum of the overlay volume and half of the sample volume. The logarithm of the corresponding rotor radius is used as the starting point of the integration procedure.
6. The natural logarithms of the rotor radii corresponding to the accumulated volumes of the fractions inside the range of interest are written in column G. Only the first fraction is treated as described under step 5.
7. The differences between the values at each successive step in column G are entered in column H, representing the dR/R values used in eqn (5.1).
8. The product of sedim values (column F) and ln (radius) difference values (column H) are entered in column I, and accumulated in column J.
9. The equivalent sedimentation coefficients of the fractions of specific interest may now be calculated by dividing the values in column J with the previously calculated force–time integral (or by multiplication with the reciprocal integral, whichever is most convenient).

Depending on the purpose of the experiment the calculation of the sedimentation coefficient which corresponds to the final mass centre can be done in different ways. For many purposes it is sufficient to pick out the fraction with

Table 5.3 AN EXAMPLE OF MANUAL CALCULATION OF SEDIMENTATION COEFFICIENTS*

Date:

	Sample 2 ml	Overlay 100 ml	Duration 299 min	Acceleration 15 min	Experiment No:
					Deceleration 10 min
	Speed 47 100 rev/min	$1/\int\!\int\omega^2 \times dt \times 2.20 \times 10^{-12}$	Particle density 1.4	Temperature 8 °C	

Fraction number A	Volume ml B	Accumulated volume C	Activity D	Sucrose conc. E	Sedim F	ln (radius) G	ln (radius) difference H	Sedim × difference I	$\sum$ Sedim × difference J	$s_{20,w} \times 10^{-13}$ K
1	13.9	13.9								
2	7.9	21.8								
3	13.4	35.2								
4	13.2	48.4								
5	12.6	61.0								
6	12.4	73.4								
7	12.5	85.9								
8	12.5	98.4				V_i 1.169 3				
$j = 9$	12.2	110.6	1 468	4.1	1.614	1.199 0	0.029 7	0.048	0.048	1.1
10	12.4	123.0	5 883	4.8	1.680	1.237 8	0.038 8	0.065	0.113	2.5
11	11.8	134.8	14 083	5.6	1.716	1.263 0	0.026 8	0.045	0.158	3.5
12	12.3	147.1	16 700	6.0	1.752	1.296 3	0.033 3	0.058	0.216	4.8
13	12.4	159.5	15 632	6.7	1.789	1.323 5	0.027 2	0.049	0.265	5.8
14	12.4	171.9	7 687	7.1	1.828	1.354 1	0.030 6	0.056	0.321	7.1
15	11.9	183.8	2 745	7.8	1.910	1.385 8	0.031 7	0.061	0.382	8.4
16	12.5	196.3	631	9.1	1.998	1.411 8	0.026 0	0.052	0.434	9.5

*Practical handling of centrifuge data for calculation of sedimentation coefficients. Simulated data corresponding to *Figure 5.3* are used. (From Funding and Steensgaard, 1973.)

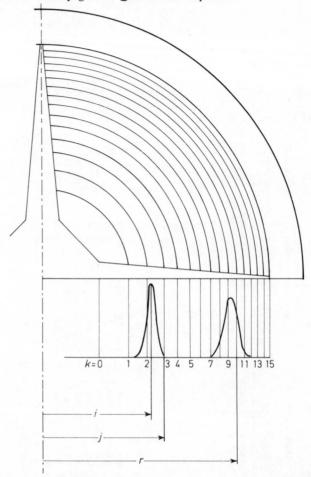

Figure 5.3 One sector of a BXIV rotor with diagrammatic representation of the initial and final sample activity distribution. The upper part is a horizontal cross-section showing the radii of 45 ml increments of volume. The lower part shows the activity versus rotor radius. Initially the sample mass centre is at i, during centrifugation it migrates to its final position at radius r. The fraction numbers (k) are shown at the end of each fraction and j depicts the fraction containing the initial mass centre. (From Funding and Steensgaard, 1973)

the highest activity and take the corresponding sedimentation coefficient as the desired value. However, if the volume of the fractions is not too large, an interpolation may be justified. The simplest way to do this is to plot the activities *versus* the calculated sedimentation coefficient, remembering that these calculations give the sedimentation coefficient corresponding to the last drop in a fraction, whereas the activities are more representative of the midpoint of a fraction. More sophisticated procedures have been described also (Ridge, 1973; Funding, 1973).

Empirical Procedures

When swing-out or fixed-angle rotors are used, sedimentation coefficients may be estimated by comparing the sedimentation pattern of test particles with the sedimentation pattern of standard markers (Martin and Ames, 1961). This method is, in principle, physically correct only when isokinetic gradients are used (Noll, 1967; Steensgaard, 1970). However, because shallow sucrose gradients which are linear with respect to radius represent a fairly good approximation to isokinetic gradients, sedimentation coefficients obtained in this way with such gradients may be regarded as quite accurate. Moreover, as overloading seems to affect the sedimentation rates of all particles in the same way (cf. *Figures 5.6* and *5.7*), it is likely that errors due to overloading will tend to be minimized in this approach.

ACCURACY OF SEDIMENTATION COEFFICIENTS CALCULATED FROM DENSITY-GRADIENT DATA USING LARGE-SCALE ZONAL ROTORS

To obtain accurate sedimentation coefficients from zonal centrifugation data, it is necessary to use a perfectly functioning zonal centrifuge system. The seals should be very carefully adjusted and polished to prevent any leakages or cross-feed during loading and unloading of the rotor contents. The temperature must be maintained at a constant and known level. The particle densities must be known within the limits of a few per cent. Sucrose concentrations and biological activities (or other measurements of particle concentration) in the collected fractions must be carefully determined, at least to within an accuracy of 1%.

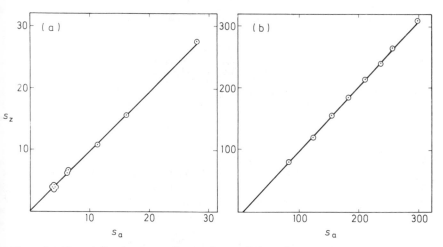

Figure 5.4 *The relation between sedimentation coefficients found by analytical ultra-centrifugation (s_a) and zonal ultracentrifugation (s_z) for proteins (a) and polysomes (b). [Adapted from Steensgaard (1974) and Norman (1971a), respectively]*

Figure 5.4 shows the results of two studies (Norman, 1971a; Steensgaard, 1974) which were done in order to investigate how accurately sedimentation coefficients could be estimated from zonal centrifugation data if all these critical factors were very carefully controlled. Test particles were ribosomes plus poly-somes and well-defined proteins. It appears from *Figure 5.4* that the correlation between sedimentation coefficients obtained by analytical ultracentrifugation (s_a) and those obtained from zonal centrifugation (s_z) data is indeed very good in these cases. This leads to the conclusions that (*a*) it is possible to estimate sedimentation coefficients accurately by zonal centrifugation and, because the deviations seem to be randomly ordered, that (*b*) proteins, ribosomes and polysomes sediment in sucrose gradients [at least below 35% (w/w) sucrose], as predicted from eqn (5.1). Hence, interactions between particles and gradient material, if any, in this case appear to be negligible for calculation purposes.

The reproducibility of sedimentation coefficients which have been obtained by zonal centrifugation is quite good. Results from a study of D-glucose-6-phosphate dehydrogenase and 6-phospho-D-gluconate dehydrogenase in bovine and human erythrocytes are shown in *Table 5.4* (Steensgaard and Funding,

Table 5.4   THE PRECISION OF $S_{20,w}$ ESTIMATION BY ZONAL CENTRIFUGATION*

	$S_{20,w}$ ± s.e.m.	(*n*)
Bovine foetal:		
D-glucose-6-phosphate dehydrogenase	7.5 ± 0.1	(10)
6-phospho-D-gluconate dehydrogenase	5.7 ± 0.1	(9)
Haemoglobin	3.6 ± 0.1	(5)
Bovine adult:		
D-glucose-6-phosphate dehydrogenase	7.2 ± 0.2	(8)
6-phospho-D-gluconate dehydrogenase	6.1 ± 0.1	(6)
Haemoglobin	3.8 ± 0.1	(7)
Normal human:		
D-glucose-6-phosphate dehydrogenase	6.4 ± 0.1	(17)
6-phospho-D-gluconate dehydrogenase	6.1 ± 0.2	(8)
Haemoglobin	4.3 ± 0.1	(3)
Deficient human:		
D-glucose-6-phosphate dehydrogenase	6.4 ± 0.3	(4)
6-phospho-D-gluconate dehydrogenase	6.2 ± 0.1	(4)
Haemoglobin	4.0 ± 0.2	(4)

*Equivalent sedimentation coefficients ($S_{20,w}$) ± standard error of mean of D-glucose-6-phosphate dehydrogenase, 6-phospho-D-gluconate dehydrogenase and haemoglobin from bovine foetal, bovine adult, normal human and D-glucose-6-phosphate dehydrogenase-deficient erythrocytes determined by zonal ultracentrifugation. The numbers in brackets denote the number of samples. (From Steensgaard and Funding, 1972.)

1972). Sedimentation coefficients which are obtained by analytical ultracentrifugation are traditionally given to two decimal places. The results from zonal centrifugation, shown in *Table 5.4*, indicate that in most cases it is only justified to give sedimentation coefficients obtained by zonal centrifugation in whole Svedberg units. If replicate determinations are done with the same sample, the average may be given to an accuracy of one decimal place when so justified by a statistical analysis.

Some Important Sources of Error

In view of the complexity of the zonal centrifugation technique it is not surprising that there are several possible sources of error. In principle, one can distinguish between three different kinds of error. First, some parameters used in the calculations may not be known with sufficient accuracy, such as the particle density and the actual rotor temperature during the centrifugation. Secondly, the necessary volumetric determinations may present some technical difficulties. Thirdly, there is the critical problem of overloading of the sample zone, as this indirectly gives rise to artificially high estimates of sedimentation coefficients.

All of these may be classified as systematic errors; they do not include random measuring errors.

Particle Density and Rotor Temperature

Zonal centrifugation is frequently chosen for determining sedimentation coefficients in cases where the particles only exist in a partly purified form, and this also includes the cases where particle densities have not been directly determined experimentally. However, several particle densities, obtained by isopycnic centrifugation, have been published; in particular, *The Handbook of Biochemistry* (Ed. H.A. Sober, 1970) and the series *Methodological Developments in Biochemistry* (Ed. E. Reid, 1971–74) may be helpful sources of information on particle densities of different macromolecules and subcellular organelles from a variety of different sources. Some particle densities in sucrose solutions are given in *Table 5.2.*

The effect of a small error in particle density on the accuracy of a calculated sedimentation coefficient can be illustrated by recalculation of the same set of data using different particle density values. An example of such a recalculation is given in *Table 5.5* (Steensgaard and Hill, 1970). This shows that within the range given (typical of proteins) the numerical value of particle density has a relatively small effect on the calculated sedimentation coefficients. However, sedimentation coefficients calculated in this way should always be accompanied by information on the particle density used. If sedimentation coefficients are calculated for two different (not too distant) particle densities, linear interpolation will be justified in most cases (Steensgaard, Funding and Jacobsen, 1973).

The rotor temperature presents a somewhat analogous problem when sedimentation coefficients are calculated from zonal centrifugation data. If the temperature used in the calculation is higher than the actual temperature of the rotor,

Table 5.5 THE INFLUENCE OF PARTICLE DENSITY ON SEDIMENTATION COEFFICIENT CALCULATIONS*

	Particle densities					
Cut No.	1.34	1.36	1.38	1.40	1.42	1.44
38	0.45	0.45	0.45	0.45	0.45	0.45
43	4.89	4.88	4.87	4.86	4.85	4.84
48	8.81	8.78	8.76	8.74	8.73	8.71
53	12.62	12.58	12.55	12.52	12.49	12.47
58	16.21	16.16	16.12	16.08	16.04	16.01
63	19.99	19.93	19.87	19.82	19.77	19.73
68	23.30	23.22	23.15	23.09	23.03	22.98
73	26.07	25.98	25.90	25.82	25.76	25.69
78	29.19	29.03	28.94	28.85	28.77	28.70
83	32.03	31.91	31.80	31.70	31.61	31.53
88	35.01	34.87	34.75	34.64	34.54	34.45
93	38.05	37.89	37.75	37.62	37.50	37.40

*The influence of particle density on computed sedimentation coefficients in selected cuts. The sample consisted of 500 μg of ^{131}I-labelled human serum albumin (alkylated and monomerized) in 2 ml of 0.05 M tris buffer, pH 8.0. Centrifugation was performed in a BXIV aluminium rotor on an isokinetic 2–9.7% sucrose gradient using 250 ml of overlay. The run time was 5 h at 30 000 rev/min. The temperature was 10 °C. (From Steensgaard and Hill, 1970.)

Table 5.6 THE INFLUENCE OF TEMPERATURE ON SEDIMENTATION COEFFICIENT CALCULATIONS*

Fraction No.	*Temperature* ($^{\circ}$C)					
	4	6	8	10	12	14
12	1.03	0.97	0.92	0.87	0.82	0.77
17	6.41	6.01	5.65	5.32	5.02	4.74
22	11.89	11.15	10.47	9.86	9.29	8.77
27	16.14	15.12	14.21	13.36	12.59	11.88
32	20.14	18.87	17.71	16.66	15.69	14.80
37	23.87	22.35	20.97	19.72	18.56	17.51
42	27.44	25.68	24.09	22.63	21.30	20.08
47	31.62	29.58	27.73	26.04	24.50	23.09
52	34.41	32.18	30.16	28.32	26.63	25.09
57	37.99	35.51	33.27	31.23	29.36	27.64
62	41.64	38.90	36.43	34.18	32.12	30.23
67	45.74	42.71	39.98	37.48	35.20	33.12

*The effects in selected cuts of the use of various temperatures in the calculation of sedimentation coefficients. Experimental data: sample, 2 ml of human serum diluted 1:22; overlay, 100 ml; isokinetic 3–23% (w/w) sucrose gradient; titanium BXIV rotor; rotor speed, 47 000 rev/min; duration, 5 h; temperature, 8 $^{\circ}$C

the calculated sedimentation coefficient will be lower than its true value (Steensgaard and Hill, 1970; Steensgaard, 1971). An example of the impact of the temperature used on the calculated sedimentation coefficients is given in *Table 5.6*. It appears from this table that the rotor temperature should be known to within 1 $^{\circ}$C. The temperature can be controlled within such limits in most modern centrifuges. However, the following practical hints may be of value in obtaining well-defined rotor temperatures.

Liquids passing through the feed head will be heated a few degrees, and therefore the gradient solutions, the sample and the overlay solution should be precooled to a temperature somewhat lower than the desired rotor temperature. The exact difference depends on the spring load in the rotor seal and on the length of the tubes in the system, but a 3 $^{\circ}$C temperature difference can be considered a good rule of thumb. High air humidity, aggravated by minor (or major) leakages from the feed head, tends to produce heavy condensation in the centrifuge bowl, which later affects the efficiency of evacuation. A poor vacuum may give rise to poor temperature control, because the presence of vapour in the bowl may be misinterpreted by the temperature control system as indicating too high a rotor temperature, leading to overcooling of the rotor. Actual rotor temperatures up to 5 $^{\circ}$C below the thermostat setting have been observed under unfavourable conditions in our laboratory (see pp. 68–72).

However, these technical difficulties can be overcome by proper consideration of all parts of the equipment. With the centrifuges and rotors available at present, the only completely accurate method of measuring the actual temperature inside the rotor is to carry out a test run, interrupt centrifugation, unscrew the rotor lid and measure the temperature of the rotor contents. This should be done occasionally to check the temperature control of the centrifuge.

Volumetric Control

Because all measurements of radii are based on volumetric measurements, the latter should be carried out as carefully as possible. It is our experience that the sample and the overlay solution can be introduced into the rotor with sufficient accuracy by the use of measuring cylinders and/or syringes. The main problem is to ensure that no liquid passes from the edge-feed connection to the centre-feed connection in the feed head during loading of the gradient, sample and overlay solutions. Cross-feed, in this sense, mainly appears if the stationary seal is insufficiently adjusted to the rotating seal. Adjustment involves centring the two seals, lining up the axes and finally finding the proper spring load. The importance of careful adjustment of the feed-head position cannot be over-emphasized. It is likely that problems with cross-feed and leakages present the largest technical source of errors in determination of sedimentation coefficients by zonal centrifugation (see pp. 68–72).

Overloading of the Sample Zone

Overloading of the sample zone occurs when too much sample material is applied to the gradient. When small amounts of sample are used the zones sediment as predicted by the Svedberg equation with correction for the effects of the gradient [as given by eqn (5.1)]. The agreement between the theoretically predicted and experimentally found sedimentation patterns is indeed so good that numerical simulation of zonal centrifugation is possible (Steensgaard, Funding and Meuwissen, 1973). However, when larger amounts of sample are used, distinct deviations from the sedimentation patterns given by the Svedberg equation are observed.

Deviations from expected sedimentation behaviour caused by overloading can be illustrated by the results of the two zonal centrifugations, as shown in *Figure 5.5*. In both cases, the macromolecular species in the sample were human serum albumin, catalase and β-galactosidase. In *Figure 5.5(a)*, the components were centrifuged alone, whereas in *Figure 5.5(b)* an extra 100 mg of human serum albumin was added to the sample before centrifugation. Two distinct effects of increased sample load are seen in these experiments: (*i*) the zones are definitely broadened in the experiment with higher load; and (*ii*) the final sample mass centres have migrated further into the gradient in the run with higher sample load.

Both of these effects have been studied further (Steensgaard, Møller and Funding, 1975). The standard sample [in *Figure 5.5(a)*] was used in all experiments, and human serum albumin was used to increase the load over a wide range. By use of a computer program, zone widths at half peak-height were estimated and sedimentation coefficients were determined by use of a procedure which estimated a precise location of the mass centres of the final zones by numerical analysis of the zone shapes.

The zone widths and the calculated sedimentation coefficients were plotted in two graphs (*Figures 5.6* and *5.7*) using the amount of added human serum albumin as abscissa. Overloading has clear-cut effects, giving rise to broadened and dislocated zones. Both of these effects are detrimental to the analytical use of zonal centrifugation, because the dislocated zones will give higher sedimentation coefficients, and the non-gaussian zone broadening makes the estimation of the final zone mass centres difficult.

Another feature should be noted in *Figures 5.6* and *5.7*, namely that the transition to the overloading of zones appears to be gradual. Therefore, it is difficult,

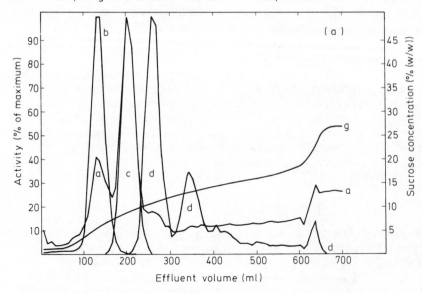

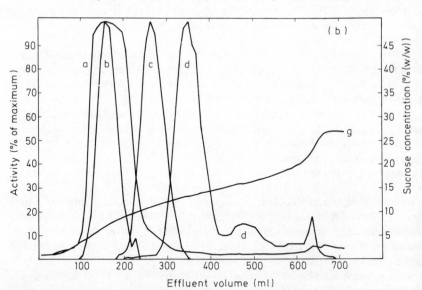

Figure 5.5 Computer-drawn effluent patterns of a zonal centrifuged multicomponent sample (a) without and (b) with overloading caused by addition of human serum albumin. The experiments were carried out in a BXIV titanium rotor on an isokinetic 3–23% (w/w) sucrose gradient with 2 ml sample and 100 ml overlay. The centrifugations were performed at 8 °C with a rotor speed of 47 000 rev/min until a force-time integral of 4.5 × 10[11] rad²/s was reached (that is, approximately after 5 h). The sample consisted of 0.3 mg of ^{131}I-human serum albumin (alkylated and monomerized), 0.5 mg of beef liver catalase and 0.1 mg of β-galactosidase. In addition, the sample in (b) contained 100 mg of unlabelled human serum albumin. The curves are standardized to show percentages of the maximum value on the left ordinate. Curve a shows the absorption at 219.5 nm, b radioactivity, c catalase activity, and d galactosidase activity. The sucrose concentration referring to the right ordinate is denoted by g. (From Steensgaard, Møller and Funding, 1975)

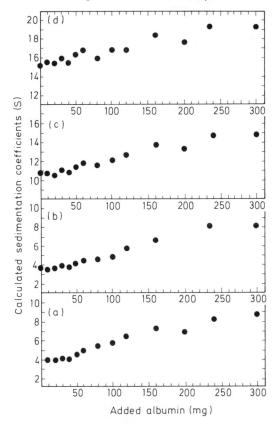

Figure 5.6 The relationship between calculated sedimentation coefficients and the amount of human serum albumin added. (a) represents human serum albumin (based on absorption measurements) and (b), (c) and (d) represent human serum albumin (radioactivity measurements), catalase and β-galactosidase (activity measurements), respectively. All values are averages of two determinations. (From Steensgaard, Møller and Funding, 1975)

or even impossible, to give a precise upper maximum limit for a non-overloaded sample zone.

The physical background of the phenomenon of overloading is not fully understood. Originally, Svensson, Hagdahl and Lerner (1957) suggested that overloading takes place when the density gradient is partially or totally destroyed by the presence of an excess of sample molecules or particles. Berman (1966) successfully developed a mathematical formula for calculating the maximum sample load which a given gradient can support at the beginning of the run. Later, Spragg and Rankin (1967) found experimental evidence that Berman's formula closely predicts density instability using deviation from gaussian zone shape as the criterion for overloading (see Chapter 2).

Recently, Meuwissen (1973a, 1973b) proposed an alternative explanation suggesting that instability can arise on two levels. Density instability of the sample zone appears when the principle of a steadily increasing density throughout the sample and the gradient is violated. However, hydrodynamic instability can arise at lower loads also (see p. 13). One important aspect of Meuwissen's concept is that the instability occurs in the descending part of an overloaded

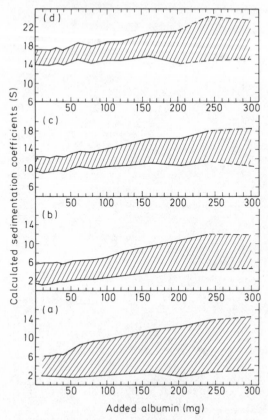

Figure 5.7 Zone broadening caused by addition of increasing amounts of human serum albumin. The zone widths in each of the points in Figure 5.6 *have been calculated in Svedberg units as the width at half peak-height on the best fitting gaussian approximation. The dashed parts of the curves represent assessed values due to biphasic effluent patterns. Same notation as in* Figure 5.6. *(From Steensgaard, Møller and Funding, 1975)*

sample zone, thus providing a tentative explanation of the experimental observation that overloaded zones are dislocated in the centrifugal direction. Also,
because the cause of hydrodynamic instability is supposed to be diffusional
movements, it can be expected that the transition from a non-overloaded sample
to an overloaded one is gradual, involving changes in activity terms as the concentration of sample molecules increases. Thus, the question of whether or not a given
zone is overloaded does not have an easy answer. In a few selected cases, for example by the use of triangular or inverted zones, Berman and Meuwissen have derived
formulae which make it possible to calculate the maximum permissible load
under these specific conditions (Berman, 1966; Meuwissen, 1973a, 1973b). As
mentioned previously, Spragg and Rankin (1967) have suggested that the shape
of the zones can be used as a criterion of overloading because zones as a general
rule assume a gaussian shape very rapidly in the rotor. Hence, deviation from a
gaussian zone shape indicates that the sample zone is overloaded. However, two
factors may restrict the use of this criterion. First, it has been found that zones
with and without gaussian zone shapes can exist in the same effluent pattern
(Steensgaard, Møller and Funding, 1975), and it seems that only the zone of

that component of the sample which causes the overloading shows a definite deviation from gaussian zone shape. Secondly, neither parametric nor non-parametric statistical distribution analysis provides a clear-cut indication as to whether a given distribution can safely be regarded as gaussian or not.

We prefer another way to assess possible overloading based on the dislocation of the final zone mass centre of a standard substance. The principle of this is to define a non-overloaded zone as a zone whose mass centre sediments in agreement with the Svedberg equation. A substance with known sedimentation coefficient (s_a) is added to the sample in trace amounts, and the sedimentation coefficient (s_z) is calculated from the experimental data. The ratio s_z/s_a then serves as an index of overloading, and only if this ratio is close, or equal, to unity, can the gradient in question be regarded as non-overloaded. This approach is justified because, as seen in *Figure 5.6*, the mass centres of zones are dislocated to an extent dependent upon the degree of overloading. In this context it should be noted also that calculation of sedimentation coefficients following gradient centrifugation can help in diagnosing possible experimental errors in gradient centrifugation experiments. Sedimentation coefficients are reduced if losses of overlay or sample have occurred, whereas sedimentation coefficients are increased with overloaded sample zones.

OTHER ANALYTICAL USES

Halsall and Schumaker (1970) found that the diffusion constant of equine cytochrome c could be estimated with an accuracy of 2% by measurements of zone widths following isopycnic centrifugation in a NaBr gradient. We have tried to calculate diffusion coefficients of different proteins after rate-zonal centrifugation. This proved to yield larger diffusion coefficients than expected, (unpublished observations), probably because the zones are broadened simply by the introduction of the sample into the rotor due to the design of feed heads and rotors (Price and Kovacs, 1969; Steensgaard and Funding, 1971).

It has been proposed (G.B. Cline, personal communication) that rate-zonal centrifugation might be used to estimate particle densities with specific reference to sucrose gradients (or, indeed, to gradients of other solutes). In principle, this could be done in the following way. The sedimentation coefficient of the particles is measured in an analytical centrifuge (that is, without a density gradient). The sedimentation rate of the particles through a sucrose gradient in a zonal or swing-out rotor is also measured and sedimentation coefficients are calculated using a wide range of particle densities. The particle density which results in complete agreement between the sedimentation coefficient as determined in the analytical centrifuge and that calculated from the sedimentation rate through the gradient is the density of the particle in aqueous sucrose.

ACKNOWLEDGEMENTS

Part of this work was done while one of the authors (J.S.) was Visiting Professor at the Department of Biology, University of Alabama in Birmingham, USA. The authors wish to thank Dr George B. Cline for many helpful discussions and suggestions during this work. The program was written and tested in co-operation with Bang Maw Liu, MA, at the University of Alabama in Birmingham. This work was supported in part by grants from Statens Lægevidenskabelige Forskningsråd and Fonden til Lægevidenskabens Fremme, both at Copenhagen, Denmark.

APPENDIX I: FORTRAN PROGRAM FOR CALCULATING SEDIMENTATION COEFFICIENTS

The following FORTRAN program is the simplest possible for calculation of sedimentation coefficients from zonal centrifugation data. The sequence of calculations is described in *Figure 5.1*, and the individual steps, explained in the text, are denoted by labels 1 to 5. The definitions of the most important symbols are given in *Table 5.1*.

The radius–volume relationship (for a BXIV rotor) is used at labels 98 and 99. When using a BXV rotor these should be changed accordingly. If other rotor types are used (including swing-out and fixed-angle rotors) appropriate formulae should be derived and inserted in the program at these labels.

The program has been tested against a more elaborate program used for several years at the Institute of Medical Biochemistry, University of Aarhus, Denmark (Steensgaard, 1974; Funding, 1973). Because the present program uses very simple formulae for calculation of radius–volume relationships, small deviations can be found in the results. However, using critical data for zonal centrifugation of proteins, the maximum errors do not exceed 0.1 S.

It should be noted that input/output statements frequently differ with the local computer systems. Hence, it is advisable to check this before use. The data used in the present example can be found in the depicted output (*Table 2*).

Table 1

```
      PROGRAM ZONAL (INPUT,OUTPUT,TAPE5=INPUT,TAPE6=OUTPUT)
C     JENS STEENSGAARD AND BANG MAW LIL
C     MINIMUM PROGRAM FOR CALCULATION OF EQUIVALENT
C     SEDIMENTATION COEFFICIENTS AFTER ZONAL CENTRIFUGATION
C     FOR BXIV ROTORS ONLY. PROGRAM IN FORTRAN IV
      REAL IVOL,MD0,MD1,MD2,MY
      INTEGER DAY,MONTH,YEAR
      READ(5,12) N,DAY,MONTH,YEAR
      WRITE(6,50)
      DO  10 I=1,4
   10 WRITE(6,70)
      WRITE (6,100)N,DAY,MONTH,YEAR
      WRITE(6,150)
 1    READ(5,15) P,G,A,O
      IP=IFIX(P)
      IA=IFIX(A)
      WRITE(6,300)IP,IA
      IO=IFIX(O)
      IQ=IFIX(O)
      WRITE(6,310)IG,IO
      C1=((3.141593*Q)**2.)*(P+(A+O)/3.)/15.
 2    READ(5,20)PDENS
      WRITE(6,320)PDENS
      C2=(PDENS-0.9982)/1.005
 3    READ(5,25)OV,SA,TE
      WRITE(6,330)OV,SA
      ITE=IFIX(TE)
      WRITE(6,340)ITE
      AV=0.
      OLDS=0.
      IVOL=OV+SA/2.
   98 OLDR=0.5947+SQRT(IVOL*0.05495+1.394)
      MD0=(-5.8513271E-6*TE+3.9680504E-5)*TE+0.0003698
      MD1=(1.2392833E-5*TE-1.0578919E-3)*TE+0.38982371
      MD2=(-8.9239737E-6*TE+4.7530081E-4)*TE+0.17097594
```

Table 1 *continued*

```
          WRITE(6,150)
          WRITE(6,360)
          WRITE(6,150)
    4     READ(5,30)NO,VOL,SC
          AV=AV+VOL
          IF (AV.GT.641.) GO TO 1000
          IF (AV.LT.IVOL) GO TO 4
   99 R=0.5947+SQRT(AV*0.05495+1.394)
          DROR=(R-OLDR)/((R+OLDR)/2.)
          OLDR=R
          DENS=(MD2*SC/100.0+MD1)*SC/100.0+MD0
          MY=SC/(SC+(100.0-SC)*18.9924)
          AD=((((((4.5921911E9*MY-1.1028981E9)*MY+1.0323349E8)*MY
        * -4.6927102E6)*MY+1.0504137E5)*MY-1.1435741E3)*MY+9.4112153)
        * *MY-1.5019327
          BD=(((((((-5.4970416E11*MY+1.3532907E11)*MY-1.2985834E10)*MY
        * +6.0654775E8)*MY-1.4184371F7)*MY+1.6911611E5)*MY+1.6077073E3)
        * *MY+2.11E9907E2
          CG=146.06E35-25.251728*SQRT(1+(MY/0.070674842)**2)
          VISC=10.0**(AC+BD/(TE+CG))
    5     Z=DROR*VISC/(PDENS-DENS)
          S=Z*C2/C1
          S=S+OLDS
          OLDS=S
          SEDCOF=S/1.E-13
          WRITE(6,370)NC,VOL,AV,SC,R,VISC,DENS,SEDCOF
          GO TO 4
   12     FORMAT(4I3)
   15     FORMAT(4F7.5)
   20     FORMAT(F7.5)
   25     FORMAT(3F7.5)
   30     FORMAT(I2,2F5.2)
   50     FORMAT(1H1)
   70     FORMAT(1H0)
  100     FORMAT(X,10X,*ZONAL CENTRIFUGATION NO. *,I4,18X,*DATE *,I2,*/*,I
        *        */19*,I2,X)
  150     FORMAT(X,10X,62(1H-))
  300     FORMAT(X,10X,*DURATION OF RUN:*,I4,* MINS*,10X,*ACCELERATION TIM
        **,I4,* MINS*)
  310     FORMAT(X,10X,*ROTOR SPEED:   *,I5,* RPM*,10X,*DECELERATION TIME
        *,I4,* MINS*)
  320     FORMAT(X,10X,*PARTICLE DENSITY:*,F7.3,* (G/CCM)*)
  330     FORMAT(X,10X,*OVERLAY VOLUME:*,F7.3,* ML*,10X,*SAMPLE VOLUME:*,3
        *F7.3,* ML*)
  340     FORMAT(X,10X,*ROTOR TEMPERATURE:*,I2,* CENTIGRADES*)
  360     FORMAT(X,10X,* NO   VOL      AV*,7X,*SC %     R*,6X,*VISC*,6X,*DEN
        **,6X,*S*)
  370     FORMAT(X,10X,I3,3X,F5.2,3X,F6.2,3X,F5.2,3X,F4.2,3X,F7.5,3X,F7.5,
        * 3X,F4.1)
 1000 STOP
          END
```

Table 2

```
ZONAL CENTRIFUGATION NC.      2                    DATE 27/10/1975
-----------------------------------------------------------------
DURATION CF RUN: 291 MINS          ACCELERATION TIME:   15 MINS
ROTOR SPEED:    47629  RPM         DECELERATION TIME:   10 MINS
PARTICLE DENSITY:   1.400 (G/CCM)
OVERLAY VCLUME:150.000 ML          SAMPLE VCLUME:      2.000 ML
ROTOR TEMPERATURE: 8 CENTIGRADES
-----------------------------------------------------------------
```

NO	VOL	AV	SC %	R	VISC	DENS	S
15	10.08	152.82	6.58	3.72	1.67885	1.02621	.2
16	10.53	163.35	7.72	3.81	1.74221	1.03085	1.2
17	10.53	173.88	8.65	3.90	1.79725	1.03467	2.2
18	10.51	184.39	9.67	3.99	1.86138	1.03890	3.2
19	10.52	194.91	10.16	4.07	1.89367	1.04094	4.2
20	10.55	205.46	11.10	4.16	1.95849	1.04488	5.2
21	10.01	215.47	11.84	4.23	2.01232	1.04800	6.1
22	10.52	225.99	12.31	4.31	2.04788	1.05000	7.1
23	10.54	236.53	12.71	4.39	2.07900	1.05170	8.0
24	10.53	247.06	13.13	4.46	2.11258	1.05349	9.0
25	10.09	257.15	13.39	4.53	2.13383	1.05461	9.8
26	10.53	267.68	13.68	4.61	2.15797	1.05585	10.7
27	10.52	278.20	14.01	4.68	2.18602	1.05727	11.6
28	10.51	288.71	14.24	4.75	2.20593	1.05826	12.5
29	10.50	299.21	14.56	4.82	2.23416	1.05965	13.3
30	10.50	309.71	14.80	4.89	2.25573	1.06069	14.1
31	10.00	319.71	15.08	4.95	2.28135	1.06190	14.9
32	10.52	330.23	15.27	5.02	2.29901	1.06273	15.7
33	10.50	340.73	15.48	5.08	2.31880	1.06364	16.5
34	10.50	351.23	15.76	5.14	2.34563	1.06487	17.3
35	10.05	361.28	15.97	5.20	2.36609	1.06579	18.0
36	10.49	371.77	16.18	5.27	2.38686	1.06671	18.8
37	10.49	382.26	16.42	5.33	2.41096	1.06776	19.5
38	10.49	392.75	16.58	5.39	2.42725	1.06846	20.3
39	10.48	403.23	16.79	5.45	2.44891	1.06939	21.0
40	10.49	413.72	16.88	5.51	2.45830	1.06978	21.7
41	10.00	423.72	17.17	5.56	2.48893	1.07106	22.4
42	10.49	434.21	17.27	5.62	2.49964	1.07151	23.1
43	10.50	444.71	17.48	5.68	2.52238	1.07244	23.8
44	10.48	455.19	17.68	5.73	2.54435	1.07332	24.5
45	10.05	465.24	17.88	5.79	2.56664	1.07421	25.2
46	10.48	475.72	17.89	5.84	2.56776	1.07426	25.8
47	10.47	486.19	18.21	5.90	2.60411	1.07568	26.5
48	10.48	496.67	18.36	5.95	2.62144	1.07635	27.2
49	10.48	507.15	18.52	6.00	2.64013	1.07706	27.8
50	10.48	517.63	18.72	6.06	2.66380	1.07796	28.5
51	9.98	527.61	18.82	6.11	2.67576	1.07840	29.1
52	10.46	538.07	18.93	6.16	2.68902	1.07890	29.7
53	9.55	547.62	19.19	6.21	2.72079	1.08006	30.3
54	10.44	558.06	19.45	6.26	2.75316	1.08123	30.9
55	10.05	568.11	19.62	6.31	2.77467	1.08200	31.5
56	10.49	578.60	19.86	6.36	2.80549	1.08308	32.1
57	10.47	589.07	20.08	6.41	2.83422	1.08407	32.8
58	10.48	599.55	20.54	6.45	2.89584	1.08616	33.4
59	10.48	610.03	20.81	6.50	2.93301	1.08738	34.0
60	10.48	620.51	21.43	6.55	3.02127	1.09021	34.7
61	9.98	630.49	22.28	6.60	3.14925	1.09410	35.3
62	10.46	640.95	23.33	6.65	3.31942	1.09895	36.0

APPENDIX II: CALCULATING RADIUS–VOLUME RELATIONSHIPS

According to Norman (1971b) the radius (R) which corresponds to a given volume (V) can be calculated from the following expressions:

$$R = \begin{cases} \dfrac{-c_2 + [c_2^2 - 4c_1(c_3 - V)]^{\frac{1}{2}}}{2c_1} & \text{for } V_c < V \leqslant V_m \\[2em] k_0 + k_1(V/100) & \text{for } V_m < V \end{cases}$$

The values of the constants are as follows:

	BXIV rotor	*BXV rotor*
V_c	70.3 ml	104.2 ml
V_m	457 ml	1208 ml
c_1	18.311	25.747
c_2	−19.88	−35.14
c_3	−27.42	−19.18
k_0	3.344 97	4.314 52
k_1	0.518 165	0.273 54

It should be noted that the design of rotors from different firms may vary slightly. Hence, it is advisable to check the proposed radius–volume relationship experimentally before use.

APPENDIX III: SEDIM VALUES FOR VARIOUS PARTICLE DENSITIES

Table 1

SEDIM VALUES AT PARTICLE DENSITY 1.1

SUCROSE CONC.	≠	4	5	CENTIGRADES 6	7	8	9	10
0	≠	1.591	1.542	1.495	1.45	1.407	1.366	1.327
.5	≠	1.645	1.594	1.545	1.499	1.454	1.411	1.37
1	≠	1.701	1.649	1.598	1.55	1.504	1.459	1.417
1.5	≠	1.761	1.707	1.654	1.604	1.556	1.51	1.465
2	≠	1.825	1.768	1.713	1.661	1.611	1.563	1.517
2.5	≠	1.892	1.833	1.776	1.722	1.669	1.619	1.572
3	≠	1.963	1.901	1.842	1.786	1.731	1.679	1.629
3.5	≠	2.039	1.974	1.912	1.853	1.797	1.742	1.69
4	≠	2.119	2.051	1.987	1.925	1.866	1.81	1.755
4.5	≠	2.204	2.134	2.067	2.002	1.94	1.881	1.824
5	≠	2.295	2.222	2.151	2.084	2.019	1.957	1.898
5.5	≠	2.392	2.315	2.241	2.171	2.103	2.038	1.976
6	≠	2.496	2.415	2.338	2.263	2.192	2.124	2.059
6.5	≠	2.607	2.522	2.44	2.363	2.288	2.217	2.148
7	≠	2.726	2.636	2.551	2.469	2.391	2.315	2.243
7.5	≠	2.853	2.759	2.669	2.583	2.5	2.421	2.346
8	≠	2.991	2.892	2.797	2.706	2.619	2.535	2.455
8.5	≠	3.139	3.034	2.934	2.838	2.746	2.658	2.573
9	≠	3.299	3.188	3.082	2.98	2.883	2.79	2.701
9.5	≠	3.472	3.355	3.242	3.134	3.032	2.933	2.838
10	≠	3.66	3.535	3.416	3.302	3.192	3.088	2.987
10.5	≠	3.865	3.732	3.605	3.483	3.367	3.256	3.149
11	≠	4.088	3.946	3.811	3.682	3.558	3.439	3.326
11.5	≠	4.332	4.181	4.036	3.898	3.766	3.639	3.518
12	≠	4.601	4.439	4.284	4.136	3.994	3.859	3.729
12.5	≠	4.897	4.723	4.556	4.397	4.246	4.1	3.961
13	≠	5.224	5.037	4.858	4.686	4.523	4.367	4.217
13.5	≠	5.588	5.385	5.192	5.007	4.831	4.662	4.5
14	≠	5.994	5.774	5.565	5.365	5.174	4.991	4.816
14.5	≠	6.45	6.211	5.983	5.765	5.557	5.359	5.169
15	≠	6.964	6.703	6.454	6.217	5.99	5.773	5.565
15.5	≠	7.549	7.263	6.989	6.729	6.48	6.242	6.014
16	≠	8.219	7.903	7.601	7.313	7.039	6.776	6.526
16.5	≠	8.992	8.641	8.306	7.987	7.682	7.391	7.113
17	≠	9.893	9.5	9.126	8.769	8.428	8.103	7.793
17.5	≠	10.96	10.51	10.09	9.688	9.305	8.939	8.589
18	≠	12.23	11.72	11.24	10.78	10.35	9.929	9.533
18.5	≠	13.77	13.18	12.63	12.1	11.6	11.12	10.67
19	≠	15.67	14.99	14.34	13.73	13.14	12.58	12.05
19.5	≠	18.08	17.27	16.5	15.77	15.07	14.41	13.78
20	≠	21.23	20.24	19.29	18.4	17.56	16.75	15.99
20.5	≠	25.49	24.24	23.05	21.93	20.87	19.86	18.91
21	≠	31.58	29.93	28.37	26.89	25.5	24.19	22.95
21.5	≠	40.98	38.63	36.42	34.35	32.41	30.59	28.88
22	≠	57.32	53.55	50.05	46.81	43.8	41	38.41

Table 1 *continued*

SEDIM VALUES AT PARTICLE DENSITY 1.1

SUCROSE CONC.	≠	4	5	CENTIGRADES 6	7	8	9	10
22.5	≠	92.64	84.94	77.99	71.71	66.02	60.86	56.1
	≠							
23	≠							
23.5	≠							
24	≠							
24.5	≠							
	≠							
25	≠							
	≠							
25.5	≠							
26	≠							
26.5	≠							
27	≠							
	≠							
27.5	≠							
	≠							
28	≠							
28.5	≠							
29	≠							
29.5	≠							
	≠							
30	≠							
	≠							
30.5	≠							
31	≠							
31.5	≠							
32	≠							
	≠							
32.5	≠							
	≠							
33	≠							
33.5	≠							
34	≠							
34.5	≠							
	≠							
35	≠							
	≠							
35.5	≠							
36	≠							
36.5	≠							
37	≠							
	≠							
37.5	≠							
	≠							
38	≠							
38.5	≠							
39	≠							
39.5	≠							
	≠							
40	≠							
	≠							
40.5	≠							
41	≠							
41.5	≠							
42	≠							
	≠							
42.5	≠							
	≠							
43	≠							
43.5	≠							
44	≠							
44.5	≠							

Table 1 *continued*

SEDIM VALUES AT PARTICLE DENSITY 1.1

SUCROSE CONC.	≠	11	12	13	14	15	16	17
0	≠	1.288	1.252	1.217	1.183	1.15	1.119	1.089
.5	≠	1.331	1.293	1.257	1.221	1.188	1.155	1.124
1	≠	1.376	1.336	1.298	1.262	1.227	1.193	1.161
1.5	≠	1.423	1.382	1.343	1.305	1.269	1.233	1.2
2	≠	1.473	1.43	1.389	1.35	1.312	1.276	1.241
2.5	≠	1.525	1.481	1.439	1.398	1.358	1.321	1.284
3	≠	1.581	1.535	1.491	1.448	1.407	1.368	1.33
3.5	≠	1.64	1.592	1.546	1.502	1.459	1.418	1.378
4	≠	1.703	1.653	1.605	1.558	1.514	1.471	1.43
4.5	≠	1.77	1.717	1.667	1.619	1.572	1.527	1.484
5	≠	1.84	1.786	1.733	1.682	1.634	1.587	1.542
5.5	≠	1.916	1.859	1.803	1.75	1.7	1.651	1.603
6	≠	1.996	1.936	1.878	1.823	1.77	1.718	1.669
6.5	≠	2.082	2.019	1.959	1.9	1.844	1.791	1.739
7	≠	2.174	2.108	2.044	1.983	1.924	1.868	1.813
7.5	≠	2.273	2.203	2.136	2.072	2.01	1.95	1.893
8	≠	2.379	2.305	2.235	2.167	2.102	2.039	1.979
8.5	≠	2.493	2.415	2.34	2.269	2.2	2.134	2.07
9	≠	2.615	2.533	2.454	2.379	2.306	2.236	2.169
9.5	≠	2.748	2.661	2.578	2.498	2.421	2.347	2.276
10	≠	2.891	2.799	2.711	2.626	2.544	2.466	2.391
10.5	≠	3.047	2.949	2.855	2.765	2.679	2.595	2.516
11	≠	3.217	3.113	3.013	2.917	2.825	2.736	2.651
11.5	≠	3.402	3.291	3.184	3.082	2.983	2.889	2.798
12	≠	3.605	3.486	3.372	3.262	3.157	3.056	2.959
12.5	≠	3.828	3.7	3.578	3.46	3.348	3.24	3.136
13	≠	4.074	3.936	3.805	3.679	3.557	3.441	3.33
13.5	≠	4.346	4.198	4.056	3.92	3.789	3.664	3.544
14	≠	4.648	4.488	4.335	4.187	4.046	3.911	3.781
14.5	≠	4.987	4.813	4.646	4.486	4.333	4.186	4.045
15	≠	5.367	5.177	4.995	4.821	4.654	4.494	4.341
15.5	≠	5.797	5.589	5.389	5.199	5.016	4.841	4.673
16	≠	6.286	6.057	5.838	5.628	5.427	5.234	5.05
16.5	≠	6.847	6.593	6.351	6.118	5.896	5.683	5.479
17	≠	7.497	7.214	6.943	6.684	6.436	6.199	5.973
17.5	≠	8.256	7.938	7.633	7.343	7.065	6.799	6.545
18	≠	9.154	8.793	8.448	8.118	7.804	7.503	7.216
18.5	≠	10.23	9.817	9.421	9.044	8.684	8.34	8.012
19	≠	11.54	11.06	10.6	10.17	9.748	9.35	8.97
19.5	≠	13.18	12.61	12.07	11.55	11.06	10.59	10.14
20	≠	15.26	14.57	13.92	13.3	12.71	12.15	11.61
20.5	≠	18.01	17.15	16.34	15.57	14.85	14.16	13.5
21	≠	21.78	20.68	19.63	18.65	17.72	16.84	16.02
21.5	≠	27.28	25.77	24.36	23.03	21.78	20.62	19.52
22	≠	36	33.76	31.68	29.75	27.95	26.27	24.72

Table 1 *continued*

SEDIM VALUES AT PARTICLE DENSITY 1.1

SUCROSE CONC.	≠	CENTIGRADES						
		11	12	13	14	15	16	17
22.5	≠	51.93	48.06	44.53	41.32	38.38	35.69	33.2
	≠							
23	≠	90.15	80.9	72.88	65.9	59.78	54.39	49.6
23.5	≠							94.2
24	≠							
24.5	≠							
	≠							
25	≠							
	≠							
25.5	≠							
26	≠							
26.5	≠							
27	≠							
	≠							
27.5	≠							
	≠							
28	≠							
28.5	≠							
29	≠							
29.5	≠							
	≠							
30	≠							
	≠							
30.5	≠							
31	≠							
31.5	≠							
32	≠							
	≠							
32.5	≠							
	≠							
33	≠							
33.5	≠							
34	≠							
34.5	≠							
	≠							
35	≠							
	≠							
35.5	≠							
36	≠							
36.5	≠							
37	≠							
	≠							
37.5	≠							
	≠							
38	≠							
38.5	≠							
39	≠							
39.5	≠							
	≠							
40	≠							
	≠							
40.5	≠							
41	≠							
41.5	≠							
42	≠							
	≠							
42.5	≠							
	≠							
43	≠							
43.5	≠							
44	≠							
44.5	≠							

Table 1 *continued*

SEDIM VALUES AT PARTICLE DENSITY 1.1

SUCROSE CONC.	≠	CENTIGRADES 18	19	20	21	22	23	24
0	≠	1.06	1.032	1.004	.9783	.953	.9286	.9051
.5	≠	1.094	1.065	1.036	1.009	.9832	.9579	.9335
1	≠	1.129	1.099	1.07	1.042	1.015	.9887	.9633
1.5	≠	1.167	1.136	1.106	1.076	1.048	1.021	.9948
2	≠	1.207	1.174	1.143	1.113	1.083	1.055	1.028
2.5	≠	1.249	1.215	1.182	1.151	1.12	1.091	1.063
3	≠	1.293	1.258	1.224	1.191	1.16	1.129	1.099
3.5	≠	1.34	1.303	1.268	1.234	1.201	1.169	1.138
4	≠	1.39	1.352	1.315	1.279	1.245	1.211	1.179
4.5	≠	1.443	1.403	1.364	1.327	1.291	1.256	1.223
5	≠	1.499	1.457	1.416	1.378	1.34	1.304	1.269
5.5	≠	1.558	1.514	1.472	1.431	1.392	1.354	1.318
6	≠	1.621	1.575	1.531	1.489	1.448	1.408	1.37
6.5	≠	1.689	1.641	1.594	1.55	1.507	1.465	1.425
7	≠	1.761	1.71	1.662	1.615	1.57	1.526	1.484
7.5	≠	1.838	1.785	1.734	1.685	1.637	1.591	1.547
8	≠	1.921	1.865	1.811	1.759	1.709	1.661	1.615
8.5	≠	2.009	1.95	1.894	1.839	1.787	1.736	1.687
9	≠	2.105	2.043	1.983	1.925	1.87	1.816	1.764
9.5	≠	2.207	2.142	2.078	2.018	1.959	1.902	1.848
10	≠	2.318	2.249	2.182	2.117	2.055	1.995	1.938
10.5	≠	2.439	2.365	2.294	2.225	2.159	2.096	2.035
11	≠	2.569	2.491	2.415	2.342	2.272	2.205	2.14
11.5	≠	2.711	2.628	2.547	2.469	2.395	2.323	2.254
12	≠	2.866	2.777	2.691	2.608	2.528	2.452	2.378
12.5	≠	3.036	2.94	2.848	2.76	2.674	2.592	2.514
13	≠	3.223	3.12	3.021	2.926	2.835	2.747	2.662
13.5	≠	3.429	3.318	3.211	3.109	3.011	2.916	2.825
14	≠	3.657	3.537	3.422	3.312	3.205	3.103	3.005
14.5	≠	3.91	3.781	3.656	3.536	3.421	3.311	3.205
15	≠	4.194	4.053	3.917	3.787	3.662	3.542	3.427
15.5	≠	4.513	4.358	4.21	4.068	3.932	3.801	3.675
16	≠	4.873	4.704	4.541	4.385	4.236	4.092	3.954
16.5	≠	5.284	5.097	4.918	4.746	4.581	4.422	4.271
17	≠	5.756	5.548	5.349	5.158	4.975	4.799	4.631
17.5	≠	6.302	6.069	5.846	5.633	5.429	5.233	5.045
18	≠	6.941	6.678	6.427	6.187	5.957	5.736	5.525
18.5	≠	7.698	7.399	7.113	6.839	6.578	6.327	6.088
19	≠	8.608	8.263	7.933	7.618	7.317	7.03	6.756
19.5	≠	9.72	9.315	8.93	8.562	8.212	7.878	7.559
20	≠	11.11	10.62	10.17	9.73	9.315	8.92	8.544
20.5	≠	12.88	12.29	11.74	11.21	10.71	10.23	9.776
21	≠	15.23	14.49	13.79	13.13	12.51	11.92	11.36
21.5	≠	18.48	17.51	16.6	15.74	14.94	14.18	13.46
22	≠	23.26	21.91	20.65	19.48	18.38	17.36	16.4

Table 1 *continued*

SUCROSE	‡	\multicolumn{6}{c}{SEDIM VALUES AT PARTICLE DENSITY 1.1 CENTIGRADES}					
CONC.	‡	18	19	20	21	22	23	2
22.5	‡	30.97	28.89	26.99	25.24	23.62	22.13	20.
	‡							
23	‡	45.41	41.64	38.28	35.26	32.55	30.1	27.
23.5	‡	82.15	72.22	63.96	57.01	51.1	46.03	41.
24	‡						93.53	79.
24.5	‡							
	‡							
25	‡							
	‡							
25.5	‡							
26	‡							
26.5	‡							
27	‡							
	‡							
27.5	‡							
	‡							
28	‡							
28.5	‡							
29	‡							
29.5	‡							
	‡							
30	‡							
	‡							
30.5	‡							
31	‡							
31.5	‡							
32	‡							
	‡							
32.5	‡							
	‡							
33	‡							
33.5	‡							
34	‡							
34.5	‡							
	‡							
35	‡							
	‡							
35.5	‡							
36	‡							
36.5	‡							
37	‡							
	‡							
37.5	‡							
	‡							
38	‡							
38.5	‡							
39	‡							
39.5	‡							
	‡							
40	‡							
	‡							
40.5	‡							
41	‡							
41.5	‡							
42	‡							
	‡							
42.5	‡							
	‡							
43	‡							
43.5	‡							
44	‡							
44.5	‡							

Table 2

```
                SEDIM VALUES AT PARTICLE DENSITY 1.2
```

SUCROSE CONC.	4	5	6	7	8	9	10
0	1.574	1.525	1.479	1.435	1.393	1.353	1.314
.5	1.611	1.561	1.514	1.469	1.425	1.384	1.344
1	1.649	1.599	1.55	1.504	1.459	1.416	1.376
1.5	1.69	1.637	1.587	1.54	1.494	1.45	1.408
2	1.731	1.678	1.626	1.577	1.53	1.485	1.442
2.5	1.775	1.72	1.667	1.616	1.568	1.522	1.478
3	1.82	1.763	1.709	1.657	1.607	1.56	1.514
3.5	1.867	1.808	1.753	1.699	1.648	1.599	1.552
4	1.916	1.856	1.798	1.743	1.691	1.64	1.592
4.5	1.967	1.905	1.846	1.789	1.735	1.683	1.633
5	2.02	1.956	1.895	1.837	1.781	1.727	1.676
5.5	2.076	2.01	1.947	1.886	1.829	1.774	1.721
6	2.133	2.065	2	1.938	1.879	1.822	1.768
6.5	2.194	2.124	2.056	1.992	1.931	1.872	1.816
7	2.257	2.184	2.115	2.049	1.985	1.925	1.867
7.5	2.323	2.248	2.176	2.108	2.042	1.98	1.92
8	2.392	2.314	2.24	2.169	2.101	2.037	1.975
8.5	2.464	2.383	2.307	2.233	2.163	2.096	2.033
9	2.539	2.456	2.376	2.301	2.228	2.159	2.093
9.5	2.618	2.532	2.449	2.371	2.296	2.224	2.156
10	2.7	2.611	2.526	2.444	2.367	2.293	2.222
10.5	2.787	2.694	2.606	2.521	2.441	2.364	2.291
11	2.877	2.781	2.69	2.602	2.519	2.439	2.363
11.5	2.972	2.873	2.778	2.687	2.6	2.518	2.439
12	3.072	2.969	2.87	2.776	2.686	2.6	2.518
12.5	3.177	3.069	2.967	2.869	2.775	2.686	2.601
13	3.287	3.175	3.068	2.967	2.87	2.777	2.689
13.5	3.403	3.286	3.175	3.07	2.969	2.872	2.78
14	3.525	3.403	3.288	3.178	3.073	2.973	2.877
14.5	3.653	3.527	3.406	3.292	3.182	3.078	2.978
15	3.788	3.657	3.531	3.412	3.298	3.189	3.085
15.5	3.931	3.793	3.663	3.538	3.419	3.306	3.198
16	4.081	3.938	3.801	3.671	3.547	3.429	3.316
16.5	4.24	4.09	3.948	3.812	3.682	3.559	3.441
17	4.408	4.251	4.102	3.96	3.825	3.696	3.573
17.5	4.586	4.422	4.266	4.117	3.976	3.841	3.713
18	4.774	4.602	4.439	4.284	4.136	3.995	3.86
18.5	4.973	4.793	4.622	4.46	4.305	4.157	4.016
19	5.184	4.996	4.817	4.646	4.484	4.329	4.181
19.5	5.409	5.211	5.023	4.844	4.673	4.511	4.357
20	5.647	5.439	5.242	5.054	4.875	4.705	4.542
20.5	5.901	5.682	5.475	5.277	5.089	4.91	4.74
21	6.171	5.941	5.722	5.515	5.317	5.129	4.95
21.5	6.459	6.216	5.986	5.768	5.56	5.362	5.173
22	6.766	6.51	6.268	6.037	5.818	5.61	5.411

Table 2 *continued*

SEDIM VALUES AT PARTICLE DENSITY 1.2

SUCROSE CONC.	≠	CENTIGRADES						
	≠	4	5	6	7	8	9	10
22.5	≠	7.094	6.824	6.569	6.325	6.094	5.874	5.66
	≠							
23	≠	7.445	7.16	6.89	6.633	6.389	6.157	5.93
23.5	≠	7.821	7.52	7.234	6.963	6.705	6.46	6.22
24	≠	8.225	7.906	7.603	7.316	7.043	6.783	6.53
24.5	≠	8.658	8.32	7.999	7.695	7.406	7.131	6.87
	≠							
25	≠	9.124	8.765	8.425	8.102	7.795	7.504	7.22
	≠							
25.5	≠	9.626	9.245	8.883	8.54	8.214	7.905	7.61
26	≠	10.17	9.762	9.377	9.012	8.666	8.337	8.02
26.5	≠	10.75	10.32	9.911	9.523	9.154	8.804	8.47
27	≠	11.39	10.93	10.49	10.07	9.681	9.308	8.95
	≠							
27.5	≠	12.08	11.58	11.12	10.67	10.25	9.853	9.47
	≠							
28	≠	12.83	12.3	11.8	11.32	10.87	10.45	10.0
28.5	≠	13.64	13.07	12.54	12.03	11.55	11.09	10.6
29	≠	14.53	13.92	13.35	12.8	12.28	11.79	11.3
29.5	≠	15.51	14.85	14.23	13.64	13.08	12.56	12.0
	≠							
30	≠	16.58	15.87	15.2	14.56	13.96	13.39	12.8
	≠							
30.5	≠	17.76	16.99	16.26	15.58	14.93	14.31	13.7
31	≠	19.06	18.23	17.44	16.7	15.99	15.33	14.7
31.5	≠	20.5	19.59	18.74	17.93	17.17	16.45	15.7
32	≠	22.1	21.11	20.18	19.3	18.47	17.68	16.9
	≠							
32.5	≠	23.88	22.8	21.79	20.83	19.92	19.06	18.2
	≠							
33	≠	25.88	24.7	23.58	22.53	21.53	20.6	19.7
33.5	≠	28.13	26.83	25.6	24.44	23.35	22.32	21.3
34	≠	30.67	29.23	27.87	26.59	25.39	24.25	23.1
34.5	≠	33.56	31.96	30.45	29.04	27.7	26.44	25.2
	≠							
35	≠	36.86	35.08	33.4	31.82	30.33	28.93	27.6
	≠							
35.5	≠	40.66	38.66	36.78	35.01	33.34	31.78	30.3
36	≠	45.06	42.8	40.68	38.69	36.82	35.06	33.4
36.5	≠	50.21	47.64	45.24	42.98	40.86	38.87	37
37	≠	56.28	53.35	50.6	48.02	45.6	43.33	41.2
	≠							
37.5	≠	63.53	60.14	56.97	54	51.22	48.62	46.1
	≠							
38	≠	72.3	68.34	64.65	61.2	57.97	54.94	52.1
38.5	≠	83.07	78.4	74.04	69.97	66.17	62.62	59.2
39	≠	96.56	90.95	85.73	80.86	76.33	72.1	68.1
39.5	≠				94.67	89.16	84.04	79.2
	≠							
40	≠						99.46	93.5
	≠							
40.5	≠							
41	≠							
41.5	≠							
42	≠							
	≠							
42.5	≠							
	≠							
43	≠							
43.5	≠							
44	≠							
44.5	≠							

Table 2 *continued*

SEDIM VALUES AT PARTICLE DENSITY 1.2

SUCROSE CONC.	CENTIGRADES 11	12	13	14	15	16	17
0	1.277	1.241	1.207	1.174	1.142	1.112	1.083
.5	1.306	1.269	1.234	1.201	1.168	1.137	1.107
1	1.337	1.299	1.263	1.228	1.195	1.163	1.132
1.5	1.368	1.33	1.293	1.257	1.223	1.19	1.159
2	1.401	1.361	1.323	1.287	1.252	1.218	1.186
2.5	1.435	1.394	1.355	1.318	1.282	1.247	1.214
3	1.471	1.429	1.389	1.35	1.313	1.277	1.243
3.5	1.507	1.464	1.423	1.383	1.345	1.309	1.274
4	1.546	1.501	1.459	1.418	1.379	1.341	1.305
4.5	1.586	1.54	1.496	1.454	1.414	1.375	1.338
5	1.627	1.58	1.535	1.492	1.45	1.41	1.372
5.5	1.67	1.622	1.575	1.531	1.488	1.447	1.408
6	1.715	1.665	1.618	1.572	1.528	1.485	1.444
6.5	1.762	1.711	1.661	1.614	1.569	1.525	1.483
7	1.811	1.758	1.707	1.658	1.611	1.566	1.523
7.5	1.862	1.807	1.755	1.704	1.656	1.609	1.565
8	1.916	1.859	1.804	1.752	1.702	1.654	1.608
8.5	1.971	1.913	1.856	1.803	1.751	1.701	1.654
9	2.029	1.969	1.911	1.855	1.801	1.75	1.701
9.5	2.09	2.027	1.967	1.91	1.854	1.801	1.75
10	2.154	2.089	2.027	1.967	1.91	1.855	1.802
10.5	2.22	2.153	2.089	2.027	1.968	1.911	1.856
11	2.29	2.22	2.153	2.089	2.028	1.969	1.913
11.5	2.363	2.291	2.221	2.155	2.091	2.03	1.972
12	2.439	2.364	2.293	2.224	2.158	2.095	2.034
12.5	2.52	2.442	2.367	2.296	2.227	2.162	2.099
13	2.604	2.523	2.446	2.371	2.3	2.232	2.167
13.5	2.692	2.608	2.528	2.451	2.377	2.306	2.238
14	2.785	2.698	2.614	2.534	2.457	2.384	2.313
14.5	2.883	2.792	2.705	2.622	2.542	2.465	2.392
15	2.986	2.891	2.8	2.714	2.631	2.551	2.475
15.5	3.094	2.995	2.901	2.811	2.724	2.641	2.562
16	3.208	3.105	3.007	2.913	2.823	2.736	2.654
16.5	3.329	3.221	3.119	3.02	2.926	2.836	2.75
17	3.456	3.344	3.236	3.134	3.036	2.942	2.852
17.5	3.59	3.473	3.361	3.254	3.151	3.053	2.959
18	3.732	3.609	3.492	3.38	3.273	3.171	3.073
18.5	3.882	3.754	3.631	3.514	3.402	3.295	3.193
19	4.041	3.906	3.778	3.656	3.539	3.426	3.319
19.5	4.209	4.068	3.934	3.806	3.683	3.566	3.453
20	4.388	4.24	4.099	3.965	3.836	3.713	3.595
20.5	4.577	4.422	4.275	4.133	3.998	3.869	3.746
21	4.779	4.616	4.461	4.313	4.171	4.035	3.906
21.5	4.994	4.822	4.659	4.503	4.354	4.212	4.076
22	5.222	5.042	4.87	4.706	4.549	4.4	4.256

Table 2 *continued*

SEDIM VALUES AT PARTICLE DENSITY 1.2

SUCROSE CONC.	≠	11	12	CENTIGRADES 13	14	15	16	17
22.5	≠	5.466	5.276	5.095	4.922	4.757	4.599	4.44
	≠							
23	≠	5.726	5.526	5.335	5.153	4.979	4.812	4.65
23.5	≠	6.004	5.793	5.591	5.399	5.215	5.04	4.87
24	≠	6.302	6.078	5.865	5.662	5.468	5.283	5.10
24.5	≠	6.621	6.384	6.159	5.944	5.739	5.543	5.35
	≠							
25	≠	6.963	6.713	6.474	6.246	6.029	5.822	5.62
	≠							
25.5	≠	7.331	7.065	6.812	6.571	6.34	6.121	5.91
26	≠	7.728	7.445	7.176	6.919	6.675	6.442	6.22
26.5	≠	8.155	7.854	7.568	7.295	7.036	6.788	6.55
27	≠	8.616	8.296	7.991	7.701	7.424	7.161	6.91
	≠							
27.5	≠	9.115	8.773	8.448	8.139	7.844	7.563	7.29
	≠							
28	≠	9.656	9.291	8.944	8.613	8.298	7.999	7.71
28.5	≠	10.24	9.853	9.481	9.128	8.791	9.471	8.16
29	≠	10.88	10.46	10.07	9.687	9.326	8.983	8.65
29.5	≠	11.58	11.13	10.7	10.3	9.909	9.541	9.19
	≠							
30	≠	12.34	11.86	11.4	10.96	10.55	10.15	9.77
	≠							
30.5	≠	13.18	12.66	12.16	11.69	11.24	10.81	10.4
31	≠	14.1	13.53	13	12.49	12	11.54	11.1
31.5	≠	15.12	14.5	13.92	13.37	12.84	12.35	11.8
32	≠	16.24	15.57	14.94	14.34	13.77	13.23	12.7
	≠							
32.5	≠	17.48	16.76	16.07	15.41	14.79	14.21	13.6
	≠							
33	≠	18.87	18.07	17.32	16.61	15.93	15.29	14.6
33.5	≠	20.42	19.55	18.72	17.94	17.2	16.5	15.8
34	≠	22.16	21.2	20.3	19.44	18.62	17.85	17.1
34.5	≠	24.13	23.17	22.07	21.12	20.22	19.37	18.5
	≠							
35	≠	26.36	25.19	24.08	23.03	22.03	21.09	20.2
	≠							
35.5	≠	28.91	27.6	26.36	25.19	24.09	23.04	22.0
36	≠	31.84	30.37	28.98	27.67	26.44	25.27	24.1
36.5	≠	35.24	33.58	32.01	30.54	29.15	27.83	26.5
37	≠	39.2	37.31	35.54	33.87	32.29	30.8	29.4
	≠							
37.5	≠	43.87	41.71	39.68	37.77	35.97	34.28	32.6
	≠							
38	≠	49.45	46.95	44.61	42.41	40.33	38.39	36.5
38.5	≠	56.18	53.27	50.53	47.97	45.56	43.3	41.1
39	≠	64.45	61	57.77	54.74	51.91	49.25	46.7
39.5	≠	74.81	70.65	66.77	63.15	59.76	56.59	53.6
	≠							
40	≠	88.07	82.97	78.22	73.8	69.67	65.82	62.2
	≠							
40.5	≠		99.16	93.2	87.66	82.51	77.73	73.2
41	≠					99.72	93.59	87.9
41.5	≠							
42	≠							
	≠							
42.5	≠							
	≠							
43	≠							
43.5	≠							
44	≠							
44.5	≠							

Table 2 *continued*

SEDIM VALUES AT PARTICLE DENSITY 1.2

SUCROSE CONC.	≠	CENTIGRADES 18	19	20	21	22	23	24
0	≠	1.055	1.028	1.002	.9764	.9522	.9288	.9063
.5	≠	1.078	1.051	1.024	.9981	.9733	.9493	.9262
1	≠	1.103	1.074	1.047	1.021	.9951	.9706	.9469
1.5	≠	1.128	1.099	1.071	1.044	1.018	.9925	.9682
2	≠	1.155	1.125	1.096	1.068	1.041	1.015	.9903
2.5	≠	1.182	1.151	1.122	1.093	1.065	1.039	1.013
3	≠	1.21	1.179	1.148	1.119	1.091	1.063	1.037
3.5	≠	1.24	1.207	1.176	1.146	1.117	1.089	1.062
4	≠	1.27	1.237	1.205	1.174	1.144	1.115	1.087
4.5	≠	1.302	1.268	1.235	1.203	1.172	1.142	1.114
5	≠	1.335	1.3	1.266	1.233	1.201	1.171	1.141
5.5	≠	1.37	1.333	1.298	1.264	1.232	1.2	1.17
6	≠	1.405	1.368	1.332	1.297	1.263	1.231	1.2
6.5	≠	1.443	1.404	1.367	1.331	1.296	1.263	1.231
7	≠	1.481	1.441	1.403	1.366	1.33	1.296	1.263
7.5	≠	1.522	1.481	1.441	1.403	1.366	1.33	1.296
8	≠	1.564	1.521	1.48	1.441	1.403	1.366	1.331
8.5	≠	1.608	1.564	1.522	1.481	1.442	1.404	1.368
9	≠	1.654	1.608	1.565	1.523	1.482	1.443	1.406
9.5	≠	1.702	1.655	1.609	1.566	1.524	1.484	1.445
10	≠	1.752	1.703	1.656	1.611	1.568	1.527	1.486
10.5	≠	1.804	1.754	1.705	1.659	1.614	1.571	1.53
11	≠	1.859	1.806	1.756	1.708	1.662	1.618	1.575
11.5	≠	1.916	1.862	1.81	1.76	1.712	1.666	1.622
12	≠	1.976	1.92	1.866	1.814	1.765	1.717	1.671
12.5	≠	2.038	1.98	1.925	1.871	1.82	1.77	1.723
13	≠	2.104	2.044	1.986	1.931	1.877	1.826	1.777
13.5	≠	2.173	2.111	2.051	1.993	1.938	1.885	1.833
14	≠	2.246	2.181	2.119	2.059	2.001	1.946	1.893
14.5	≠	2.322	2.254	2.19	2.127	2.068	2.01	1.955
15	≠	2.402	2.332	2.264	2.2	2.137	2.078	2.02
15.5	≠	2.486	2.413	2.343	2.275	2.211	2.149	2.089
16	≠	2.574	2.498	2.425	2.355	2.288	2.223	2.161
16.5	≠	2.668	2.588	2.512	2.439	2.369	2.302	2.237
17	≠	2.766	2.683	2.604	2.528	2.455	2.385	2.317
17.5	≠	2.869	2.783	2.7	2.621	2.545	2.472	2.402
18	≠	2.979	2.889	2.802	2.72	2.64	2.564	2.49
18.5	≠	3.094	3	2.91	2.824	2.74	2.661	2.584
19	≠	3.216	3.118	3.024	2.933	2.847	2.763	2.683
19.5	≠	3.346	3.243	3.144	3.049	2.959	2.872	2.788
20	≠	3.483	3.375	3.271	3.172	3.077	2.986	2.899
20.5	≠	3.628	3.515	3.406	3.303	3.203	3.107	3.016
21	≠	3.782	3.663	3.55	3.441	3.336	3.236	3.14
21.5	≠	3.946	3.821	3.702	3.587	3.478	3.373	3.272
22	≠	4.12	3.989	3.863	3.743	3.628	3.517	3.412

Table 2 *continued*

SEDIM VALUES AT PARTICLE DENSITY 1.2

SUCROSE CONC.	≠	CENTIGRADES						
	≠	18	19	20	21	22	23	24
22.5	≠	4.305	4.167	4.035	3.909	3.788	3.671	3.56
	≠							
23	≠	4.502	4.357	4.218	4.085	3.957	3.835	3.71
23.5	≠	4.712	4.559	4.413	4.273	4.139	4.01	3.88
24	≠	4.937	4.776	4.621	4.473	4.332	4.196	4.06
24.5	≠	5.178	5.007	4.844	4.688	4.538	4.395	4.25
	≠							
25	≠	5.435	5.255	5.082	4.917	4.759	4.608	4.46
	≠							
25.5	≠	5.711	5.52	5.337	5.163	4.995	4.835	4.68
26	≠	6.008	5.805	5.611	5.426	5.249	5.079	4.91
26.5	≠	6.327	6.111	5.906	5.709	5.521	5.341	5.16
27	≠	6.67	6.441	6.222	6.014	5.814	5.623	5.44
	≠							
27.5	≠	7.04	6.797	6.564	6.342	6.129	5.926	5.73
	≠							
28	≠	7.441	7.181	6.933	6.696	6.47	6.253	6.04
28.5	≠	7.874	7.597	7.332	7.079	6.838	6.607	6.38
29	≠	8.345	8.048	7.765	7.494	7.236	6.989	6.75
29.5	≠	8.856	8.538	8.235	7.945	7.668	7.404	7.15
	≠							
30	≠	9.414	9.072	8.746	8.435	8.139	7.855	7.58
	≠							
30.5	≠	10.02	9.655	9.304	8.97	8.651	8.347	8.05
31	≠	10.69	10.29	9.915	9.555	9.211	8.884	8.57
31.5	≠	11.42	10.99	10.58	10.2	9.825	9.471	9.13
32	≠	12.23	11.76	11.32	10.9	10.5	10.12	9.75
	≠							
32.5	≠	13.12	12.61	12.13	11.68	11.24	10.83	10.4
	≠							
33	≠	14.1	13.56	13.03	12.54	12.06	11.61	11.1
33.5	≠	15.2	14.6	14.03	13.49	12.97	12.48	12.0
34	≠	16.43	15.77	15.15	14.55	13.99	13.45	12.9
34.5	≠	17.81	17.08	16.39	15.74	15.12	14.53	13.9
	≠							
35	≠	19.36	18.56	17.8	17.08	16.39	15.74	15.1
	≠							
35.5	≠	21.11	20.23	19.38	18.59	17.83	17.11	16.4
36	≠	23.12	22.13	21.19	20.3	19.46	18.66	17.9
36.5	≠	25.42	24.31	23.26	22.26	21.32	20.43	19.5
37	≠	28.08	26.82	25.64	24.52	23.46	22.46	21.5
	≠							
37.5	≠	31.18	29.75	28.41	27.14	25.94	24.8	23.7
	≠							
38	≠	34.82	33.2	31.56	30.21	28.84	27.55	26.3
38.5	≠	39.17	37.29	35.51	33.84	32.26	30.78	29.3
39	≠	44.41	42.21	40.14	38.19	36.36	34.63	33
39.5	≠	50.84	48.23	45.78	43.48	41.32	39.29	37.3
	≠							
40	≠	58.86	55.72	52.78	50.03	47.44	45.02	42.7
	≠							
40.5	≠	69.13	65.26	61.66	58.29	55.15	52.21	49.6
41	≠	82.65	77.77	73.23	69.01	65.09	61.43	58.0
41.5	≠		94.79	88.38	83.41	78.36	73.68	69.
42	≠					96.86	90.6	84.8
	≠							
42.5	≠							
	≠							
43	≠							
43.5	≠							
44	≠							
44.5	≠							

Table 3

SEDIM VALUES AT PARTICLE DENSITY 1.3

SUCROSE CONC.	≠	4	5	CENTIGRADES 6	7	8	9	10
0	≠	1.568	1.52	1.474	1.43	1.388	1.348	1.31
.5	≠	1.6	1.551	1.504	1.459	1.416	1.375	1.336
1	≠	1.633	1.582	1.534	1.488	1.445	1.403	1.362
1.5	≠	1.667	1.615	1.566	1.519	1.474	1.431	1.39
2	≠	1.702	1.649	1.599	1.551	1.505	1.461	1.419
2.5	≠	1.738	1.684	1.633	1.584	1.537	1.492	1.448
3	≠	1.776	1.721	1.668	1.618	1.569	1.523	1.479
3.5	≠	1.815	1.759	1.705	1.653	1.603	1.556	1.511
4	≠	1.856	1.798	1.742	1.689	1.639	1.59	1.544
4.5	≠	1.898	1.838	1.781	1.727	1.675	1.625	1.578
5	≠	1.942	1.88	1.822	1.766	1.713	1.662	1.613
5.5	≠	1.987	1.924	1.864	1.807	1.752	1.699	1.649
6	≠	2.034	1.969	1.908	1.849	1.792	1.738	1.687
6.5	≠	2.083	2.016	1.953	1.892	1.834	1.779	1.726
7	≠	2.133	2.065	2	1.937	1.878	1.821	1.767
7.5	≠	2.186	2.115	2.048	1.984	1.923	1.865	1.809
8	≠	2.24	2.168	2.099	2.033	1.97	1.91	1.853
8.5	≠	2.297	2.222	2.151	2.084	2.019	1.957	1.898
9	≠	2.356	2.279	2.206	2.136	2.07	2.006	1.945
9.5	≠	2.417	2.338	2.263	2.191	2.122	2.057	1.994
10	≠	2.481	2.399	2.322	2.248	2.177	2.109	2.045
10.5	≠	2.547	2.463	2.383	2.307	2.234	2.164	2.098
11	≠	2.616	2.529	2.447	2.368	2.293	2.221	2.153
11.5	≠	2.688	2.598	2.513	2.432	2.354	2.281	2.21
12	≠	2.763	2.67	2.582	2.499	2.419	2.342	2.269
12.5	≠	2.841	2.745	2.654	2.568	2.485	2.406	2.331
13	≠	2.922	2.823	2.729	2.64	2.555	2.473	2.396
13.5	≠	3.006	2.905	2.808	2.715	2.627	2.543	2.463
14	≠	3.095	2.99	2.889	2.794	2.703	2.616	2.533
14.5	≠	3.187	3.078	2.974	2.876	2.781	2.692	2.606
15	≠	3.283	3.171	3.063	2.961	2.864	2.771	2.682
15.5	≠	3.384	3.267	3.156	3.05	2.949	2.853	2.761
16	≠	3.489	3.368	3.253	3.143	3.039	2.939	2.844
16.5	≠	3.599	3.473	3.354	3.241	3.132	3.029	2.931
17	≠	3.714	3.584	3.46	3.342	3.23	3.123	3.021
17.5	≠	3.834	3.699	3.571	3.449	3.332	3.222	3.116
18	≠	3.96	3.82	3.687	3.56	3.439	3.324	3.215
18.5	≠	4.092	3.946	3.808	3.677	3.551	3.432	3.318
19	≠	4.23	4.079	3.935	3.799	3.669	3.545	3.427
19.5	≠	4.375	4.218	4.068	3.926	3.791	3.663	3.54
20	≠	4.527	4.363	4.208	4.061	3.92	3.786	3.659
20.5	≠	4.686	4.516	4.355	4.201	4.055	3.916	3.784
21	≠	4.854	4.677	4.509	4.349	4.197	4.052	3.914
21.5	≠	5.03	4.845	4.67	4.504	4.346	4.195	4.052
22	≠	5.215	5.022	4.84	4.667	4.502	4.345	4.196

Table 3 *continued*

SEDIM VALUES AT PARTICLE DENSITY 1.3

SUCROSE CONC.	≠	CENTIGRADES						
	≠	4	5	6	7	8	9	10
22.5	≠	5.409	5.209	5.019	4.838	4.666	4.503	4.34
	≠							
23	≠	5.614	5.405	5.206	5.018	4.839	4.668	4.50
23.5	≠	5.83	5.612	5.404	5.207	5.02	4.843	4.67
24	≠	6.058	5.829	5.613	5.407	5.212	5.026	4.85
24.5	≠	6.297	6.059	5.832	5.617	5.413	5.22	5.03
	≠							
25	≠	6.551	6.301	6.064	5.839	5.626	5.423	5.23
	≠							
25.5	≠	6.818	6.557	6.309	6.074	5.85	5.638	5.43
26	≠	7.101	6.827	6.567	6.321	6.087	5.865	5.65
26.5	≠	7.4	7.112	6.84	6.582	6.337	6.105	5.88
27	≠	7.716	7.415	7.129	6.858	6.602	6.358	6.12
	≠							
27.5	≠	8.051	7.735	7.435	7.151	6.881	6.626	6.38
	≠							
28	≠	8.406	8.074	7.759	7.461	7.178	6.909	6.65
28.5	≠	8.783	8.433	8.102	7.789	7.492	7.21	6.94
29	≠	9.184	8.815	8.467	8.137	7.824	7.528	7.24
29.5	≠	9.609	9.221	8.854	8.507	8.178	7.866	7.57
	≠							
30	≠	10.06	9.652	9.266	8.9	8.553	8.225	7.91
	≠							
30.5	≠	10.54	10.11	9.704	9.318	8.952	8.606	8.27
31	≠	11.06	10.6	10.17	9.762	9.377	9.011	8.66
31.5	≠	11.6	11.12	10.67	10.24	9.829	9.443	9.07
32	≠	12.19	11.68	11.2	10.74	10.31	9.903	9.51
	≠							
32.5	≠	12.81	12.27	11.76	11.28	10.83	10.39	9.98
	≠							
33	≠	13.48	12.91	12.37	11.85	11.38	10.92	10.4
33.5	≠	14.2	13.59	13.02	12.48	11.96	11.48	11.0
34	≠	14.97	14.32	13.71	13.14	12.59	12.08	11.5
34.5	≠	15.79	15.1	14.46	13.84	13.27	12.72	12.2
	≠							
35	≠	16.68	15.95	15.26	14.6	13.99	13.41	12.8
	≠							
35.5	≠	17.63	16.85	16.12	15.42	14.77	14.15	13.5
36	≠	18.66	17.83	17.04	16.3	15.6	14.94	14.3
36.5	≠	19.77	18.88	18.04	17.25	16.5	15.8	15.1
37	≠	20.97	20.01	19.12	18.27	17.47	16.72	16.0
	≠							
37.5	≠	22.26	21.24	20.28	19.37	18.52	17.72	16.9
	≠							
38	≠	23.67	22.57	21.54	20.57	19.65	18.79	17.9
38.5	≠	25.19	24.01	22.9	21.86	20.88	19.95	19.0
39	≠	26.85	25.58	24.38	23.26	22.21	21.21	20.2
39.5	≠	28.65	27.28	25.99	24.78	23.65	22.58	21.5
	≠							
40	≠	30.61	29.13	27.74	26.44	25.22	24.06	22.9
	≠							
40.5	≠	32.75	31.15	29.65	28.24	26.92	25.68	24.5
41	≠	35.09	33.36	31.74	30.21	28.79	27.44	26.1
41.5	≠	37.65	35.78	34.02	32.37	30.82	29.37	28
42	≠	40.47	38.43	36.52	34.73	33.05	31.47	29.9
	≠							
42.5	≠	43.56	41.34	39.26	37.31	35.49	33.78	32.1
	≠							
43	≠	46.96	44.54	42.27	40.15	38.17	36.31	34.56
43.5	≠	50.72	48.07	45.6	43.28	41.12	39.09	37.1
44	≠	54.87	51.97	49.26	46.74	44.37	42.15	40.0
44.5	≠	59.47	56.29	53.32	50.55	47.96	45.54	43.2

Table 3 *continued*

SEDIM VALUES AT PARTICLE DENSITY 1.3

SUCROSE CONC.	≠	CENTIGRADES						
	≠	11	12	13	14	15	16	17
0	≠	1.273	1.237	1.203	1.171	1.14	1.11	1.081
	≠							
.5	≠	1.298	1.262	1.227	1.194	1.162	1.131	1.102
1	≠	1.324	1.287	1.251	1.217	1.185	1.153	1.123
1.5	≠	1.351	1.313	1.277	1.242	1.208	1.176	1.145
2	≠	1.378	1.34	1.303	1.267	1.233	1.2	1.168
	≠							
2.5	≠	1.407	1.367	1.329	1.293	1.258	1.224	1.192
	≠							
3	≠	1.437	1.396	1.357	1.32	1.284	1.25	1.217
3.5	≠	1.467	1.426	1.386	1.348	1.311	1.276	1.242
4	≠	1.499	1.456	1.416	1.376	1.339	1.303	1.268
4.5	≠	1.532	1.488	1.446	1.406	1.368	1.331	1.295
	≠							
5	≠	1.566	1.521	1.478	1.437	1.397	1.359	1.323
	≠							
5.5	≠	1.601	1.555	1.511	1.469	1.428	1.389	1.352
6	≠	1.638	1.59	1.545	1.502	1.46	1.42	1.382
6.5	≠	1.675	1.627	1.581	1.536	1.493	1.452	1.413
7	≠	1.715	1.665	1.617	1.571	1.528	1.485	1.445
	≠							
7.5	≠	1.755	1.704	1.655	1.608	1.563	1.52	1.478
	≠							
8	≠	1.798	1.745	1.695	1.645	1.6	1.535	1.513
8.5	≠	1.841	1.787	1.735	1.686	1.638	1.592	1.548
9	≠	1.887	1.831	1.778	1.727	1.678	1.631	1.586
9.5	≠	1.934	1.877	1.822	1.769	1.719	1.67	1.624
	≠							
10	≠	1.983	1.924	1.868	1.813	1.762	1.712	1.664
	≠							
10.5	≠	2.034	1.973	1.915	1.859	1.806	1.755	1.705
11	≠	2.087	2.024	1.964	1.907	1.852	1.799	1.749
11.5	≠	2.142	2.078	2.016	1.957	1.9	1.845	1.793
12	≠	2.2	2.133	2.069	2.008	1.95	1.894	1.84
	≠							
12.5	≠	2.259	2.191	2.125	2.062	2.001	1.944	1.888
	≠							
13	≠	2.321	2.25	2.183	2.118	2.055	1.996	1.939
13.5	≠	2.386	2.313	2.243	2.176	2.111	2.05	1.991
14	≠	2.454	2.378	2.306	2.236	2.17	2.106	2.045
14.5	≠	2.524	2.446	2.371	2.299	2.231	2.165	2.102
	≠							
15	≠	2.597	2.516	2.439	2.365	2.294	2.226	2.161
	≠							
15.5	≠	2.674	2.59	2.51	2.434	2.36	2.29	2.223
16	≠	2.754	2.667	2.584	2.505	2.429	2.357	2.288
16.5	≠	2.837	2.747	2.662	2.58	2.501	2.426	2.355
17	≠	2.924	2.831	2.743	2.658	2.577	2.499	2.425
	≠							
17.5	≠	3.015	2.919	2.827	2.739	2.655	2.575	2.498
	≠							
18	≠	3.11	3.011	2.915	2.824	2.737	2.654	2.574
18.5	≠	3.21	3.107	3.008	2.913	2.823	2.737	2.654
19	≠	3.314	3.207	3.104	3.006	2.913	2.823	2.738
19.5	≠	3.423	3.312	3.205	3.104	3.007	2.914	2.825
	≠							
20	≠	3.538	3.422	3.311	3.206	3.105	3.009	2.916
	≠							
20.5	≠	3.658	3.537	3.422	3.313	3.208	3.108	3.012
21	≠	3.783	3.658	3.539	3.425	3.316	3.212	3.112
21.5	≠	3.915	3.785	3.661	3.542	3.429	3.321	3.217
22	≠	4.053	3.918	3.789	3.665	3.548	3.435	3.327

Table 3 *continued*

SEDIM VALUES AT PARTICLE DENSITY 1.3

SUCROSE CONC.	≠	11	12	CENTIGRADES 13	14	15	16	17
22.5	≠	4.199	4.058	3.923	3.795	3.672	3.555	3.443
23	≠	4.352	4.205	4.064	3.931	3.803	3.681	3.564
23.5	≠	4.512	4.359	4.213	4.073	3.94	3.813	3.692
24	≠	4.682	4.522	4.369	4.224	4.085	3.952	3.826
24.5	≠	4.86	4.693	4.533	4.382	4.237	4.099	3.967
25	≠	5.047	4.873	4.706	4.548	4.397	4.252	4.115
25.5	≠	5.245	5.062	4.889	4.723	4.565	4.414	4.271
26	≠	5.454	5.263	5.081	4.908	4.743	4.585	4.435
26.5	≠	5.674	5.474	5.284	5.103	4.93	4.765	4.608
27	≠	5.906	5.697	5.498	5.308	5.127	4.955	4.791
27.5	≠	6.152	5.933	5.724	5.525	5.336	5.155	4.983
28	≠	6.412	6.182	5.963	5.755	5.556	5.367	5.187
28.5	≠	6.688	6.446	6.216	5.998	5.789	5.591	5.402
29	≠	6.98	6.726	6.484	6.255	6.036	5.828	5.629
29.5	≠	7.289	7.022	6.768	6.527	6.297	6.079	5.87
30	≠	7.617	7.336	7.069	6.816	6.574	6.344	6.125
30.5	≠	7.966	7.67	7.389	7.122	6.868	6.626	6.395
31	≠	8.337	8.025	7.729	7.447	7.179	6.925	6.682
31.5	≠	8.731	8.402	8.09	7.793	7.511	7.242	6.987
32	≠	9.151	8.803	8.474	8.161	7.863	7.58	7.31
32.5	≠	9.599	9.231	8.883	8.552	8.238	7.939	7.654
33	≠	10.08	9.688	9.319	8.969	8.637	8.321	8.021
33.5	≠	10.59	10.17	9.785	9.414	9.063	8.729	8.411
34	≠	11.13	10.7	10.28	9.89	9.517	9.164	8.828
34.5	≠	11.72	11.25	10.81	10.4	10	9.628	9.272
35	≠	12.34	11.85	11.38	10.94	10.52	10.13	9.747
35.5	≠	13.01	12.49	11.99	11.52	11.08	10.66	10.26
36	≠	13.73	13.18	12.65	12.15	11.68	11.23	10.8
36.5	≠	14.51	13.91	13.35	12.82	12.32	11.84	11.38
37	≠	15.34	14.71	14.11	13.54	13	12.49	12.01
37.5	≠	16.24	15.56	14.92	14.32	13.74	13.2	12.69
38	≠	17.21	16.49	15.8	15.15	14.54	13.96	13.41
38.5	≠	18.26	17.48	16.75	16.06	15.4	14.78	14.19
39	≠	19.39	18.56	17.78	17.03	16.33	15.67	15.04
39.5	≠	20.63	19.73	18.89	18.09	17.33	16.62	15.95
40	≠	21.96	21	20.09	19.23	18.42	17.66	16.93
40.5	≠	23.41	22.38	21.4	20.48	19.6	18.78	18
41	≠	24.99	23.87	22.82	21.83	20.89	20	19.16
41.5	≠	26.72	25.51	24.37	23.29	22.28	21.32	20.42
42	≠	28.6	27.29	26.06	24.9	23.8	22.77	21.79
42.5	≠	30.66	29.24	27.91	26.65	25.46	24.34	23.29
43	≠	32.92	31.38	29.93	28.56	27.28	26.07	24.92
43.5	≠	35.4	33.72	32.15	30.66	29.27	27.95	26.71
44	≠	38.13	36.3	34.58	32.97	31.45	30.02	28.67
44.5	≠	41.14	39.14	37.27	35.51	33.85	32.29	30.83

Table 3 *continued*

SEDIM VALUES AT PARTICLE DENSITY 1.3

SUCROSE CONC.	≠	18	19	20	21	22	23	24
				CENTIGRADES				
0	≠	1.053	1.026	1.001	.9758	.9519	.9289	.9067
	≠							
.5	≠	1.073	1.046	1.02	.9944	.97	.9465	.9238
1	≠	1.094	1.066	1.039	1.014	.9886	.9646	.9414
1.5	≠	1.116	1.087	1.06	1.033	1.008	.9833	.9596
2	≠	1.138	1.109	1.081	1.054	1.028	1.003	.9783
	≠							
2.5	≠	1.161	1.131	1.102	1.075	1.048	1.022	.9976
	≠							
3	≠	1.185	1.154	1.125	1.096	1.069	1.043	1.018
3.5	≠	1.209	1.178	1.148	1.119	1.091	1.064	1.038
4	≠	1.235	1.203	1.172	1.142	1.113	1.086	1.059
4.5	≠	1.261	1.228	1.196	1.166	1.137	1.108	1.081
	≠							
5	≠	1.288	1.254	1.222	1.191	1.161	1.132	1.104
	≠							
5.5	≠	1.316	1.281	1.248	1.216	1.185	1.156	1.127
6	≠	1.345	1.31	1.275	1.243	1.211	1.181	1.151
6.5	≠	1.375	1.339	1.304	1.27	1.238	1.206	1.176
7	≠	1.406	1.369	1.333	1.298	1.265	1.233	1.202
	≠							
7.5	≠	1.438	1.4	1.363	1.328	1.294	1.261	1.229
	≠							
8	≠	1.472	1.432	1.395	1.358	1.323	1.289	1.257
8.5	≠	1.506	1.466	1.427	1.39	1.354	1.319	1.286
9	≠	1.542	1.501	1.461	1.422	1.385	1.35	1.315
9.5	≠	1.58	1.537	1.496	1.456	1.418	1.381	1.346
	≠							
10	≠	1.618	1.574	1.532	1.491	1.452	1.414	1.378
	≠							
10.5	≠	1.658	1.613	1.569	1.528	1.487	1.449	1.411
11	≠	1.7	1.653	1.609	1.565	1.524	1.484	1.446
11.5	≠	1.743	1.695	1.649	1.605	1.562	1.521	1.482
12	≠	1.788	1.739	1.691	1.646	1.602	1.559	1.519
	≠							
12.5	≠	1.835	1.784	1.735	1.688	1.643	1.599	1.557
	≠							
13	≠	1.884	1.831	1.781	1.732	1.685	1.641	1.598
13.5	≠	1.934	1.88	1.828	1.778	1.73	1.684	1.639
14	≠	1.987	1.931	1.877	1.826	1.776	1.728	1.683
14.5	≠	2.042	1.984	1.929	1.875	1.824	1.775	1.728
	≠							
15	≠	2.099	2.039	1.982	1.927	1.874	1.824	1.775
	≠							
15.5	≠	2.159	2.097	2.038	1.981	1.926	1.874	1.824
16	≠	2.221	2.157	2.096	2.037	1.981	1.927	1.875
16.5	≠	2.286	2.22	2.157	2.096	2.038	1.982	1.928
17	≠	2.353	2.285	2.22	2.157	2.097	2.039	1.983
	≠							
17.5	≠	2.424	2.353	2.286	2.221	2.158	2.098	2.041
	≠							
18	≠	2.498	2.425	2.355	2.287	2.223	2.161	2.101
18.5	≠	2.575	2.499	2.427	2.357	2.29	2.226	2.164
19	≠	2.656	2.577	2.502	2.429	2.36	2.294	2.23
19.5	≠	2.74	2.658	2.58	2.505	2.434	2.365	2.299
	≠							
20	≠	2.828	2.743	2.662	2.585	2.51	2.439	2.37
	≠							
20.5	≠	2.92	2.833	2.749	2.668	2.591	2.517	2.445
21	≠	3.017	2.926	2.839	2.755	2.675	2.598	2.524
21.5	≠	3.118	3.024	2.933	2.846	2.763	2.683	2.606
22	≠	3.225	3.126	3.032	2.942	2.855	2.772	2.692

Table 3 *continued*

SEDIM VALUES AT PARTICLE DENSITY 1.3

SUCROSE CONC.	≠	18	19	20	21	22	23	24
				CENTIGRADES				
22.5	≠	3.336	3.234	3.136	3.042	2.952	2.865	2.783
	≠							
23	≠	3.453	3.346	3.244	3.147	3.053	2.963	2.877
23.5	≠	3.576	3.465	3.359	3.257	3.159	3.066	2.977
24	≠	3.705	3.589	3.478	3.373	3.271	3.174	3.081
24.5	≠	3.84	3.72	3.604	3.494	3.388	3.287	3.19
	≠							
25	≠	3.983	3.857	3.737	3.622	3.512	3.406	3.305
	≠							
25.5	≠	4.133	4.002	3.876	3.756	3.641	3.531	3.426
26	≠	4.291	4.154	4.023	3.898	3.778	3.663	3.553
26.5	≠	4.458	4.315	4.178	4.047	3.921	3.801	3.686
27	≠	4.634	4.484	4.34	4.203	4.072	3.947	3.827
	≠							
27.5	≠	4.819	4.662	4.512	4.369	4.232	4.101	3.975
	≠							
28	≠	5.015	4.85	4.693	4.543	4.4	4.263	4.131
28.5	≠	5.221	5.049	4.885	4.728	4.577	4.434	4.296
29	≠	5.44	5.259	5.087	4.922	4.765	4.614	4.47
29.5	≠	5.671	5.482	5.301	5.128	4.963	4.805	4.654
	≠							
30	≠	5.916	5.717	5.527	5.345	5.172	5.006	4.848
	≠							
30.5	≠	6.176	5.966	5.767	5.576	5.394	5.22	5.053
31	≠	6.451	6.231	6.021	5.82	5.629	5.446	5.271
31.5	≠	6.743	6.511	6.29	6.079	5.877	5.685	5.501
32	≠	7.054	6.809	6.576	6.354	6.142	5.939	5.745
	≠							
32.5	≠	7.384	7.126	6.88	6.646	6.422	6.208	6.004
	≠							
33	≠	7.735	7.463	7.203	6.956	6.72	6.495	6.28
33.5	≠	8.109	7.821	7.547	7.286	7.037	6.799	6.572
34	≠	8.508	8.203	7.914	7.637	7.374	7.123	6.883
34.5	≠	8.933	8.611	8.304	8.012	7.734	7.468	7.215
	≠							
35	≠	9.388	9.047	8.722	8.412	8.117	7.836	7.568
	≠							
35.5	≠	9.875	9.512	9.167	8.839	8.527	8.229	7.945
36	≠	10.4	10.01	9.644	9.296	8.964	8.649	8.347
36.5	≠	10.95	10.54	10.16	9.785	9.433	9.097	8.778
37	≠	11.55	11.12	10.7	10.31	9.935	9.578	9.238
	≠							
37.5	≠	12.2	11.73	11.29	10.87	10.47	10.09	9.732
	≠							
38	≠	12.89	12.39	11.92	11.48	11.05	10.65	10.26
38.5	≠	13.63	13.1	12.6	12.13	11.67	11.24	10.83
39	≠	14.44	13.87	13.34	12.83	12.34	11.88	11.44
39.5	≠	15.31	14.7	14.13	13.58	13.06	12.57	12.1
	≠							
40	≠	16.25	15.6	14.98	14.4	13.84	13.31	12.81
	≠							
40.5	≠	17.26	16.57	15.9	15.28	14.68	14.12	13.58
41	≠	18.37	17.62	16.91	16.23	15.59	14.99	14.41
41.5	≠	19.57	18.76	17.99	17.27	16.58	15.93	15.31
42	≠	20.87	20	19.17	18.39	17.65	16.95	16.28
	≠							
42.5	≠	22.29	21.35	20.46	19.62	18.82	18.06	17.34
	≠							
43	≠	23.84	22.83	21.86	20.95	20.09	19.27	18.5
43.5	≠	25.54	24.44	23.39	22.41	21.47	20.59	19.75
44	≠	27.4	26.2	25.07	24	22.99	22.03	21.12
44.5	≠	29.44	28.14	26.91	25.74	24.65	23.61	22.62

Table 4

SEDIM VALUES AT PARTICLE DENSITY 1.4

SUCROSE CONC.	≠	4	5	6	7	8	9	10
				CENTIGRADES				
0	≠	1.565	1.517	1.471	1.428	1.386	1.346	1.307
.5	≠	1.594	1.545	1.499	1.454	1.411	1.37	1.331
1	≠	1.624	1.574	1.527	1.481	1.437	1.396	1.356
1.5	≠	1.655	1.604	1.556	1.503	1.464	1.422	1.381
2	≠	1.688	1.635	1.586	1.538	1.492	1.449	1.407
2.5	≠	1.721	1.667	1.616	1.568	1.521	1.477	1.434
3	≠	1.755	1.7	1.648	1.599	1.551	1.506	1.462
3.5	≠	1.791	1.735	1.681	1.63	1.582	1.535	1.491
4	≠	1.827	1.77	1.716	1.663	1.614	1.556	1.52
4.5	≠	1.865	1.807	1.751	1.697	1.646	1.598	1.551
5	≠	1.905	1.845	1.787	1.733	1.68	1.63	1.583
5.5	≠	1.945	1.884	1.825	1.769	1.716	1.664	1.615
6	≠	1.987	1.924	1.864	1.807	1.752	1.699	1.649
6.5	≠	2.031	1.966	1.905	1.846	1.789	1.736	1.684
7	≠	2.076	2.01	1.946	1.886	1.828	1.773	1.72
7.5	≠	2.123	2.055	1.99	1.928	1.869	1.812	1.758
8	≠	2.171	2.101	2.035	1.971	1.91	1.852	1.797
8.5	≠	2.221	2.15	2.081	2.016	1.953	1.894	1.837
9	≠	2.273	2.2	2.129	2.062	1.998	1.937	1.879
9.5	≠	2.327	2.252	2.179	2.11	2.045	1.982	1.922
10	≠	2.383	2.305	2.231	2.16	2.093	2.028	1.966
10.5	≠	2.442	2.361	2.285	2.212	2.142	2.076	2.013
11	≠	2.502	2.419	2.341	2.265	2.194	2.126	2.061
11.5	≠	2.564	2.48	2.399	2.321	2.248	2.178	2.11
12	≠	2.63	2.542	2.459	2.379	2.303	2.231	2.162
12.5	≠	2.697	2.607	2.521	2.439	2.361	2.287	2.216
13	≠	2.767	2.674	2.586	2.502	2.421	2.345	2.271
13.5	≠	2.84	2.745	2.653	2.567	2.484	2.405	2.329
14	≠	2.916	2.817	2.723	2.634	2.549	2.467	2.389
14.5	≠	2.995	2.893	2.796	2.704	2.616	2.532	2.452
15	≠	3.077	2.972	2.872	2.777	2.686	2.599	2.517
15.5	≠	3.163	3.054	2.951	2.853	2.759	2.67	2.584
16	≠	3.252	3.14	3.033	2.932	2.835	2.743	2.655
16.5	≠	3.345	3.229	3.119	3.014	2.914	2.819	2.728
17	≠	3.441	3.322	3.208	3.099	2.996	2.898	2.804
17.5	≠	3.542	3.418	3.301	3.189	3.082	2.98	2.883
18	≠	3.648	3.519	3.398	3.282	3.171	3.066	2.966
18.5	≠	3.757	3.625	3.499	3.379	3.264	3.156	3.052
19	≠	3.872	3.734	3.604	3.48	3.362	3.249	3.142
19.5	≠	3.991	3.849	3.714	3.585	3.463	3.347	3.236
20	≠	4.116	3.969	3.829	3.696	3.569	3.448	3.334
20.5	≠	4.247	4.094	3.949	3.811	3.68	3.555	3.436
21	≠	4.384	4.225	4.075	3.931	3.795	3.666	3.542
21.5	≠	4.527	4.362	4.206	4.057	3.916	3.782	3.654
22	≠	4.676	4.505	4.343	4.189	4.042	3.903	3.77

Table 4 *continued*

SEDIM VALUES AT PARTICLE DENSITY 1.4

SUCROSE CONC.	≠	CENTIGRADES 4	5	6	7	8	9	10
22.5	≠	4.833	4.655	4.487	4.327	4.175	4.03	3.89
	≠							
23	≠	4.997	4.813	4.637	4.471	4.313	4.163	4.02
23.5	≠	5.169	4.977	4.795	4.622	4.458	4.302	4.15
24	≠	5.35	5.15	4.961	4.781	4.61	4.448	4.29
24.5	≠	5.539	5.331	5.134	4.947	4.769	4.6	4.44
	≠							
25	≠	5.738	5.521	5.316	5.121	4.936	4.76	4.59
	≠							
25.5	≠	5.947	5.721	5.507	5.304	5.111	4.928	4.75
26	≠	6.166	5.931	5.708	5.496	5.296	5.105	4.92
26.5	≠	6.397	6.152	5.919	5.698	5.489	5.29	5.10
27	≠	6.64	6.384	6.141	5.911	5.692	5.485	5.28
	≠							
27.5	≠	6.896	6.628	6.375	6.134	5.906	5.69	5.48
	≠							
28	≠	7.166	6.886	6.621	6.37	6.131	5.905	5.69
28.5	≠	7.451	7.158	6.88	6.617	6.368	6.132	5.90
29	≠	7.751	7.444	7.154	6.879	6.618	6.371	6.13
29.5	≠	8.068	7.746	7.442	7.154	6.882	6.623	6.37
	≠							
30	≠	8.402	8.066	7.747	7.445	7.16	6.889	6.63
	≠							
30.5	≠	8.757	8.403	8.069	7.753	7.453	7.17	6.90
31	≠	9.131	8.76	8.409	8.078	7.764	7.466	7.18
31.5	≠	9.528	9.138	8.77	8.421	8.092	7.78	7.48
32	≠	9.949	9.539	9.151	8.785	8.439	8.111	7.8
	≠							
32.5	≠	10.39	9.963	9.556	9.171	8.807	8.462	8.13
	≠							
33	≠	10.87	10.41	9.985	9.58	9.196	8.834	8.49
33.5	≠	11.37	10.89	10.44	10.01	9.61	9.228	8.86
34	≠	11.91	11.4	10.92	10.47	10.05	9.647	9.26
34.5	≠	12.48	11.94	11.44	10.96	10.52	10.09	9.69
	≠							
35	≠	13.08	12.52	11.99	11.48	11.01	10.56	10.1
	≠							
35.5	≠	13.73	13.13	12.57	12.04	11.54	11.07	10.6
36	≠	14.42	13.79	13.19	12.63	12.1	11.6	11.1
36.5	≠	15.16	14.49	13.86	13.26	12.7	12.17	11.6
37	≠	15.94	15.23	14.56	13.93	13.34	12.78	12.2
	≠							
37.5	≠	16.79	16.03	15.32	14.65	14.02	13.43	12.8
	≠							
38	≠	17.69	16.89	16.13	15.42	14.75	14.12	13.5
38.5	≠	18.66	17.81	17	16.25	15.54	14.87	14.2
39	≠	19.7	18.79	17.94	17.13	16.38	15.56	14.9
39.5	≠	20.82	19.85	18.94	18.08	17.27	16.51	15.8
	≠							
40	≠	22.02	20.99	20.01	19.1	18.24	17.43	16.6
	≠							
40.5	≠	23.32	22.21	21.17	20.19	19.28	18.41	17.6
41	≠	24.72	23.53	22.42	21.37	20.39	19.47	18.6
41.5	≠	26.23	24.96	23.76	22.64	21.59	20.61	19.6
42	≠	27.86	26.49	25.21	24.02	22.89	21.83	20.8
	≠							
42.5	≠	29.62	28.16	26.78	25.5	24.29	23.16	22.0
	≠							
43	≠	31.53	29.96	28.48	27.1	25.8	24.59	23.4
43.5	≠	33.6	31.91	30.32	28.83	27.44	26.13	24.9
44	≠	35.85	34.02	32.31	30.71	29.21	27.8	26.4
44.5	≠	38.3	36.32	34.47	32.75	31.13	29.62	28.2

Table 4 *continued*

SEDIM VALUES AT PARTICLE DENSITY 1.4

SUCROSE CONC.	≠	CENTIGRADES						
	≠	11	12	13	14	15	16	17
0	≠	1.271	1.236	1.202	1.169	1.138	1.108	1.08
	≠							
.5	≠	1.294	1.258	1.223	1.19	1.159	1.128	1.099
1	≠	1.318	1.281	1.246	1.212	1.18	1.148	1.119
1.5	≠	1.342	1.305	1.269	1.234	1.201	1.169	1.139
2	≠	1.367	1.329	1.292	1.257	1.223	1.191	1.16
	≠							
2.5	≠	1.393	1.354	1.317	1.281	1.246	1.213	1.181
	≠							
3	≠	1.42	1.38	1.342	1.305	1.27	1.236	1.204
3.5	≠	1.448	1.407	1.368	1.33	1.294	1.26	1.226
4	≠	1.477	1.435	1.395	1.356	1.319	1.284	1.25
4.5	≠	1.506	1.463	1.422	1.383	1.345	1.309	1.274
	≠							
5	≠	1.537	1.493	1.451	1.411	1.372	1.335	1.3
	≠							
5.5	≠	1.569	1.524	1.481	1.439	1.4	1.362	1.326
6	≠	1.601	1.555	1.511	1.469	1.429	1.39	1.352
6.5	≠	1.635	1.588	1.543	1.5	1.458	1.418	1.38
7	≠	1.67	1.622	1.575	1.531	1.489	1.448	1.409
	≠							
7.5	≠	1.706	1.657	1.609	1.564	1.52	1.478	1.438
	≠							
8	≠	1.744	1.693	1.644	1.598	1.553	1.51	1.469
8.5	≠	1.782	1.73	1.68	1.633	1.587	1.543	1.501
9	≠	1.823	1.769	1.718	1.669	1.622	1.577	1.533
9.5	≠	1.864	1.809	1.757	1.706	1.658	1.612	1.567
	≠							
10	≠	1.907	1.851	1.797	1.745	1.695	1.648	1.602
	≠							
10.5	≠	1.952	1.894	1.838	1.785	1.734	1.685	1.639
11	≠	1.998	1.939	1.882	1.827	1.775	1.724	1.676
11.5	≠	2.046	1.985	1.926	1.87	1.816	1.765	1.715
12	≠	2.096	2.033	1.973	1.915	1.86	1.807	1.756
	≠							
12.5	≠	2.148	2.083	2.021	1.961	1.904	1.85	1.798
	≠							
13	≠	2.202	2.135	2.071	2.01	1.951	1.895	1.841
13.5	≠	2.257	2.188	2.123	2.06	1.999	1.942	1.886
14	≠	2.315	2.244	2.176	2.112	2.05	1.99	1.933
14.5	≠	2.375	2.302	2.232	2.166	2.102	2.04	1.982
	≠							
15	≠	2.438	2.363	2.291	2.222	2.156	2.093	2.032
	≠							
15.5	≠	2.503	2.425	2.351	2.28	2.212	2.147	2.085
15	≠	2.571	2.491	2.414	2.341	2.271	2.204	2.139
16.5	≠	2.641	2.558	2.479	2.404	2.331	2.262	2.196
17	≠	2.714	2.629	2.547	2.469	2.395	2.323	2.255
	≠							
17.5	≠	2.791	2.703	2.618	2.538	2.461	2.387	2.316
	≠							
18	≠	2.87	2.779	2.692	2.609	2.529	2.453	2.38
18.5	≠	2.953	2.859	2.769	2.683	2.601	2.522	2.447
19	≠	3.04	2.942	2.849	2.76	2.675	2.594	2.516
19.5	≠	3.13	3.029	2.933	2.841	2.753	2.669	2.588
	≠							
20	≠	3.224	3.12	3.02	2.925	2.834	2.747	2.664
	≠							
20.5	≠	3.322	3.214	3.111	3.012	2.918	2.828	2.742
21	≠	3.425	3.313	3.206	3.104	3.006	2.913	2.824
21.5	≠	3.532	3.416	3.305	3.199	3.098	3.002	2.91
22	≠	3.644	3.523	3.409	3.299	3.194	3.094	2.999

Table 4 *continued*

SEDIM VALUES AT PARTICLE DENSITY 1.4

SUCROSE CONC.	≠	CENTIGRADES 11	12	13	14	15	16	17
22.5	≠	3.761	3.636	3.517	3.403	3.295	3.191	3.092
	≠							
23	≠	3.884	3.754	3.63	3.512	3.4	3.292	3.189
23.5	≠	4.012	3.877	3.749	3.626	3.509	3.398	3.291
24	≠	4.146	4.006	3.873	3.746	3.624	3.508	3.398
24.5	≠	4.287	4.141	4.003	3.871	3.745	3.624	3.509
	≠							
25	≠	4.434	4.283	4.139	4.001	3.87	3.745	3.626
	≠							
25.5	≠	4.589	4.432	4.282	4.139	4.002	3.872	3.748
26	≠	4.751	4.587	4.431	4.282	4.141	4.005	3.876
26.5	≠	4.922	4.751	4.588	4.433	4.286	4.145	4.01
27	≠	5.101	4.923	4.753	4.592	4.438	4.291	4.151
	≠							
27.5	≠	5.289	5.103	4.926	4.758	4.598	4.445	4.299
	≠							
28	≠	5.487	5.293	5.108	4.933	4.766	4.606	4.454
28.5	≠	5.695	5.492	5.3	5.117	4.942	4.776	4.617
29	≠	5.914	5.703	5.501	5.31	5.128	4.954	4.78
29.5	≠	6.145	5.924	5.714	5.513	5.323	5.142	4.96
	≠							
30	≠	6.389	6.157	5.937	5.728	5.529	5.339	5.158
	≠							
30.5	≠	6.645	6.403	6.173	5.954	5.745	5.547	5.358
31	≠	6.917	6.663	6.421	6.192	5.974	5.766	5.568
31.5	≠	7.203	6.937	6.684	6.443	6.215	5.997	5.79
32	≠	7.506	7.226	6.961	6.709	6.469	6.241	6.024
	≠							
32.5	≠	7.826	7.533	7.254	6.989	6.738	6.499	6.271
	≠							
33	≠	8.165	7.857	7.564	7.285	7.022	6.771	6.532
33.5	≠	8.525	8.2	7.892	7.6	7.323	7.059	6.80
34	≠	8.906	8.564	8.24	7.933	7.641	7.364	7.1
34.5	≠	9.31	8.95	8.609	8.286	7.978	7.687	7.40
	≠							
35	≠	9.739	9.36	9.001	8.66	8.336	8.029	7.737
	≠							
35.5	≠	10.2	9.796	9.416	9.057	8.716	8.392	8.084
36	≠	10.68	10.26	9.858	9.479	9.119	8.777	8.45
36.5	≠	11.2	10.75	10.33	9.927	9.547	9.187	8.84
37	≠	11.75	11.28	10.83	10.41	10	9.622	9.26
	≠							
37.5	≠	12.34	11.84	11.36	10.91	10.49	10.09	9.70
	≠							
38	≠	12.97	12.43	11.93	11.46	11.01	10.58	10.1
38.5	≠	13.64	13.07	12.54	12.04	11.56	11.11	10.68
39	≠	14.36	13.76	13.19	12.65	12.15	11.67	11.2
39.5	≠	15.12	14.49	13.89	13.32	12.78	12.27	11.7
	≠							
40	≠	15.95	15.27	14.63	14.03	13.45	12.91	12.4
	≠							
40.5	≠	16.83	16.11	15.43	14.78	14.18	13.6	13.0
41	≠	17.78	17.01	16.29	15.6	14.95	14.34	13.7
41.5	≠	18.81	17.98	17.21	16.47	15.78	15.13	14.5
42	≠	19.91	19.03	18.2	17.41	16.68	15.98	15.3
	≠							
42.5	≠	21.09	20.15	19.26	18.43	17.64	16.89	16.1
	≠							
43	≠	22.37	21.36	20.41	19.51	18.67	17.87	17.1
43.5	≠	23.75	22.67	21.65	20.69	19.79	18.93	18.1
44	≠	25.25	24.08	22.99	21.96	20.99	20.07	19.2
44.5	≠	26.86	25.61	24.44	23.33	22.29	21.31	20.3

Table 4 *continued*

SEDIM VALUES AT PARTICLE DENSITY 1.4

SUCROSE CONC.	≠	18	19	CENTIGRADES 20	21	22	23	24
0	≠	1.052	1.026	1	.9755	.9518	.9289	.9069
.5	≠	1.071	1.044	1.018	.9925	.9683	.945	.9226
1	≠	1.09	1.062	1.036	1.01	.9854	.9616	.9387
1.5	≠	1.11	1.081	1.054	1.028	1.003	.9787	.9553
2	≠	1.13	1.101	1.073	1.047	1.021	.9962	.9724
2.5	≠	1.151	1.121	1.093	1.066	1.04	1.014	.9899
3	≠	1.172	1.142	1.113	1.085	1.059	1.033	1.008
3.5	≠	1.194	1.164	1.134	1.106	1.078	1.052	1.027
4	≠	1.217	1.186	1.156	1.127	1.099	1.072	1.046
4.5	≠	1.241	1.209	1.178	1.148	1.12	1.092	1.066
5	≠	1.265	1.233	1.201	1.171	1.141	1.113	1.086
5.5	≠	1.291	1.257	1.225	1.193	1.164	1.135	1.107
6	≠	1.317	1.282	1.249	1.217	1.187	1.157	1.129
6.5	≠	1.343	1.308	1.274	1.242	1.21	1.18	1.151
7	≠	1.371	1.335	1.3	1.267	1.235	1.204	1.174
7.5	≠	1.4	1.363	1.327	1.293	1.26	1.228	1.198
8	≠	1.429	1.391	1.355	1.32	1.286	1.254	1.223
8.5	≠	1.46	1.421	1.384	1.348	1.313	1.28	1.248
9	≠	1.492	1.452	1.414	1.377	1.341	1.307	1.274
9.5	≠	1.525	1.484	1.444	1.407	1.37	1.335	1.302
10	≠	1.559	1.517	1.476	1.437	1.4	1.364	1.33
10.5	≠	1.594	1.551	1.509	1.469	1.431	1.394	1.359
11	≠	1.63	1.586	1.543	1.502	1.463	1.425	1.389
11.5	≠	1.668	1.622	1.579	1.537	1.496	1.457	1.42
12	≠	1.707	1.66	1.615	1.572	1.531	1.491	1.452
12.5	≠	1.748	1.699	1.653	1.609	1.566	1.525	1.486
13	≠	1.79	1.74	1.693	1.647	1.603	1.561	1.521
13.5	≠	1.833	1.782	1.733	1.687	1.641	1.598	1.556
14	≠	1.878	1.826	1.776	1.728	1.681	1.637	1.594
14.5	≠	1.925	1.872	1.82	1.77	1.722	1.677	1.632
15	≠	1.974	1.919	1.865	1.814	1.765	1.718	1.673
15.5	≠	2.025	1.968	1.913	1.86	1.81	1.761	1.714
16	≠	2.078	2.019	1.962	1.908	1.856	1.806	1.758
16.5	≠	2.133	2.072	2.013	1.957	1.904	1.852	1.802
17	≠	2.19	2.127	2.067	2.009	1.954	1.9	1.849
17.5	≠	2.249	2.184	2.122	2.062	2.005	1.95	1.898
18	≠	2.311	2.244	2.18	2.118	2.059	2.003	1.948
18.5	≠	2.375	2.306	2.24	2.176	2.115	2.057	2.001
19	≠	2.442	2.37	2.302	2.237	2.174	2.113	2.055
19.5	≠	2.511	2.438	2.367	2.299	2.234	2.172	2.112
20	≠	2.584	2.508	2.435	2.365	2.298	2.233	2.172
20.5	≠	2.66	2.581	2.506	2.433	2.364	2.297	2.233
21	≠	2.739	2.657	2.579	2.504	2.433	2.364	2.298
21.5	≠	2.821	2.737	2.656	2.579	2.504	2.433	2.365
22	≠	2.907	2.82	2.736	2.656	2.579	2.505	2.435

Table 4 *continued*

SEDIM VALUES AT PARTICLE DENSITY 1.4

SUCROSE CONC.	≠	CENTIGRADES						
	≠	18	19	20	21	22	23	24
22.5	≠	2.997	2.907	2.82	2.737	2.657	2.581	2.50
	≠							
23	≠	3.091	2.997	2.907	2.821	2.739	2.66	2.58
23.5	≠	3.189	3.092	2.999	2.909	2.824	2.742	2.66
24	≠	3.292	3.191	3.094	3.002	2.913	2.828	2.74
24.5	≠	3.399	3.294	3.194	3.098	3.006	2.918	2.83
	≠							
25	≠	3.512	3.403	3.299	3.199	3.103	3.012	2.92
	≠							
25.5	≠	3.63	3.516	3.408	3.304	3.205	3.11	3.01
26	≠	3.753	3.635	3.522	3.415	3.311	3.213	3.11
26.5	≠	3.882	3.76	3.642	3.53	3.423	3.32	3.22
27	≠	4.018	3.89	3.768	3.651	3.54	3.433	3.33
	≠							
27.5	≠	4.16	4.027	3.9	3.778	3.662	3.551	3.44
	≠							
28	≠	4.309	4.171	4.038	3.912	3.791	3.675	3.56
28.5	≠	4.466	4.321	4.183	4.052	3.926	3.805	3.69
29	≠	4.631	4.48	4.336	4.198	4.067	3.941	3.82
29.5	≠	4.804	4.646	4.496	4.353	4.216	4.085	3.95
	≠							
30	≠	4.986	4.822	4.665	4.515	4.372	4.235	4.10
	≠							
30.5	≠	5.178	5.006	4.842	4.686	4.536	4.393	4.25
31	≠	5.38	5.2	5.029	4.865	4.709	4.559	4.41
31.5	≠	5.593	5.405	5.225	5.054	4.891	4.734	4.58
32	≠	5.817	5.62	5.433	5.253	5.082	4.919	4.76
	≠							
32.5	≠	6.054	5.848	5.651	5.463	5.284	5.113	4.95
	≠							
33	≠	6.305	6.088	5.882	5.685	5.497	5.318	5.14
33.5	≠	6.569	6.342	6.125	5.919	5.722	5.534	5.35
34	≠	6.849	6.61	6.383	6.166	5.959	5.762	5.57
34.5	≠	7.145	6.894	6.655	6.427	6.21	6.003	5.80
	≠							
35	≠	7.459	7.195	6.943	6.704	6.476	6.258	6.05
	≠							
35.5	≠	7.792	7.513	7.249	6.997	6.756	6.527	6.30
36	≠	8.144	7.851	7.572	7.307	7.054	6.813	6.58
36.5	≠	8.519	8.21	7.916	7.636	7.369	7.116	6.87
37	≠	8.917	8.591	8.28	7.985	7.704	7.436	7.18
	≠							
37.5	≠	9.34	8.996	8.668	8.356	8.06	7.777	7.50
	≠							
38	≠	9.791	9.426	9.08	8.751	8.437	8.139	7.85
38.5	≠	10.27	9.885	9.519	9.17	8.839	8.524	8.22
39	≠	10.78	10.37	9.986	9.618	9.267	8.933	8.61
39.5	≠	11.33	10.9	10.49	10.09	9.723	9.37	9.03
	≠							
40	≠	11.91	11.45	11.02	10.6	10.21	9.835	9.47
	≠							
40.5	≠	12.54	12.05	11.59	11.15	10.73	10.33	9.95
41	≠	13.21	12.69	12.2	11.73	11.28	10.86	10.4
41.5	≠	13.93	13.37	12.85	12.35	11.88	11.43	11
42	≠	14.69	14.1	13.55	13.02	12.51	12.04	11.5
	≠							
42.5	≠	15.52	14.89	14.29	13.73	13.19	12.69	12.2
	≠							
43	≠	16.41	15.74	15.1	14.5	13.93	13.38	12.8
43.5	≠	17.37	16.65	15.97	15.32	14.71	14.13	13.5
44	≠	18.4	17.63	16.9	16.21	15.56	14.94	14.3
44.5	≠	19.51	18.68	17.9	17.16	16.47	15.81	15.1

APPENDIX IV: CALCULATING THE NATURAL LOGARITHM OF THE ROTOR RADIUS

Table 1

B-XIV ROTOR. THE NATURAL LOGARITM OF THE ROTOR RADIUS

VOLUME: (ML) :	0	2	4	6	8
30 :	0.8587	0.8699	0.8810	0.8920	0.9029
40 :	0.9137	0.9246	0.9353	0.9459	0.9564
50 :	0.9668	0.9771	0.9873	0.9974	1.0074
60 :	1.0173	1.0270	1.0367	1.0444	1.0519
70 :	1.0595	1.0669	1.0743	1.0817	1.0890
80 :	1.0962	1.1034	1.1105	1.1176	1.1247
90 :	1.1317	1.1386	1.1455	1.1523	1.1591
100 :	1.1659	1.1726	1.1793	1.1859	1.1925
110 :	1.1990	1.2055	1.2119	1.2172	1.2224
120 :	1.2275	1.2327	1.2378	1.2429	1.2480
130 :	1.2530	1.2580	1.2630	1.2680	1.2729
140 :	1.2777	1.2824	1.2870	1.2917	1.2963
150 :	1.3008	1.3054	1.3100	1.3145	1.3190
160 :	1.3235	1.3279	1.3324	1.3378	1.3433
170 :	1.3487	1.3541	1.3594	1.3648	1.3701
180 :	1.3753	1.3806	1.3858	1.3910	1.3961
190 :	1.4006	1.4043	1.4081	1.4118	1.4155
200 :	1.4191	1.4228	1.4265	1.4301	1.4338
210 :	1.4374	1.4410	1.4446	1.4480	1.4513
220 :	1.4547	1.4581	1.4614	1.4647	1.4681
230 :	1.4714	1.4747	1.4780	1.4812	1.4845
240 :	1.4877	1.4907	1.4938	1.4969	1.4999
250 :	1.5029	1.5059	1.5090	1.5120	1.5150
260 :	1.5179	1.5209	1.5239	1.5268	1.5298
270 :	1.5327	1.5357	1.5386	1.5415	1.5444
280 :	1.5473	1.5502	1.5531	1.5560	1.5588
290 :	1.5616	1.5643	1.5669	1.5696	1.5723
300 :	1.5749	1.5776	1.5802	1.5828	1.5855
310 :	1.5881	1.5907	1.5933	1.5957	1.5982
320 :	1.6006	1.6030	1.6054	1.6078	1.6102
330 :	1.6126	1.6150	1.6174	1.6198	1.6222
340 :	1.6245	1.6269	1.6292	1.6316	1.6339
350 :	1.6363	1.6386	1.6409	1.6433	1.6456
360 :	1.6479	1.6502	1.6525	1.6546	1.6568
370 :	1.6589	1.6610	1.6632	1.6653	1.6674
380 :	1.6695	1.6716	1.6737	1.6758	1.6779
390 :	1.6800	1.6821	1.6842	1.6863	1.6883
400 :	1.6904	1.6925	1.6945	1.6966	1.6986
410 :	1.7007	1.7027	1.7047	1.7075	1.7103
420 :	1.7130	1.7157	1.7185	1.7212	1.7239
430 :	1.7266	1.7293	1.7320	1.7347	1.7374
440 :	1.7393	1.7406	1.7419	1.7431	1.7444
450 :	1.7456	1.7469	1.7482	1.7494	1.7507
460 :	1.7519	1.7532	1.7544	1.7561	1.7577
470 :	1.7594	1.7610	1.7627	1.7643	1.7660

Table 1 *continued*

B-XIV ROTOR. THE NATURAL LOGARITM OF THE ROTOR RADIUS

VOLUME: (ML) :	0	2	4	6	8
480 :	1.7676	1.7692	1.7709	1.7725	1.7741
490 :	1.7758	1.7776	1.7793	1.7811	1.7829
500 :	1.7846	1.7863	1.7881	1.7898	1.7916
510 :	1.7933	1.7950	1.7967	1.7985	1.8002
520 :	1.8019	1.8036	1.8053	1.8070	1.8087
530 :	1.8105	1.8122	1.8138	1.8155	1.8172
540 :	1.8189	1.8204	1.8220	1.8235	1.8251
550 :	1.8266	1.8282	1.8297	1.8312	1.8328
560 :	1.8343	1.8358	1.8374	1.8390	1.8407
570 :	1.8423	1.8440	1.8456	1.8473	1.8489
580 :	1.8505	1.8522	1.8538	1.8554	1.8570
590 :	1.8587	1.8603	1.8619	1.8635	1.8651
600 :	1.8667	1.8684	1.8700	1.8716	1.8732
610 :	1.8748	1.8763	1.8779	1.8795	1.8810
620 :	1.8825	1.8840	1.8856	1.8871	1.8886
630 :	1.8901	1.8916	1.8931	1.8946	1.8961
640 :	1.8976				

Table 2

B-XV ROTOR. THE NATURAL LOGARITM OF THE ROTOR RADIUS

VOLUME: (ML) :	0	5	10	15	20
75 :		0.9821	0.9983	1.0143	1.0301
100 :	1.0456	1.0608	1.0759	1.0907	1.1053
125 :	1.1180	1.1306	1.1430	1.1553	1.1674
150 :	1.1794	1.1913	1.2030	1.2134	1.2238
175 :	1.2340	1.2442	1.2542	1.2641	1.2740
200 :	1.2837	1.2923	1.3009	1.3093	1.3177
225 :	1.3261	1.3343	1.3425	1.3507	1.3584
250 :	1.3661	1.3737	1.3813	1.3888	1.3962
275 :	1.4036	1.4110	1.4177	1.4243	1.4309
300 :	1.4375	1.4440	1.4504	1.4569	1.4633
325 :	1.4693	1.4753	1.4813	1.4873	1.4932
350 :	1.4991	1.5049	1.5107	1.5160	1.5212
375 :	1.5263	1.5315	1.5366	1.5417	1.5468
400 :	1.5518	1.5568	1.5618	1.5668	1.5717
425 :	1.5767	1.5816	1.5864	1.5913	1.5958
450 :	1.6004	1.6049	1.6094	1.6139	1.6184
475 :	1.6228	1.6273	1.6314	1.6356	1.6397
500 :	1.6438	1.6479	1.6520	1.6561	1.6601
525 :	1.6642	1.6682	1.6722	1.6762	1.6801
550 :	1.6841	1.6880	1.6919	1.6956	1.6993
575 :	1.7029	1.7066	1.7102	1.7138	1.7174

Table 2 *continued*

```
        B-XV ROTOR. THE NATURAL LOGARITM OF THE ROTOR RADIUS
------------------------------------------------------------------------
```

VOLUME: (ML) :	0	5	10	15	20
600 :	1.7210	1.7246	1.7281	1.7317	1.7352
625 :	1.7387	1.7422	1.7457	1.7492	1.7523
650 :	1.7555	1.7586	1.7617	1.7648	1.7679
675 :	1.7710	1.7741	1.7772	1.7802	1.7833
700 :	1.7863	1.7894	1.7924	1.7954	1.7984
725 :	1.8014	1.8044	1.8074	1.8103	1.8133
750 :	1.8162	1.8192	1.8221	1.8251	1.8280
775 :	1.8309	1.8338	1.8367	1.8396	1.8424
800 :	1.8453	1.8479	1.8504	1.8530	1.8555
825 :	1.8581	1.8606	1.8631	1.8656	1.8681
850 :	1.8706	1.8731	1.8756	1.8781	1.8806
875 :	1.8831	1.8856	1.8880	1.8905	1.8929
900 :	1.8954	1.8978	1.9002	1.9027	1.9051
925 :	1.9075	1.9099	1.9123	1.9147	1.9171
950 :	1.9195	1.9219	1.9242	1.9265	1.9288
975 :	1.9311	1.9333	1.9356	1.9378	1.9401
1000 :	1.9423	1.9446	1.9468	1.9490	1.9513
1025 :	1.9535	1.9557	1.9579	1.9601	1.9622
1050 :	1.9643	1.9664	1.9685	1.9706	1.9727
1075 :	1.9748	1.9769	1.9789	1.9810	1.9831
1100 :	1.9851	1.9872	1.9892	1.9913	1.9933
1125 :	1.9953	1.9972	1.9992	2.0011	2.0031
1150 :	2.0050	2.0070	2.0089	2.0108	2.0127
1175 :	2.0147	2.0166	2.0185	2.0204	2.0223
1200 :	2.0242	2.0260	2.0278	2.0296	2.0314
1225 :	2.0332	2.0350	2.0368	2.0386	2.0404
1250 :	2.0422	2.0440	2.0458	2.0475	2.0493
1275 :	2.0511	2.0528	2.0546	2.0564	2.0581
1300 :	2.0599	2.0616	2.0634	2.0651	2.0669
1325 :	2.0686	2.0703	2.0721	2.0738	2.0755
1350 :	2.0773	2.0790	2.0807	2.0823	2.0840
1375 :	2.0856	2.0872	2.0889	2.0905	2.0921
1400 :	2.0937	2.0953	2.0969	2.0986	2.1002
1425 :	2.1018	2.1034	2.1050	2.1066	2.1082
1450 :	2.1099	2.1116	2.1132	2.1149	2.1166
1475 :	2.1182	2.1199	2.1215	2.1232	2.1248
1500 :	2.1264	2.1281	2.1297	2.1314	2.1330
1525 :	2.1346	2.1362	2.1379	2.1395	2.1411
1550 :	2.1427	2.1443	2.1459	2.1475	2.1491
1575 :	2.1507	2.1523	2.1539	2.1555	2.1571
1600 :	2.1587	2.1608	2.1628	2.1648	2.1669
1625 :	2.1689	2.1709	2.1729	2.1749	2.1769
1650 :	2.1789	2.1809	2.1829	2.1849	

REFERENCES

ANDERSON, N.G. (1966). *Natn. Cancer Inst. Monogr.*, No. 21
BARBER, E.J. (1966). *Natn. Cancer Inst. Monogr.*, No. 21, 219
BERMAN, A.S. (1966). *Natn. Cancer Inst. Monogr.*, No. 21, 41
BIRNIE, G.D. (1973). In *Methodological Developments in Biochemistry*, vol. 3, p. 17. Ed. E. Reid. London; Longmans
BISHOP, B.S. (1966). *Natn. Cancer Inst. Monogr.*, No. 21, 175
CLINE, G.B. and RYEL, R.B. (1971). In *Methods in Enzymology*, vol. XXII, p. 168. Eds S.P. Colowick and N.O. Kaplan. New York and London; Academic Press
DE DUVE, C. (1965). *Harvey Lect.*, **59**, 49
FRITSCH, A. (1973). *Analyt. Biochem.*, **55**, 57
FUNDING, L. (1973). In European Symposium of Zonal Centrifugation in Density Gradient, *Spectra 2000*, vol. 4, p. 45. Ed. J.-C. Cherman. Paris; Editions Cité Nouvelle
FUNDING, L. and STEENSGAARD, J. (1973). MSE Application Information, A8/6/73. Crawley, Sussex, UK; MSE Instruments Ltd
HALSALL, H.B. and SCHUMAKER, V.N. (1969). *Analyt. Biochem.*, **30**, 368
HALSALL, H.B. and SCHUMAKER, V.N. (1970). *Biochem. biophys. Res. Commun.*, **39**, 479
HINTON, R.H. (1971). In *Separations with Zonal Rotors*, p. Z5.1. Ed. E. Reid. Guildford, UK; Univ. of Surrey Press
LEACH, J.M. (1971). In *Separations with Zonal Rotors*, p. Z4.1. Ed. E. Reid. Guildford, UK; Univ. of Surrey Press
McEWEN, C.R. (1967). *Analyt. Biochem.*, **20**, 114
MARTIN, R.G. and AMES, B.N. (1961). *J. biol. Chem.*, **236**, 1372
MEUWISSEN, J.A.T.P. (1973a). In *Methodological Developments in Biochemistry*, vol. 3, p. 29. Ed. E. Reid. London; Longmans
MEUWISSEN, J.A.T.P. (1973b). In European Symposium of Zonal Centrifugation in Density Gradient, *Spectra 2000*, vol. 4, p. 21. Ed. J.-C. Cherman. Paris; Editions Cité Nouvelle
NOLL, H. (1967). *Nature, Lond.*, **215**, 360
NORMAN, M. (1971a). In *Separations with Zonal Rotors*, p. P6.1. Ed. E. Reid. Guildford, UK; Univ. of Surrey Press
NORMAN, M.R. (1971b). In *Separations with Zonal Rotors*, p. Z3.1. Ed. E. Reid. Guildford, UK; Univ. of Surrey Press
PRICE, C.A. and KOVACS, A. (1969). *Analyt. Biochem.*, **28**, 460
REID, E. (Ed.) (1971–74). *Methodological Developments in Biochemistry*, vols. 1–4, Guildford, UK; Univ. of Surrey Press/London; Longmans
RIDGE, D. (1973). In European Symposium of Zonal Centrifugation in Density Gradient, *Spectra 2000*, vol. 4, p. 39. Ed. J.-C. Cherman. Paris; Editions Cité Nouvelle
SOBER, H.A. (Ed.) (1970). *Handbook of Biochemistry.* Cleveland, Ohio; The Chemical Rubber Co.
SPRAGG, S.P. and RANKIN, C.T. Jr. (1967). *Biochim. biophys. Acta*, **141**, 164
STECK, T.L., STRAUS, J.H. and WALLACH, D.F.H. (1970). *Biochim. biophys. Acta*, **203**, 385
STEENSGAARD, J. (1970). *Eur. J. Biochem.*, **16**, 66

STEENSGAARD, J. (1971). In *Separations with Zonal Rotors,* p. M3.1. Ed. E. Reid. Guildford, UK; Univ. of Surrey Press

STEENSGAARD, J. (1974). 'Separation og analyse af macromolekyler ved s-zoneultracentrifugering', Aarhus University, Denmark

STEENSGAARD, J. and FUNDING, L. (1971). *Acta path. microbiol. scand.,* Sect. B, **79**, 19

STEENSGAARD, J. and FUNDING, L. (1972). *Acta vet. scand.,* **13**, 305

STEENSGAARD, J. and HILL, R.J. (1970). *Analyt. Biochem.,* **34**, 485

STEENSGAARD, J., FUNDING, L. and JACOBSEN, C. (1973). In *Methodological Developments in Biochemistry,* vol. 3, p. 91. Ed. E. Reid. London; Longmans

STEENSGAARD, J., FUNDING, L. and MEUWISSEN, J.A.T.P. (1973). *Eur. J. Biochem.,* **39**, 481

STEENSGAARD, J., MØLLER, N.P.H. and FUNDING, L. (1975). *Eur. J. Biochem.,* **51**, 483

SVEDBERG, T. and PEDERSEN, K.O. (1940). *The Ultracentrifuge,* p. 5. Oxford; Oxford Univ. Press

SVENSSON, H., HAGDAHL, L. and LERNER, K.D. (1957). *Sci. Tools,* **4**, 1

WATTIAUX, R., WATTIAUX-DE CONINCK, S. and RONVEAUX-DUPAL, M.F. (1971). *Eur. J. Biochem.,* **22**, 31

WATTIAUX, R., WATTIAUX-DE CONINCK, S., COLLOT, M. and RONVEAUX-DUPAL, M.F. (1973). In European Symposium of Zonal Centrifugation in Density Gradient, *Spectra 2000,* vol. 4, p. 63. Ed. J.-C. Cherman. Paris; Editions Cité Nouvelle

WEAST, R.C. (Ed.) (1966). *Handbook of Chemistry and Physics,* 46th edn, Cleveland, Ohio; The Chemical Rubber Co.

YOUNG, B.D. (1972). MSE Application Information, A6/6/72. Crawley, Sussex, UK; MSE Instruments Ltd

6 Isopycnic Centrifugation in Ionic Media

G.D. BIRNIE
The Beatson Institute for Cancer Research, Glasgow

Isopycnic centrifugation (more correctly termed equilibrium density-gradient centrifugation) has proved to be an enormously powerful technique in molecular and cell biology. Not only has its use provided valuable clues to the structures and chemical compositions of biological materials, but it has also been invaluable for separating, fractionating and purifying a host of biological macromolecules and particles. Fractionation of biological materials by buoyant density-gradient centrifugation in preparative ultracentrifuges is one of the simplest of techniques, and requires no sophisticated equipment apart from the ultracentrifuge and rotor. In essence, it consists of centrifuging the mixture which is to be fractionated in a density gradient formed from a solution of a suitable substance (almost invariably in water) until each species of particle in the mixture reaches its isopycnic point, that is, the point in the gradient at which the density of the solution equals the buoyant density of that species of particle. Thus, when a mixture of particles of different buoyant densities is centrifuged to equilibrium in a density gradient, each component of the mixture migrates to its own characteristic isopycnic point and is thereby separated from all the other components of the mixture. It is an equilibrium method, which means that, once the particles have formed equilibrium bands at their isopycnic points, there is no change in the distribution of the particles in the gradient no matter for how much longer centrifugation is prolonged.

The technique is now used for such a variety of purposes that no single contribution can cover all its many facets. This chapter concentrates, therefore, on three main aspects of the use of ionic buoyant density-gradient solutes, considering each of them strictly from a practitioner's point of view. First, the basic technique is described, and some indication of the advantages and disadvantages of different methods is given. Secondly, those parameters which have a direct bearing on the formation of ionic density gradients, and of isopycnic zones of particles within them, are discussed. Particular attention is given to the application of simple equilibrium gradient theory to separations in preparative ultracentrifuges to show how this can be used as a guide when the conditions for a fractionation experiment are being chosen. Thirdly, the factors which affect the accuracy of analytical data obtained from isopycnic centrifugation experiments are discussed, and some steps which can be taken to minimize errors are described. Finally, a few examples of fractionations are given to indicate the scope of the technique, and its use for the isolation of DNA, RNA and protein on a preparative scale is described.

BASIC PROCEDURES

There are two methods of setting up a buoyant density-gradient fractionation. The first involves forming the density gradient *in situ*, and consists of preparing a uniform mixture of the sample which is to be fractionated with a solution of the salt which is to form the density gradient in a suitable buffer (such as HEPES, tris-HCl or phosphate); the mixture is then centrifuged until (*a*) an equilibrium gradient is formed by redistribution of the salt in the centrifugal field, and (*b*) the components of the sample have reached their isopycnic positions. The second method is similar to that used for rate-zonal fractionations (see Chapter 3); it consists of preparing a gradient from solutions of the salt in a suitable buffer, layering the sample to be fractionated on top of the gradient, and centrifuging until the components of the sample have reached their isopycnic positions.

Gradients Formed in situ

Step 1. A mixture of sample, buffer and salt solutions is prepared such that the concentration of salt in the mixture gives a solution of the appropriate density, that is, one which will form a gradient which covers the desired range in density. Since most density gradients formed *in situ* are relatively shallow, it is important that the initial mixture is prepared accurately so far as salt concentration is concerned. This is most conveniently done by checking the density of the initial mixture (for example, by measuring its refractive index; see p. 196) and adjusting it to the correct value by addition of concentrated salt solution or buffer. An alternative procedure which is particularly useful when the sample is dilute and, consequently, has a large volume, is to dissolve salt crystals directly in the buffered sample solution until a solution of the desired density is obtained. Whichever procedure is used, it must be remembered that the final mixture must be homogeneous; to ensure this, vigorous mixing (for example, with a vortex mixer) is essential.

The mass of salt which should be dissolved in a given volume of solvent to give a solution of a particular density is obtained from tables of density *versus* weight fraction (percentage composition by weight). The mass (g) of solute which is required to prepare V (ml) of solution of density ρ (g/cm^3) is calculated from

$$m_{solute} = V_{solution} \times \rho \times F \tag{6.1}$$

where F is the weight fraction of the solute in a solution of this density. The mass (g) of solvent required to prepare this solution is given by

$$m_{solvent} = V_{solution} \times \rho(1 - F) \tag{6.2}$$

Fine adjustments to the density of a solution are made by trial and error, but some guidance as to the amounts of buffer or salt to be used is given by

$$\Delta V_{solvent} = \frac{\rho_1 - \rho_2}{\rho_2 - 1} \tag{6.3}$$

$$\Delta m_{solute} = \rho_2 F_2 - \rho_1 F_1 \tag{6.4}$$

where $\Delta V_{solvent}$ is the volume (ml) of buffer (assuming its density to be 1.00 g/cm^3) required to decrease the density of 1 ml of solution from ρ_1 to ρ_2

(g/cm^3), and Δm_{solute} is the mass (g) of solute needed to increase the density of 1 ml from ρ_1 to ρ_2 (g/cm^3).

Step 2. The appropriate volume of the mixture is put into the centrifuge tube. Frequently, the gradient does not occupy the full capacity of the tube, in which case, to prevent the unsupported part of the tube collapsing, it must be filled by the addition of a clean, inert liquid which is immiscible with, and less dense than, water. A mineral oil such as 'medicinal' paraffin or 'light white' paraffin is suitable. Of equal importance is precise balancing of the tubes, both in respect of overall weight and of the position of the centre of gravity. Any significant imbalance leads to excessive precession of the rotor during deceleration which increases the amount of swirling and, hence, of disturbance, in the gradient. Ideally, tubes should be balanced exactly; however, the extent of imbalance which can be tolerated depends on the size and weight of the rotor to be used. In our experience, this should be less than 0.1 g for gradients to be centrifuged in small swing-out and fixed-angle rotors (see p. 293). To achieve this standard, (*a*) pairs (or triplets) of tubes of exactly the same weight when empty should be selected; (*b*) pairs of caps (when these are required) should be selected likewise; (*c*) the tubes should be balanced after the gradient mixture has been put into them; (*d*) the tubes should be rebalanced after they have been topped-up with oil. The centre of gravity of the contents of the tube changes as the gradient forms in the centrifugal field; thus, if a tube containing a sample has to be balanced with a 'blank' tube, the latter *must* contain precisely the same salt solution as the 'sample' tube.

On occasion, the material from which the centrifuge tubes are made has to be chosen with some care (see Chapter 9). Problems arise particularly with nucleic acids, especially denatured DNA and RNA, which bind tenaciously to some plastics surfaces as well as to the surfaces of glass tubes in which gradient fractions are collected. Mostly the problem is acute only with low concentrations of nucleic acids (less than 10 μg/ml). The avidity with which nucleic acids are absorbed to plastics and glass surfaces depends on the nature of the material and the composition of the solution in the tube. Nitrocellulose and most glasses absorb nucleic acids particularly avidly and polypropylene is prone to do so at high pH. Polyallomer is usually quite satisfactory, although significant losses of nucleic acids have been observed when the tubes are used for the first time. The problem with glass surfaces is easily overcome by treatment with a siliconizing fluid such as 'Repelcote' (Hopkins and Williams, Ltd) to make them water repellent. This solution is not applicable to plastics. Where losses by absorption to these surfaces are unacceptably high they can be reduced by (*a*) adding sufficient carrier DNA to the solution, or (*b*) including a detergent such as sodium lauroyl sarcosinate or sodium dodecyl sulphate in the solution, or (*c*) soaking the tubes in 0.1 M EDTA or a solution of denatured DNA for 24 h prior to use (Szybalski and Szybalski, 1971).

Step 3. The tubes are capped tightly and loaded into position in the rotor. The tubes should fit snugly but easily, otherwise it might prove difficult to remove them without disturbing the gradient at the end of the run. Occasionally, tubes which have been used several times are distorted and fit too tightly; they should be discarded.

Step 4. The rotor is put into the centrifuge and accelerated to the desired speed. The only limitation is on the maximum speed of the rotor if (*a*) the average density of the gradient exceeds 1.2 g/cm^3 (see p. 301), or (*b*) the concentration of the salt at the bottom of the gradient which is formed exceeds its solubility so that excess salt crystallizes at the bottom of the tube. Some authors argue that there is no need to derate a rotor in cases where the tubes are less than half-filled with dense gradient solution (the remaining volume being occupied by light mineral oil). However, this ignores the fact that the centre of gravity of tubes filled in this way is much lower than that of tubes filled normally. Consequently, it is best to derate the rotor even although the tubes are only partly filled with dense solution.

Step 5. The rotor is decelerated to rest and the tubes removed. It is at this stage that there is the greatest danger of disturbing the gradient. To minimize such disturbances (they cannot entirely be eliminated) the rotor should be allowed to come to rest without application of mechanical or electrical brakes, and it must be eased off the spindle and lifted out of the centrifuge gently and smoothly. In this regard, the proper use (see Chapter 9) of the rotor release key in MSE centrifuges has much to commend it. Similar care must be taken when removing the tubes from the rotor and transferring them to the fractionation apparatus.

Step 6. The gradient is fractionated (see p. 188). At this stage also, gradients are easily disturbed and, consequently, must be handled very gently. If there is paraffin oil on top of the gradient, most of it may be removed by pasteur pipette without disturbing the gradient. The gradient should be displaced slowly from the tubes in order to prevent streaming in, and disturbance of, the gradient. In this regard, it is worth noting that gradients are stable for considerable periods of time so long as they are kept on a vibration-free bench at the same temperature as during centrifugation. Disturbance of the gradient can also be minimized by (*a*) not passing it through long lengths of tubing, and (*b*) not inverting it at any point in the fractionation apparatus.

Preformed Gradients

Step 1. Solutions of the salt in an appropriate buffer are prepared, the concentrations being such that the entire range of densities required in the gradient are encompassed. A gradient is then prepared in the centrifuge tube, either by use of gradient-making equipment (see Chapter 3) or by superimposing a series of layers of solution of decreasing density in the tube, and allowing the gradient to form by diffusion. If the latter method is used each layer should be 1 cm or (at most) 2 cm thick, and diffusion should take place on a vibration-free bench, preferably at the temperature at which the gradient is to be centrifuged. The time required for the formation of a smooth gradient is normally between 8 and 24 h, but since it depends on the thickness of each layer, the differences in density, the temperature and the viscosities of the solutions, it is worth while determining the optimum time for formation of the desired gradient in a preliminary experiment. Correct balancing of these tubes, both with respect to overall weight and centre of gravity, is essential for good resolution; the same procedure as described for gradients formed *in situ* should be followed.

Step 2. The sample solution is layered on top of the gradient. The density of the sample must be less than that of the top of the gradient; unless its volume is very small in relation to that of the gradient, the density of the sample should be adjusted to within 0.01 g/cm³ of that of the top of the gradient. The tubes are then topped-up with oil if necessary, balanced again, and placed in the rotor. *Steps 3–6.* These are entirely similar to the corresponding steps for gradients formed *in situ.*

OPTIMUM DESIGN OF BUOYANT DENSITY-GRADIENT SEPARATIONS

Many successful separations of biological particles and macromolecules have been obtained using one or other of the basic procedures outlined in the previous section. However, many of the experiments have been done on a purely empirical basis, with the result that conditions to achieve acceptable separations are often arrived at by the process of trial and error. In many cases, marked improvements in the separations or decreases in the time spent in centrifugation could be achieved by proper design of the experiments. The factors which affect the separations obtainable by centrifugation of materials in density gradients are well known (see Chapters 2 and 8), and the relative importance of these factors under a variety of conditions is easily determined. Consequently, it is a simple matter to take them into consideration in order to predict the best set of conditions for any particular fractionation by buoyant density-gradient centrifugation. This approach is, of course, commonly used for analytical ultracentrifugation studies, but less so for those done in preparative ultracentrifuges. In the following sections, each of these factors is discussed in some detail from the point of view of separations done in preparative ultracentrifuges, with the object of showing how knowledge of their effects can be used when experiments are being designed, thus often saving both time and materials.

Equilibrium Gradients

The most usual way to generate a density gradient in a salt solution for isopycnic separations, particularly those involving nucleic acids, is by centrifugation. When a solution of a salt is centrifuged a time is reached at which the rate of sedimentation of the salt in the centrifugal direction is equal to the rate of diffusion of the salt in the opposite direction, that is, an equilibrium gradient of the salt has been formed. The formation of equilibrium gradients in centrifugal fields has been studied in great detail using the methods of equilibrium thermodynamics (see Chapter 2), with the result that a number of important predictions about an equilibrium gradient can be made, often with great accuracy. These are (i) the steepness of the slope ($d\rho/dr$) at each point in the gradient; (ii) the overall range of density encompassed by the gradient; (iii) the position in the gradient at which the density is the same as that of the initial solution (the isoconcentration point); and (iv) the time necessary for an equilibrium gradient to be formed.

Similar considerations lead to predictions of the time required for particles to migrate to their isopycnic positions. Predictions of this kind are clearly of enormous value in designing isopycnic experiments in preparative ultracentrifuges as well as in analytical ultracentrifuges. Consequently, it is important to realize that, although the mathematical theory underlying the predictions may appear somewhat daunting to non-mathematicians, the predictions themselves are easily made by applying a few simple equations to some parameters whose values are easily obtained, and that the only mathematics required is a little simple arithmetic.

The predictions which can be made about equilibrium density gradients all have their basis in the general equation of equilibrium centrifugation (Svedberg and Pederson, 1940; Meselson, Stahl and Vinograd, 1957; Vinograd and Hearst, 1962), which can be expressed as

$$\frac{d\rho}{dr} = \frac{M(1 - \bar{V})}{RT} \times \frac{d\rho}{d\ln a} \times \omega^2 r \qquad (6.5)$$

where M is the molecular weight, $\bar{V}$ the partial specific volume and a the thermodynamic activity coefficient of the gradient solute, ρ is the initial density of the salt solution, R is the gas constant, T is the temperature ($^{\circ}$K), ω is the angular velocity of the rotor, and r is the distance from the axis of rotation to the point in the gradient at which the slope is $d\rho/dr$.

For our purposes, this equation can be simplified very considerably by combining all the terms which depend on the concentration of the gradient solute, and on its thermodynamic properties in a given solvent, with R and T, when it becomes

$$\frac{d\rho}{dr} = \frac{\omega^2 r}{\beta^{\circ}} \qquad (6.6)$$

The term β° is called the density-gradient proportionality constant for the salt in the specified solvent, the superscript indicating that the values are only rigorously valid at 1 atm (Ifft, Martin and Kinzie, 1970). β°-values for a variety of concentrations of a number of solutes in aqueous solution have been calculated (Ifft *et al.*, 1961, 1970; Hu, Bock and Halvorson, 1962; Ludlum and Warner, 1965; McEwen, 1967); the method by which β°-values for other salts can be calculated has also been described (Ifft, Martin and Kinzie, 1970).

The Slope of the Density Gradient

From eqn (6.6) it is clear that the gradient formed during centrifugation depends on three factors; the rotational speed of the rotor, the distance of the point in the gradient from the axis of rotation, and the β°-value for the solute. Inspection of a table of β°-values, and of eqn (6.5), indicates that β°-values vary widely from one salt to another, and depend heavily on the concentration of the salt and, to a much smaller extent, on the temperature of the solution. Representative β°-values for a few salts are given in *Table 6.1*, and from these and eqn (6.6) a number of general rules become obvious. These are:

1. $d\rho/dr$ increases with the square of the speed of rotation.
2. $d\rho/dr$ increases linearly with increasing distance from the axis of rotation.
3. $d\rho/dr$ decreases (to a small extent) with increasing temperature.
4. $d\rho/dr$ depends to a very large degree on the solute.
5. $d\rho/dr$ increases with decreasing β°-value.
6. $d\rho/dr$ increases (with a very few exceptions) with increasing salt concentration.

Table 6.1 DENSITY-GRADIENT PROPORTIONALITY CONSTANTS (β°-VALUES) FOR SOME SALTS*

Density at 25 °C /cm³	Cation	β°-value $\times 10^{-9}$ Anion			
		Cl⁻	Br⁻	I⁻	SO₄²⁻
.20	Cs⁺	2.04	1.46	1.17	1.06
.30		1.55	1.06	0.81	0.76
.40		1.33	0.89	0.65	0.67
.50		1.22	0.80	0.55	0.64
.60		1.17	0.73		0.66
.70		1.14			0.69
.80		1.12			0.74
.90		1.12			
.20	Rb⁺	3.42	2.15	1.58	
.30		2.76	1.56	1.18	
.40		2.25	1.34	0.99	
.50			1.22	0.90	
.60				0.83	
.05	Na⁺	28.7	12.8	8.7	
.10		18.7	7.7	5.1	
.15		15.8	5.9	3.80	
.20		14.3	5.2	3.19	
.25			4.5	2.86	
.30				2.82	
.35				2.95	
.05	K⁺	22.1	11.3	7.9	
.10		13.1	6.2	4.28	
.15		10.1	4.6	3.44	
.20			3.80	2.55	
.25			3.36	2.21	
.30			3.05	1.96	
.40				1.73	
.30	Li⁺	42.3	8.9		
.40			9.1		
.50			10.5		
.60			11.4		
.70			11.4		
.80			10.0		

*β°-values for Cs_2SO_4 from Ludlum and Warner (1965); all others from Ifft, Martin and Kinzie (1970).

For the sake of strict accuracy, two addenda to these general rules must be stated, although they have little practical effect on separations in preparative ultracentrifuges. First, the β°-values for salts are only strictly valid at 1 atm, and eqns (6.5) and (6.6) do not take into consideration the difference in pressure between top and bottom of a gradient, nor do they allow for differences in compressibility among components of the solution. Secondly, the density gradients formed in preparative ultracentrifuges are measured after the rotor has come to rest, and an unavoidable redistribution of salt takes place during the deceleration of the rotor. This effect is most noticeable at the two extremes of the gradient, particularly in gradients in fixed-angle rotors, and results in deviation of the observed gradient from that predicted by rule (2). Consequently, the general rules derived from eqns (6.5) and (6.6) are only applicable to the middle three-quarters of the gradient as measured after the rotor has come to rest. Thus, for this and other reasons, in preparative density-gradient experiments it is advisable to design the gradient so that the particles which are to be separated will band in the middle portion of the gradient.

The Isoconcentration Point

Ifft, Martin and Kinzie (1970) have shown that the distance from the axis of rotation to the isoconcentration point (r_c) in a cylindrical tube (that is, a tube in a swing-out rotor) is given by the expression

$$r_c = [\tfrac{1}{3}(r_t^2 + r_t r_b + r_b^2)]^{\frac{1}{2}} \tag{6.7}$$

where r_t and r_b are the distances from the axis of rotation to the top and bottom, respectively, of the gradient. For sector-shaped cells and, thus, for zonal rotors, the corresponding expression is

$$r_c = [\tfrac{1}{2}(r_t^2 + r_b^2)]^{\frac{1}{2}} \tag{6.8}$$

Estimating r_c for gradients formed in fixed-angle rotors is more difficult because the geometry of a liquid column subjected to a centrifugal field in a fixed-angle rotor is complex (see *Figure 6.1*, p. 187) and, as yet, no mathematical analysis of this situation has been published. However, many experiments with equilibrium gradients of CsCl, Cs_2SO_4 and NaI, which occupied one-third to one-half of the capacity of tubes in fixed-angle rotors, suggest that eqn (6.7) gives a reasonable estimate of the isoconcentration point of these gradients. Although it must be remembered that less reliance can be placed on estimates of r_c made in this way for gradients in fixed-angle rotors than on those for gradients in swing-out and zonal rotors, nevertheless they are accurate enough for most (if not all) experiments done in preparative ultracentrifuges.

These expressions indicate that r_c is independent of the initial concentration of the salt solution and of rotor speed. This is not entirely true; for some salts (for example, CsCl) r_c increases with increasing initial concentration and increasing rotor speed (Fritsch, 1975). However, the dependence of r_c on these parameters is slight over the range of salt concentrations and rotor speeds normally used in isopycnic density-gradient experiments in preparative ultracentrifuges, and no corrections to r_c need be made. Thus, calculation of r_c is extremely simple; it requires only that the distances from the axis of rotation to the top and bottom of the gradient be known, and these can be obtained for all commonly used rotors from the data on pp. 304–312.

However, when estimating r_t, it must be remembered that centrifuge tubes are never filled completely with solution; indeed, frequently the density gradient will occupy only one-half, or less, of the length of the tube. For example, it is usual to put no more than 6 ml of solution into each tube of the MSE 3 × 6.5 Ti swing-out rotor. In this case, r_t is about 0.4 cm greater than r_{min} (p. 304). Thus:

$$r_c = [\tfrac{1}{3}(4.7^2 + 4.7 \times 10.4 + 10.4^2)]^{\frac{1}{2}} \tag{6.9}$$

that is, 7.73 cm.

Similarly, estimates can be made of r_c for equilibrium gradients formed in partly filled tubes in fixed-angle rotors, given that the assumption stated previously in regard to this situation is valid. Thus, a reasonable estimate of r_c for a 4.5 ml gradient in a tube of the MSE 10 × 10 Ti fixed-angle rotor, where r_t and r_b are about 6.5 cm and 8.5 cm, respectively, is given by

$$r_c = [\tfrac{1}{3}(6.5^2 + 6.5 \times 8.5 + 8.5^2)]^{\frac{1}{2}} \tag{6.10}$$

that is, 7.52 cm.

In virtually all of the commonly used swing-out and fixed-angle rotors, the radius of the isoconcentration point of an equilibrium gradient is only slightly greater than that of the mid-point of the *length* of the liquid column in the direction of the centrifugal field. However, centrifuge tubes have hemispherical bottoms so that the mid-point of the length of a column in a tube does not coincide with the mid-point of the volume of liquid in the column. Consequently, the isoconcentration point occurs below the half-way mark (that is, towards the dense end) of a gradient fractionated by volume. The exact position of the iso-concentration point in relation to gradient volume (and, indeed, of all other points in a gradient which are calculated in terms of distance from the rotor's axis) depends upon the geometry of the centrifuge tube, specifically on the ratio of the diameter of the tube to the length occupied by the gradient. Thus, for short gradients (length ≤ 2 × diameter), the proportion of the *volume* of the gradient lying below the isoconcentration point is significantly less than that of the *length* of the gradient below this point, while the difference is negligible for long gradients in narrow tubes (length ≥ 8 × diameter). In zonal rotors, however, the compartments are sector-shaped so that the isoconcentration point is always in the top half of the gradient volume. For example, the isoconcentration point in a BXIV rotor is at a radius of about 4.8 cm, which is about 300 ml from the top of the 660 ml gradient.

The Range in Densities

The third prediction which can be made about an equilibrium gradient, its over-all range in density, is important for two reasons. First, it ensures that the con-ditions chosen are such that the difference between the densities at the top and bottom of the gradient is sufficient to encompass the buoyant densities of the particles to be separated. Secondly, it prevents inadvertent choice of conditions which lead to such high concentrations of salt at the bottom of the gradient that crystallization of excess salt occurs. This condition can be disastrous since the density of the crystallized salt (4 g/cm^3 in the case of CsCl) may be so high that it produces stresses far in excess of the design limits of the rotor. The equation used is obtained by integrating eqn (6.6) to give

$$\rho_2 - \rho_1 = \frac{\omega^2}{2\beta^\circ}(r_2^2 - r_1^2) \tag{6.11}$$

in which ρ_1 is the density (g/cm^3) at the point in the gradient r_1 cm from the axis of rotation, and ρ_2 the density at r_2 cm.

To calculate the overall density range of a gradient formed at a given rotational speed in a particular rotor, it is only necessary to substitute the appropriate values for the angular velocity (ω) and the radii of the top and bottom of the gradient, together with the β°-value for the initial concentration of the salt from which the gradient is formed. For example, if 6 ml of CsCl of initial density 1.60 g/cm^3 is centrifuged to equilibrium at 30 000 rev/min in an MSE 3 × 6.5 Ti swing-out rotor at 25 °C, the overall density range of the gradient ($\Delta\rho$) is given by

$$\Delta\rho = \rho_b - \rho_t = \frac{9.87 \times 10^6}{2 \times 1.17 \times 10^9} \times (10.4^2 - 4.7^2) \text{ g/cm}^3 \qquad (6.12)$$

that is, 0.36 g/cm^3, where ρ_b and ρ_t are the densities at the bottom and top, respectively, of the gradient, ω was calculated from the rotational speed (see Chapter 1), the $\beta°$-value for CsCl was taken from Ifft, Martin and Kinzie (1970) and r_t and r_b from the tables of rotor dimensions in Chapter 9. This, of course, is the difference between the densities at top and bottom while the rotor is at speed. The difference measured after the rotor has come to rest is smaller because of the redistribution which takes place during deceleration (see p. 175).

Calculating the absolute densities at the top and, more important, at the bottom of an equilibrium gradient is only slightly more complicated because the position of the isoconcentration point of the gradient must first be estimated. As we have seen, for the 3 × 6.5 Ti rotor the isoconcentration point is about 7.7 cm from the axis of rotation, eqn (6.9). Substituting this radius and the density at the isoconcentration point for r_1 and ρ_1 in eqn (6.11), for a gradient formed at 30 000 rev/min:

$$\rho_b - 1.60 = \frac{9.87 \times 10^6}{2 \times 1.17 \times 10^9} \times (10.4^2 - 7.7^2) \text{ g/cm}^3 \qquad (6.13)$$

Thus, $\rho_b = 1.81$ g/cm^3.

Since the overall density range of the gradient formed under these conditions is 0.36 g/cm^3, eqn (6.12), the density at the top of the gradient (ρ_t) while at speed is 1.45 g/cm^3. Note that the density of the solution at the bottom of this gradient is close to, but does not exceed, that of a saturated solution of CsCl (1.92 g/cm^3) so that there is no danger of CsCl crystals forming at the bottom of the tube. However, if this solution of CsCl were centrifuged at 40 000 rev/min to obtain a steeper gradient, the density at the bottom would be

$$\rho_b = \frac{1.75 \times 10^7}{2 \times 1.17 \times 10^9} \times (10.4^2 - 7.7^2) + 1.60 \text{ g/cm}^3 \qquad (6.14)$$

that is, 1.97 g/cm^3. Thus the solubility of CsCl would be exceeded at the bottom of this gradient, with the result that CsCl crystals would form and the rotor would be overstressed, possibly with catastrophic consequences. To preclude any possibility of crystals forming at the bottom of the tube, the maximum density of a CsCl gradient must not exceed 1.90 g/cm^3 at 25 °C. Substituting this value in eqn (6.14) and using it to calculate ω shows that the maximum speed at which 6 ml of a solution of CsCl of initial density 1.60 g/cm^3 may be centrifuged to equilibrium at 25 °C in a tube of this rotor is 36 000 rev/min.

Time to Equilibrate the Gradient

The fourth prediction which can be made is the time necessary for the gradient to reach equilibrium. When a solution of uniform solute concentration is centrifuged, the concentration first changes rapidly at the two extremes of the liquid column, whereas it remains uniform and stable in the middle portion. As the time for which the solution is exposed to a centrifugal field increases, so the proportion of the solution which remains unchanged in concentration decreases. Eventually, only the solution at the isoconcentration point has the same concentration of solute as the initial solution, at which time a smooth, continuous equilibrium gradient has been formed from top to bottom of the centrifuge tube. The changes which occur with time in the concentration of a solute at different points in a centrifugal field are illustrated by the studies on the rates of formation of metrizamide gradients (see Chapter 7). The time required to form a density gradient which is within 1% of the true equilibrium gradient is given by

$$t = k(r_b - r_t)^2 \tag{6.15}$$

where t is the time in hours, r_t and r_b are the distances (cm) of the top and bottom, respectively, of the density gradient from the axis of rotation (Van Holde and Baldwin, 1958), and k is a constant which is inversely proportional to the diffusion coefficient of the solute which forms the gradient. Thus the time required for an equilibrium gradient to form depends on the gradient solute. Moreover, the diffusion coefficient is directly proportional to the viscosity of the solution and inversely proportional to the temperature and, thus, the time required for formation of an equilibrium gradient is decreased with increase in temperature and increased with increase in the viscosity of the initial solution. However, it should be noted that the time necessary for equilibrium to be reached is (essentially) independent of the speed of the rotor, and that by far the most important factor in determining this time is the length of the liquid column in the centrifugal field.

Contrary to popular belief, the time necessary to form an equilibrium gradient can be surprisingly long. For example, consider a solution of CsCl (ρ_i, 1.5 g/cm^3) being centrifuged to equilibrium at 20 °C (Van Holde and Baldwin, 1958). Taking the diffusion coefficient to be 220 Ficks (Baldwin and Shooter, 1963), eqn (6.15) becomes

$$t = 5.6(r_b - r_t)^2 \tag{6.16}$$

In the MSE 3 × 6.5 Ti rotor (a typical small swing-out rotor) with 6 ml of solution per tube, r_t is 4.7 cm and r_b 10.4 cm, so the time necessary to form an equilibrium gradient is

$$t = 5.6(10.4 - 4.7)^2 \text{ h} \tag{6.17}$$

that is, about 180 h. In a swing-out rotor with long tubes, such as the Beckman SW 41 Ti which contains 13 ml of solution per tube, r_t is 7.2 cm and r_b 15.9 cm (see Chapter 9), and formation of an equilibrium gradient requires centrifugation for about 420 h.

Clearly, therefore, it is important to keep the length of the liquid column as short as possible. This can be done by partly filling the tubes with the gradient solution (the remaining volume being occupied with paraffin oil to prevent collapse of the centrifuge tube), which can result in very considerable reductions in the time needed for equilibrium gradients to be formed. For example, in a Beckman SW 65 Ti rotor the time to reach equilibrium is about 100 h with a

4.5 ml gradient but only 40 h with a 3 ml gradient. Even more substantial savings in time can be achieved by using a fixed-angle rotor, particularly if the tubes are only partly filled with gradient solution. For any given volume of liquid in the normal type of centrifuge tube (that is, with length greater than diameter), the length of the liquid column in the direction of the centrifugal field decreases progressively as the angle of the tube relative to the centrifugal field increases from 0° (in a swing-out rotor) to 90° (in a vertical-tube rotor). For example, in the MSE 10 X 10 Ti fixed-angle rotor the tubes are held at an angle of 35° to the axis of rotation (55° to the centrifugal field); consequently, for a 4.5 ml gradient r_t and r_b are about 6.5 cm and 8.5 cm, respectively, and the time required to form an equilibrium CsCl gradient is about

$$t = 5.6(8.5 - 6.5)^2 \text{ h} \tag{6.18}$$

that is, about 22 h.

Even more startling savings in time are achieved with vertical-tube rotors because the gradient is formed over a very short distance, namely, the width of the tube. Predictions for these, and for other fixed-angle and swing-out rotors, can easily be made from the dimensions of the rotors given in Chapter 9. Fixed-angle rotors have other advantages over swing-out rotors for isopycnic centrifugation and these are discussed on p. 187.

Another way in which very substantial reductions in centrifugation time can be achieved is to begin the experiment by partly preforming the gradient by superimposing two or three layers of CsCl solution of the appropriate densities in the centrifuge tube. Say, for example, an equilibrium gradient is to be formed at 25 °C by centrifuging 6 ml of CsCl solution at 40 000 rev/min in the MSE 3 X 6.5 Ti rotor, and that the density at the mid-point of the gradient is to be 1.50 g/cm³. As we have seen already, eqn (6.17), if the centrifuge tube contains initially a homogeneous solution of CsCl of density 1.50 g/cm³ it takes about 180 h for the gradient to reach equilibrium. However, if the tube contains initially three 2-ml layers of CsCl, each layer occupies only one-third of the total length of the gradient and the time required to form the equilibrium gradients in all three layers is calculated from eqn (6.16) as

$$t = 5.6 \left[\frac{10.4 - 4.7}{3} \right]^2 \text{ h} \tag{6.19}$$

that is, about 20 h.

If the initial density of CsCl in each of the three layers is correctly chosen, the three gradients form a single, continuous equilibrium gradient encompassing the whole 6 ml in the tube. Since r_c is close to the mid-point of a 6 ml gradient in this rotor, it is reasonable to choose a solution of density 1.50 g/cm³ for the middle layer. From the dimensions of the rotor, the radii of the mid-points of the top, middle and bottom layers are found to be about 5.6, 7.5 and 9.4 cm, respectively, and application of eqn (6.11) indicates that the initial density of the top layer should be

$$\rho_{5.6 \text{ cm}} = 1.50 - \frac{1.75 \times 10^7}{2 \times 1.22 \times 10^9} \times (7.5^2 - 5.6^2) \text{ g/cm}^3 \tag{6.20}$$

that is, 1.32 g/cm³; and that of the bottom layer

$$\rho_{9.4\,cm} = 1.50 + \frac{1.75 \times 10^7}{2 \times 1.22 \times 10^9} \times (9.4^2 - 7.5^2)\, g/cm^3 \qquad (6.21)$$

that is, $1.73\, g/cm^3$.

Time to Equilibrate the Particles

In most experiments the sample to be fractionated in an equilibrium gradient is introduced as part of the initial homogeneous mixture in the centrifuge tube. During centrifugation the particles migrate to form zones or bands, each of which contains particles which are homogeneous with respect to buoyant density in the particular gradient medium being used. Provided that centrifugation is for a long enough time, an equilibrium position will be reached by each zone with respect to buoyant density and to zone width. Although large particles will attain equilibrium more quickly than an equilibrium gradient can be formed, small particles and macromolecules may take a considerable time to form equilibrium zones. Consequently, in order to predict the minimum time for which any particular mixture must be centrifuged to achieve equilibrium with respect to both sample particles and gradient, the time required for the smallest particles in the sample to reach their equilibrium position must be estimated. Such estimates can be obtained from eqn (6.22) (Fritsch, 1975) which has been derived from an equation of Meselson, Stahl and Vinograd (1957):

$$t = \frac{\beta^{\circ}(\rho_p - \rho_m)}{\omega^4 r_p^2 s} \times \left[1.26 + \ln \frac{r_b - r_t}{\sigma}\right] \qquad (6.22)$$

In this equation, t is the time in seconds, β° is the β°-value of the salt forming the density gradient and ρ_p is the buoyant density (g/cm^3) of the particle in that salt. The angular velocity (rad/s) of the rotor is ω, r_p is the distance (cm) from the axis of rotation to the position occupied by the particles at equilibrium, and r_t and r_b the distances (cm) to the top and bottom, respectively, of the gradient. As we have seen previously, the values of all these parameters are readily obtained from tables or by calculation. However, values for three terms (s, ρ_m and σ) are not so readily obtained: s, the sedimentation rate of the particle in a medium (of density ρ_m) consisting of a solution of the salt used to form the equilibrium gradient and σ, the standard deviation of the distribution (presumed to be gaussian) of the particles in the zone formed at equilibrium. Consequently, it is necessary to make a number of assumptions and approximations to obtain an equation which can be used easily to give reasonably close estimates of centrifugation times.

First, for s we substitute $S_{20,w} \times 10^{-13}$, which holds for non-viscous media such as CsCl; if the equilibrium gradient is formed in a solution of relative viscosity (η_{rel}) greater than 1, the corresponding substitution for s is $S_{20,w} \times 10^{-13} \times 1/\eta_{rel}$. Secondly, the value of $(r_b - r_t)/\sigma$ depends on the length of the gradient and on the sharpness of the zone formed by the particles at equilibrium. If the particles are homogeneous and of high molecular weight, σ is small and, consequently, in long gradients (5 cm), a reasonable value for $(r_b - r_t)/\sigma$ is 50. However, with short gradients and low-molecular-weight particles, $(r_b - r_t)/\sigma$ may be as low as 10. Fortunately, large differences in $(r_b - r_t)/\sigma$ have relatively little effect on t as estimated by eqn (6.22) and, since this parameter is more likely to have a low value in the normal experimental situation in preparative ultracentrifuges, it seems reasonable to give it a value of 20 in eqn (6.22). Making these substitutions, and recalculating ω as rev/min (N), gives an equation of practical use:

$$t = \frac{1.13 \times 10^{14} \times \beta° \times (\rho_p - 1)}{N^4 r_p{}^2 S_{20,w}} \tag{6.23}$$

from which t, in hours, can easily be calculated. For example, consider a sample of DNA of buoyant density 1.70 g/cm³ and minimum size 10 S ($\sim 1 \times 10^6$ daltons) which is to be centrifuged to equilibrium at 25 °C in a CsCl solution of initial density 1.70 g/cm³. If the volume of the gradient is 4.5 ml, and it is centrifuged in the MSE 10 × 10 Ti fixed-angle rotor, we have already seen that the radius of the isoconcentration point is about 7.5 cm, eqn (6.10). Thus, if the rotor is spun at 40 000 rev/min, the time required for all the DNA molecules to reach equilibrium is given by

$$t = \frac{1.13 \times 10^{14} \times 1.14 \times 10^9 \times 0.7}{(4 \times 10^4)^4 \times 7.5^2 \times 10} \text{ h} \tag{6.24}$$

that is, about 63 h. This is considerably longer than the 22 h it takes to form the equilibrium gradient, eqn (6.18). If the mixture is centrifuged for less than 63 h, the larger DNA molecules will reach their equilibrium positions but some of the smaller ones will fail to do so, and the width of the zone will be greater than that of the equilibrium zone.

It has already been shown that the length of the gradient column has a large effect on the rate at which the equilibrium gradient is formed, eqn (6.16), but only a small effect on the rate at which the sample particles attain their equilibrium position, eqn (6.23). In contrast, the time required for an equilibrium gradient to form is independent of rotor speed, eqn (6.16), whereas that needed for the sample particles to reach equilibrium is inversely proportional to the fourth power of the rotor speed, eqn (6.23). These observations form the rational behind the use of the 'gradient relaxation' technique (Anet and Strayer, 1969a) in fixed-angle rotors to shorten centrifugation times. This method consists of centrifuging first at high speed, then at a lower speed. Consider, for example, a 30 000 rev/min equilibrium gradient is required for the sake of resolution and, at this speed, it requires perhaps 90 h to centrifuge the sample particles to equilibrium. If, instead, the rotor is first spun at 45 000 rev/min for 18 h, then at 30 000 rev/min for 24 h, the particles will reach their equilibrium positions during the high-speed spin and redistribute during the subsequent low-speed spin at the same time as the new equilibrium gradient is formed. In this way, by taking advantage of the short length of the gradient column in a fixed-angle rotor and of the profound effect of rotor speed on the rate at which sample zones attain their equilibrium positions, more than half the centrifugation time can be saved without any sacrifice of the resolving power of the low-speed gradient. A note of caution is appropriate, however. The speed of the rotor during the high-speed spin must not be so great that excess salt will crystallize at the bottom of the gradient (see p. 177).

Non-equilibrium Preformed Gradients

Time for Which the Gradient is Stable

In certain experiments the time required for the sample particles to reach their isopycnic points is very considerably less than that for an equilibrium gradient to be formed, despite the use of one or other of the methods for shortening the latter. This is the case when large particles (ribosomes, viruses and the like) are sedimented in media of low viscosity such as CsCl, or even when smaller particles are centrifuged in more viscous media like aqueous caesium formate and potassium acetate. In these cases it may be necessary to limit the centrifugation time because the particles are labile (or because the centrifuge is required for other experiments). Although it would be possible to use gradients which are formed by centrifugation, but curtailing the time and thus obtaining pre-equilibrium gradients, in many situations these gradients would be too shallow to be useful. A more usual approach is to preform the gradient, either by diffusion of a series of layers of solution of different density or by using gradient-making apparatus (see p. 44). In this way the slope of the density gradient can be chosen to give the resolution and overall range of density required.

However, when planning such a gradient it is necessary to remember that it will be unstable (unless it is an equilibrium gradient) because the salt will redistribute in the centrifugal field to which it is exposed. But, just as it is possible to calculate the time necessary to form an equilibrium gradient, eqn (6.15), the time for which the initial shape of a preformed gradient is maintained can be calculated from the same parameters. For example, Baldwin and Shooter (1963) have shown that, for preformed CsCl gradients, $d\rho/dr$ at the centre of the gradient column is maintained for

$$t = 0.3(r_b - r_t)^2 \tag{6.25}$$

where t is the time in h, and r_t and r_b are the distances (cm) of the top and bottom, respectively, of the density gradient from the rotor's axis, cf. eqn (6.16). For about half of this time the initial gradient is maintained in the middle third of the tube. Since the major factor affecting this time is again the length of the gradient column, it is best to use as long a gradient as possible when centrifuging particles to equilibrium in a preformed gradient. When gradient media with relative viscosities of 2 or more are used, the time for which the gradient is stable can be prolonged significantly by centrifuging at low temperatures.

Time to Equilibrate Particles

With preformed gradients the sample is usually introduced as a narrow zone at the top of the gradient, in which case Baldwin and Shooter (1963) have shown that the particle zones reach equilibrium with respect to position and zone width in a time given by

$$t = \frac{(\rho_p - \rho_m) \times [\log(r_p - r_t)/r_t + 4.61]}{\omega^2 r_p s \times d\rho/dr} \tag{6.26}$$

where $d\rho/dr$ is the slope of the density gradient (g/cm^3 per cm) and the other symbols are as in eqn (6.22). Making the same assumptions and approximations as was done to obtain eqn (6.23) gives the corresponding practical equation

$$t = \frac{2.53 \times 10^{11} \times (\rho_p - 1)[\log(r_p - r_t)/r_t + 4.61]}{N^2 r_p S_{20,w} \times d\rho/dr} \tag{6.27}$$

where ρ_p is the buoyant density (g/cm^3) of the particles in the solution forming the gradient, N, the rotor speed (rev/min), $S_{20,w}$, the sedimentation coefficient (Svedberg units), r_t, the distance (cm) from the rotor's axis to the top of the gradient, and r_p, the distance to the position occupied by the particles at equilibrium. This applies to gradients of solutions, such as aqueous CsCl, whose relative viscosities are close to 1; for more viscous solutions the appropriate correction to $S_{20,w}$ must be made (see p. 181). Thus, the time required for the particle zones to reach equilibrium depends mainly on the sedimentation coefficient of the particles and on the rotor speed, but little on the length of the gradient (cf. p. 181).

By comparing eqn (6.27) with eqn (6.23), it can be seen that increasing the rotor's speed has a less dramatic (although still very significant) effect on the rate at which the particles attain equilibrium. However, under otherwise similar conditions, less time is required for particles to reach equilibrium if the sample is layered on top of a preformed gradient than if it is initially a component of a homogeneous mixture with the salt solution. For example, it has already been calculated that a 10 S DNA would require about 63 h of centrifugation to reach equilibrium in a 4.5 ml CsCl gradient formed by centrifugation at 40 000 rev/min in an MSE 10 × 10 Ti fixed-angle rotor, eqn (6.24). If this gradient had been preformed and the sample layered on top of it, the time required for the smallest DNA molecules to reach equilibrium would have been

$$\frac{2.53 \times 10^{11} \times (1.7 - 1) \times [\log(7.6 - 6.5)/6.5 + 4.61]}{(4 \times 10^4)^2 \times 7.5 \times 10 \times 0.116} \text{ h} \tag{6.28}$$

that is, 48 h. The slope of the gradient $(d\rho/dr)$ has been taken to be 0.116 g/cm^3 per cm, that is, the slope at the isoconcentration point as calculated from eqn (6.6

Resolving Power of Gradients

The general, practical rule is — two zones of particles will be well separated if the volume between the two peaks is twice that of their width. If this volume is less there will be overlap, although the separation may be acceptable for some purposes, particularly if the overlapped region is discarded. Whether a gradient has sufficient resolving power to give such a separation between zones of particles which differ in buoyant density by any given amount is dependent on (a) the mass of the particles in each zone; (b) the homogeneity of the particles in each zone; (c) the nature of any particle–solvent interactions; (d) the slope $(d\rho/dr)$ of the gradient over the range in densities in which the particles band; (e) the speed of the rotor; and (f) the type of rotor used. The complexity of the interaction between some of these factors and the departure from ideal behaviour of others, particularly in the kind of experiment usually done in preparative ultracentrifuges, precludes any possibility of making accurate predictions to fit all cases.

However, consideration of each of them does give some guidance (although in general terms only) as to the type of gradient which should result in acceptable separations of particles.

Gradient Capacity

The nature of the factors which determine the capacity of a gradient (that is, the mass of material which can be supported in a zone without distortion) has yet to be resolved (see Chapter 3, and Fritsch, 1975). In practice, it has been found that the actual capacity of a gradient is less, sometimes much less, than the predicted capacity. For example, the equilibrium gradient formed by centrifuging 4.5 ml of CsCl (ρ_i, 1.70 g/cm^3) at 40 000 rev/min in the MSE 10 X 10 Ti fixed-angle rotor has a slope at the isoconcentration point of 0.116 g/cm^3 per cm, eqn (6.6). Thus, this gradient should be able to support at this point at least 1 mg of material in a zone only 0.1 mm thick. In practice, however, such a gradient would be grossly overloaded by this amount of material. For example, Flamm, Birnstiel and Walker (1972) found that, to obtain a good separation between two DNAs differing in density by 0.005 g/cm^3, the maximum load on a 3 ml gradient in a swing-out rotor is 15 μg, whereas that on a 4.5 ml gradient in a fixed-angle rotor is 45 μg. The loading can be increased substantially if the materials to be separated differ widely in buoyant density. For example, when DNA is to be separated from RNA and proteins (see p. 209), a 10 ml gradient in a fixed-angle rotor has the capacity for 5 mg of DNA. As a general rule, it is better to assume that the actual capacity of a gradient for any particular species is no more than 10% of the predicted capacity.

Homogeneity of Particles

With very few exceptions, the particles and macromolecules which are dealt with in buoyant density-gradient experiments in preparative ultracentrifuges are heterogeneous with respect to density, or molecular weight, or both. Consequently, zones formed at equilibrium are often much wider than those formed by homogeneous species, and the resolving power of a gradient is much less than that estimated from average molecular weights and/or buoyant densities. This is particularly true for nucleic acids, nucleoproteins and proteins. For example, the buoyant densities of the sequences in a mammalian DNA vary from about 1.68 to 1.72 g/cm^3 in CsCl, whereas in a randomly sheared preparation the molecular weights of the fragments can vary by a factor of 10. Thus, when such a preparation is centrifuged to equilibrium, the zone occupied by the DNA may constitute a substantial proportion of the length of the gradient (see *Figure 6.2*).

Particle–Solvent Interactions

Interactions between the particles and one or other component of the salt solution from which the gradient is formed are very common, and they can have a very profound effect on the separation obtained. For example, the buoyant density of proteins is very similar in all salt solutions, whereas that of DNA varies widely because of the wide variation in water activity among aqueous solutions of salts. Consequently, a much greater separation between proteins and DNA is obtained in CsCl gradients than in gradients of Cs_2SO_4 or NaI. Another factor which can be of great importance is the presence of traces of impurities in the gradient solution. The concentration of salt used to form an equilibrium gradient is high so that even a trace impurity may be at a very significant concentration, particularly when small amounts of material have been put on to the gradient in order to obtain good resolution. Nucleic acids, in particular, very readily form complexes with heavy metal ions with consequent changes in their buoyant densities. Moreover, the affinity of DNA sequences for certain ions is dependent on their base composition, which is the basis of the fractionation of DNA by centrifugation in Cs_2SO_4 containing Ag^+ or Hg^{2+} ions (see p. 204). This type of interaction is very dependent on the nature and concentration of the heavy metal ions; consequently, it is essential to prevent the formation of any unplanned complexes by (*a*) forming the gradient from as pure a preparation of the salt as possible, and (*b*) ensuring that no other component of the gradient solution, *including the sample itself*, contains heavy metal ions (or anything else, such as actinomycin D, which might interact with the DNA in the gradient).

The Slope of the Density Gradient and Rotor Speed

The slope of an equilibrium gradient and the speed of the rotor in which it is formed are intimately interconnected in their influence on resolution. The factors affecting the slope ($d\rho/dr$) are listed on p. 174, from which it can be seen that $d\rho/dr$ is proportional to the square of the angular velocity (ω). However, although $d\rho/dr$ is low for equilibrium gradients formed at low speeds, use of such gradients does not increase resolution since the width of a zone of particles at equilibrium is inversely proportional to ω^2. Thus, in low-speed equilibrium gradients the peaks are far apart but zones are wide; in high-speed gradients the zones are narrow but the peaks are close together. Consequently, there is no advantage in centrifuging to equilibrium at low speed, except in those cases where a preparative gradient is being used as a purification procedure, and discarding the inter-band regions in which there is overlap between components does not result in an unacceptable reduction in yield. In these cases, the 'gradient relaxation' technique (p. 182) can be particularly useful when one or more of the species to be separated are of low molecular weight.

With preformed gradients, $d\rho/dr$ is, of course, predetermined so long as the time required for the particles to reach equilibrium is shorter than that for which the gradient is stable (p. 183). Thus, preformed shallow gradients in long tubes can be used with great effect to separate species of particles which are large enough to reach their equilibrium positions rapidly. Since the time for which a preformed gradient is stable is independent of rotor speed, eqn (6.25), it is clearly advantageous to centrifuge such gradients at high speed.

Choice of Rotors

So far, only one method has been found for increasing the resolving power of the equilibrium gradient formed by centrifuging a salt solution at a given speed, that is, by using a fixed-angle rotor instead of a swing-out rotor (see Flamm, Birnstiel and Walker, 1972). This is because the difference in density between the top and bottom of a gradient formed by centrifugation depends on the *horizontal* distance between these points *during centrifugation*, eqn (6.11). When the rotor is brought to rest after centrifugation is complete, and the tube is reoriented to the vertical position, the distance between the top and bottom of a gradient formed in a swing-out rotor is unchanged, whereas the corresponding distance in a gradient formed in a fixed-angle rotor is increased. In the latter,

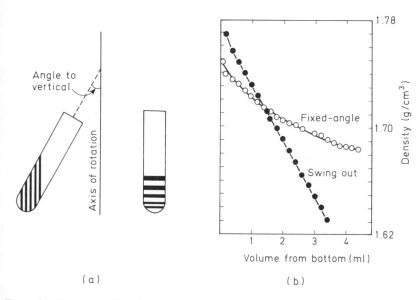

(a)   (b)

Figure 6.1 Density gradients in fixed-angle and swing-out rotors. (a) Diagram of a gradient in a tube: (left), at speed in a fixed-angle rotor and (right), at rest after reorientation to the vertical position; (b) density gradients formed by centrifuging a CsCl solution (ρ_i, 1.72 g/cm^3) in a fixed-angle rotor (o——o) and a swing-out rotor (•——•). (From Flamm et al., 1972)

this means that the gradient is now stretched over a longer distance so that $d\rho/dr$ is decreased and, consequently, resolution is increased. This is illustrated diagrammatically by *Figure 6.1*, which shows how zones (and inter-zone regions) are expanded near the top, and constricted near the bottom, of a gradient formed in a fixed-angle rotor. The practical advantage of using a fixed-angle rotor to increase the resolving power of an equilibrium gradient is shown by *Figure 6.2*. The extent to which the resolving power of an equilibrium gradient can be increased by this means increases as the angle of the tube to the direction of the centrifugal field increases from 0° (swing-out rotors) to 90° (vertical-tube rotors). Thus, the use of fixed-angle rotors for equilibrium density-gradient centrifugation offers three advantages: (*i*) a decrease in the time required for formation of an equilibrium gradient, eqn (6.18); (*ii*) an increase in resolving power; and (*iii*) a larger number of tubes and greater volume per rotor (Flamm, Birnstiel and Walker, 1972).

It should be noted that the resolving power of isopycnic gradients in zonal rotors is the same as in swing-out rotors and, thus, less than in fixed-angle rotors.

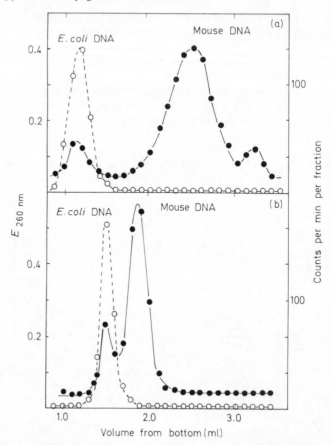

Figure 6.2 Fractionation of a mixture of mouse and E. coli *DNA by centrifugation in* CsCl: *(a) in a fixed-angle rotor and (b) in a swing-out rotor. A mixture of 3μg of* [^{14}C] *DNA from* E. coli *and 50 μg of unlabelled DNA from mouse liver was centrifuged to equilibrium in a CsCl solution (*ρ_i, *1.71 g/cm^3);* ●——●, E_{260}; ○-----○, *radioactivity. (From* Flamm *et al., 1972*

The features of a zonal rotor which increase the resolution of rate-zonal separations (see Chapter 4) do not affect the resolving power of an isopycnic gradient.

Fractionation of Gradients

After centrifugation has been completed the density gradient must be fractionated in some way to permit analysis and/or recovery of the separated zones of particles. It is essential that the proper method is chosen and that it is used correctly, since even excellent separations can be rendered less than acceptable by lack of attention and care at this stage. In the first place, the tubes must be removed from the rotor very gently and without any jarring or tilting movements. Before and during fractionation, the centrifuge tubes must be kept on a vibration-free bench; moreover, they must be maintained at the same temperature as they were during centrifugation otherwise convection currents will disrupt the gradient. For these reasons, it is usually best to leave each tube in the rotor until the gradient in it is to be fractionated.

All the useful techniques for fractionating gradients are variations of three basic methods by which fractions are collected: (*i*) from the bottom after puncturing the tube; (*ii*) from the top by displacing the gradient upwards; and (*iii*) directly from a zone. Esosteric methods, such as freezing the gradient and slicing the tube, are not worth considering except as a last resort (see Chapter 7), since they are rarely (if ever) satisfactory.

Collection from the Bottom

This was the method originally used, and consists simply of puncturing the bottom of the tube with a needle and collecting fractions by counting the drops falling from the hole made by the needle, or from the needle if the latter is hollow and is left in place. Fractionation by this method is not very satisfactory; the size of the hole formed, and that of the drops of liquid falling from it, are variable, and the flow rate is not easily controlled. In more sophisticated versions, control over flow rate is achieved by sealing the top of the centrifuge tube and connecting the air space within it to a simple manometer (Fritsch, 1975).

An alternative procedure which allows good control over flow rate and the volume of each fraction has been described by Flamm, Birnstiel and Walker (1972), and is shown in *Figure 6.3(a)*. In this method, the top of the tube, the reservoir and the syringe are filled with paraffin oil. The tube is punctured and the gradient is displaced by oil introduced from the syringe; thus the volume of each fraction is precisely controlled and measured by the volume of oil used to displace it. Alternatively, the tube may be connected to a burette filled with oil and the volume of each fraction controlled by the stopcock on the burette. Whichever method is used, it is essential to unload the gradient slowly (0.5–1 ml/min). This is particularly important if any of the zones are viscous, since such zones will drag along the walls of the tube and become displaced from their original positions in the gradient. For example, in *Figure 6.2*, the separation between mouse satellite and main-band DNA is diminished, and that between the latter and *E. coli* DNA enhanced, by displacement of the viscous zone formed by the high-molecular-weight main-band DNA.

The major difficulties of all methods involving collection from the bottom are (*a*) ensuring that there is proper laminar flow through the hole or the needle, and (*b*) avoiding contamination of gradient fractions by any material forming a pellet at (or near in the case of fixed-angle tubes) the bottom of the tube. Also, there is considerable danger of introducing a bubble of air, which travels upwards and disrupts the gradient, at the moment the tube is punctured. Finally, although puncturing the thin-walled tubes used in swing-out rotors is easy, pushing a needle through the much thicker plastic of fixed-angle rotor tubes is a difficult proposition.

Collection from the Top

Because of these difficulties, fractionation of gradients by upwards displacement is becoming used more commonly. In this method, a dense fluid is introduced

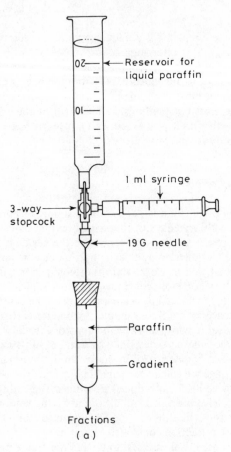

Reservoir for
liquid paraffin

1 ml syringe

3-way
stopcock

19 G needle

Paraffin

Gradient

Fractions
(a)

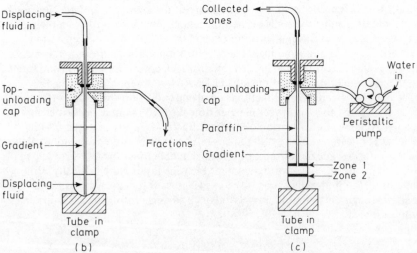

Displacing
fluid in

Top-
unloading
cap

Gradient

Displacing
fluid

Fractions

Tube in
clamp

(b)

Collected
zones

Top-unloading
cap

Paraffin

Gradient

Water
in

Peristaltic
pump

Zone 1
Zone 2

Tube in
clamp

(c)

Figure 6.3 Methods for fractionating density gradients: (a) collection of fractions from the bottom of the tube (from Flamm et al., 1972); (b) collection of fractions from the top by upwards displacement of the gradient (the MSE top-unloading cap: see Young, 1976); (c) collection of individual zones. See text for details of each method

into the bottom of the tube, thus displacing the gradient upwards into a cone-shaped collecting head and thence into a narrow tube from which fractions are collected. Several of these devices are now available commercially; at least one (manufactured by Isco) is equipped with a spectrophotometer flow cell mounted immediately above the cone, so allowing the ultraviolet absorbance of the zones in the gradient to be measured before the gradient is fractionated (see, for example, Chapter 3, *Figure 3.7*). The displacing fluid can be introduced through a hollow needle pushed through the bottom of the tube; however, as stated previously, puncturing the tubes of fixed-angle rotors is difficult. Alternatively, a narrow, rigid, hollow needle can be pushed down through the gradient to the bottom of the tube, *Figure 6.3(b)*. This method works surprisingly well; even a viscous zone of DNA remains undisturbed if the needle is introduced by a slow, smooth movement.

The most convenient reservoir for the displacing fluid is a burette, since the volume of each fraction can then be controlled very accurately simply by the volume of fluid introduced to the bottom of the gradient. Obviously, all air bubbles must be removed from the column of displacing fluid before unloading is started. The fluid used to displace the gradient may be a concentrated solution of salt; however, it is essential for such displacing solutions to be composed of the same salt as that which forms the gradient, otherwise extensive and very rapid mixing takes place because of the very large *concentration* gradients set up. It is much more convenient to use a dense liquid which is immiscible with water for displacement. Although tetrabromomethane (density 2.96 g/cm^3) has been used to displace gradients, this liquid has a rather obnoxious smell, and an oil such as Fluorochemical FC43 (supplied by the 3M Company), which has a density of about 2 g/cm^3, is to be preferred. The latter has the added advantages of a very low viscosity and complete chemical inactivity; although it is expensive it is easily recovered, and can be used many times over.

Although this method has fewer problems than the previous one, difficulties can still arise if there is a pellicle of insoluble material at the top of the gradient, although such pellicles are often easily removed before fractionation is begun. The major problem with the method is that it is difficult to avoid having the gradient inverted at some point in the fractionation apparatus, *Figure 6.3(b)*. Since inversion causes mixing and, hence, loss of resolution, it must be avoided as much as possible; where it is unavoidable, mixing should be minimized by using short lengths of narrow, non-wettable tubing (silicone rubber is most suitable). With this method also, unloading should be done slowly, particularly if any zone is viscous.

Direct Collection of Zones

It is not always necessary to fractionate a whole gradient when the object is simply to collect a zone of purified particles. So long as the required zone is visible (or can be made so) it is possible to collect it by puncturing the side of the tube with a hypodermic needle immediately below the zone, and *slowly* withdrawing the zone into a syringe. This approach is not ideal, since it is not always easy to puncture the centrifuge tube or, once this is done, to prevent leakage around the needle. Moreover, ensuring that the zone flows correctly into the tip of the needle requires extreme care in the rate at which it is withdrawn, and considerable mixing of the zone with liquid from above it usually occurs, particularly if the zone is narrow.

The top-unloading device shown in *Figure 6.3(b)* is easily adapted to allow several zones to be withdrawn sequentially from a gradient. The method involves inserting the central needle (the tip of which has been cut at right angles to its axis) to a point just below the zone to be collected, and pumping water slowly into the top of the tube, *Figure 6.3(c)*. The zone is thus displaced up through the central needle. Since the gradient is inverted in the central tube, considerable mixing takes place and, in order to recover all of a zone, it is necessary to collect about five times the volume it originally occupied.

This method has been very successful for collecting plasmid DNA which has been separated from the host *E. coli* DNA by isopycnic centrifugation in CsCl and ethidium bromide (see *Figure 6.5*, p. 203). Because of the mixing in the central tube it is necessary first to remove the upper band of *E. coli* DNA by placing the end of the needle between the two zones. However, although the separation between the zones is small (2–4 mm), the collected plasmid zone contains no more than 2% of DNA molecules (perhaps 0.5% by weight) derived from the *E. coli* genome (L. Coggins, personal communication).

Buoyant Density-gradient Solutes

The nature of the substance used to form the density gradient is of paramount importance in determining whether a particular fractionation will be successful, and in the 20 years since buoyant density-gradient centrifugation was introduced the suitability of many compounds as density-gradient solutes has been explored. Some of the compounds upon which reports have been published have never reappeared in the literature; others have enjoyed limited popularity, mainly for some special purposes; only a few have been used extensively. The properties of the ideal gradient solute have been summarized in Chapter 3 (see p. 35). Clearly, none of the salts used in buoyant density-gradient centrifugation satisfies all of these criteria. In particular, the high osmolality of the dense salts solutions causes difficulties with many cell organelles, whereas the high ionic strength of these solutions results in the disruption of most nucleoprotein complexes. It is the latter property of dense solutions of salts which limits their suitability as buoyant density-gradient media for the isopycnic sedimentation of nucleoproteins such as ribosomes and chromatin, and which led to the search for non-ionic gradient solutes (see Chapter 7). Also, although the thermodynamic activity of salts in aqueous solutions varies widely from one salt to another, and is dependent on the concentration of the salt, it is always high so that the water activity of these solutions is correspondingly low. In practical terms, this means that the hydration of particles and macromolecules in a salt gradient is very strongly dependent on the nature and concentration of the salt, and is usually low as compared to their degree of hydration in water alone. This effect is particularly notable with the nucleic acids which, consequently, have high buoyant densities in solutions of salts. Thus, for many purposes, only salts which are very soluble in water are capable of forming solutions dense enough to permit isopycnic banding of biological particles, particularly nucleic acids and nucleoproteins.

Other properties of salts in solution require attention when their suitability as buoyant density-gradient solutes is considered. The fact that the density gradient is composed of a salt gradient means that the experimenter has only limited control over the ionic conditions under which an isopycnic centrifugal separation is done, and no control over the ionic strength or osmolarity of any part of the density gradient. In addition, some of the salts (notably iodides) absorb ultraviolet light strongly, so causing difficulty with detection of materials in these gradients, whereas others (for example, iodides and sulphates) interact with commonly used liquid scintillation fluids, so causing difficulties with measurements of radioactivity in gradient fractions. Moreover, the viscosity of dense solutions of some salts (for example, caesium acetate and potassium tartrate) is high enough to cause significant retardation of the sedimentation of small particles and macromolecules. Finally, the slope $(d\rho/dr)$ of an equilibrium gradient, and hence its resolving power, depends both on the salt forming the gradient and on its concentration, and varies very widely.

Table 6.2 MAXIMUM DENSITIES OF THE SOLUTIONS OF SOME SALTS AT 20 °C

Salt	Density g/cm³	Salt	Density g/cm³
CsCl	1.91, 1.98*	NaBr	1.53
Cs$_2$SO$_4$	2.01	NaI	1.90
CsBr	1.72	Na$^+$ formate	1.32, 1.40*
CsI	1.65	KBr	1.37
Cs$^+$ formate	2.10	KI	1.72
Cs$^+$ acetate	2.00	K$^+$ formate	1.57, 1.63*
RbCl	1.49	K$^+$ acetate	1.41
RbBr	1.68	LiBr	1.83
RbI	1.95		

*These solutions are in D$_2$O; the remainder are in water.

These constraints obviously place considerable limitations on the choice of salt for any particular buoyant density-gradient separation. They, together with considerations of cost and availability, are the reasons for two salts of caesium (CsCl and Cs$_2$SO$_4$) reigning supreme as ionic buoyant density-gradient solutes. Some of the properties of these salts, and those of some others which have been used, are summarized in the following sections (see also *Table 6.2*). Other properties of these salts which have proved important in particular separations are noted in the sections in which these separations are discussed (see pp. 199–209).

Caesium chloride is by far the most commonly used salt for buoyant density-gradient experiments. This is partly for historic reasons: aqueous CsCl was the medium selected for the first studies of the analytical use of the method (Meselson, Stahl and Vinograd, 1957), and its thermodynamic parameters have been examined most thoroughly (see Ifft, Martin and Kinzie, 1970). CsCl is very soluble in water, giving very dense solutions (up to 1.91 g/cm³ at 20 °C); a significant increase in density can be obtained by using D$_2$O as solvent (1.98 g/cm³). The relative viscosity of these solutions is close to unity. Very pure (optical grade) CsCl can be purchased, and solutions of this material do not absorb ultraviolet light; however, this grade of salt is very costly, which limits its use to experiments in the analytical ultracentrifuge. Lower grades of CsCl are considerably less expensive, although still far from cheap; their solutions have a measurable absorption in the ultraviolet, but this is usually not so great as to preclude the use of these grades of CsCl for experiments in preparative ultracentrifuges. These lower grades of CsCl are frequently contaminated with traces of heavy metal ions. Although this contamination is slight in absolute terms, the concentration of heavy metal ions in the dense CsCl solutions used for density gradients can still be very significant, particularly in relation to the concentration of the particles being fractionated in the gradient. For this reason it is common practice to include EDTA (1–10 mM) in CsCl solutions. However, it is more satisfactory to remove these ions completely, and this can easily be done by passing a concentrated CsCl solution slowly through a bed of chelating resin (for example, Dowex Chelating Resin, obtainable from Sigma Chemical Corporation, St. Louis, Mo.); in our experience, a 5 ml bed of resin is sufficient to treat 100 ml of saturated CsCl. Biological particles and macromolecules can easily be separated from CsCl by, for example, gel filtration or dialysis. The salt is soluble in $HClO_4$ and is not precipitated from a 2.3 M (30% w/w) solution by addition of 2 vol of 95% ethanol, so that proteins and nucleic acids can be recovered from CsCl gradient fractions by precipitation.

Dilute solutions of CsCl do not quench the usual liquid scintillation fluids. However, the thermodynamic properties of CsCl are not ideal in all situations. The water activity of CsCl solutions is low with the result that the buoyant density of nucleic acids in CsCl is very high (see p. 201), that of RNA being greater than the maximum density obtainable with a CsCl solution at 25 °C. In addition, the β°-values for CsCl are low, so that the slope ($d\rho/dr$) of an equilibrium CsCl gradient is relatively steep (see p. 174), and its resolving power poorer than that of some other salts.

Caesium sulphate is the next most widely used salt. For the most part, its properties are similar to those of CsCl, but they do differ in a number of important respects. Although the maximum density obtainable with Cs_2SO_4 in water (2.01 g/cm³ at 25 °C) is only slightly greater than that with CsCl, the water activity in Cs_2SO_4 solutions is greater, and the buoyant densities of the nucleic acids are considerably lower in Cs_2SO_4 than in CsCl solutions (see p. 201). However, most species of RNA molecules either aggregate and/or precipitate in Cs_2SO_4 (see p. 205) so reducing the usefulness of this solute for separations of RNA. The β°-values for Cs_2SO_4 are lower than for CsCl, so that solutions of Cs_2SO_4 form even steeper gradients with poorer resolution. Moreover, sulphate ions precipitate the scintillants used in most liquid scintillation fluids so that the radioactivity of fractions cut from a Cs_2SO_4 gradient cannot be measured directly by liquid scintillation spectrometry unless diluted extensively.

Other caesium salts have also been used, although to a very limited extent. Caesium acetate, formate, bromide and iodide all give dense aqueous solutions. At equilibrium, CsBr and CsI gradients are steeper than the corresponding CsCl gradients, whereas the slope of caesium formate gradients is about the same. Caesium acetate forms a shallower gradient than CsCl at equilibrium, although this advantage is discounted for some purposes by the greater viscosity of dense caesium acetate solutions. In addition, CsI solutions absorb strongly in the ultraviolet, and are prone to oxidation by atmospheric oxygen and when mixed with liquid scintillation fluids.

Rhubidium salts have been used on occasion in place of the caesium salts, although for most purposes they seem to offer little significant advantage. The β°-values for RbCl are about twice those for CsCl so that the salt forms shallower equilibrium gradients, with resolution about twice that of CsCl gradients (Hu, Bock and Halvorson, 1962). However, the maximum density attainable with RbCl is only 1.49 g/cm³. The β°-values for RbBr and RbI are also greater than those for the corresponding caesium salts; the densities of saturated solutions of the bromides are about the same, while a saturated RbI solution (1.95 g/cm³) is considerably denser than one of aqueous CsI (1.65 g/cm³). RbI gives rise to the same difficulties as CsI in detection of materials banded in them.

Sodium and potassium halides are attractive propositions as buoyant density-gradient solutes from the point of view of their cost, which is much lower than that of CsCl. The densities of the solutions formed by NaCl and KCl are too low to be of any practical use. However, the bromides and the iodides of both sodium and potassium form dense, non-viscous solutions in water; moreover, their water activity is higher than that of CsCl solutions, so that the buoyant densities of the nucleic acids in particular are significantly lower in these solutions than in CsCl. Both single- and double-stranded RNAs can be banded isopycnically in KI and NaI density gradients without the aggregation or precipitation of the RNAs that occurs in Cs_2SO_4. The β°-values for KBr, KI, NaBr and NaI are considerably greater than those for the corresponding caesium salts and CsCl, so that they form shallower density gradients. NaBr has been used for fractionating lipoproteins (see p. 214), and both KI and NaI have proved useful for separating nucleic acids (see p. 199).

However, there are disadvantages to the use of solutions of the iodides as buoyant density-gradient solutes. First, they are very prone to oxidation, and a reducing agent ($NaHSO_3$ or dithiothreitol) must be present during density-gradient centrifugation to prevent precipitation of free iodine. They are also oxidized when mixed with liquid scintillation fluids, and the elementary iodine released quenches both ^{3}H and ^{14}C completely; however, this can readily be overcome by adding β-mercaptoethanol to the mixture prepared for liquid scintillation spectrometry. Secondly, they absorb strongly in the ultraviolet which precludes the detection of materials banded in the gradients by measurement of absorption at 260 nm or 280 nm. Purified KI can be purchased cheaply, but comparable grades of NaI are not readily available. However, the major contaminants of commercial NaI are the ions of heavy metals, and these are easily removed by passage of dense NaI solutions through chelating resin; in our experience, a 10 ml bed of Dowex Chelating Resin is sufficient to treat 100 ml of saturated NaI solution.

Other sodium and potassium salts have not been used as density-gradient solutes to any significant extent. The densities obtainable are quite high (for example, saturated potassium formate in D_2O has a density of 1.63 g/cm³), but many of these solutions are rather viscous. Moreover, the equilibrium density gradients formed by these salts tend to be too shallow to be of value for most fractionations. Despite these limitations, preformed potassium tartrate gradients have proved useful for the isopycnic banding of viruses (see p. 208) mainly because some viruses appear to retain their viability better in solutions of this salt than in the salts more often used as buoyant density-gradient solutes.

Lithium bromide is extremely soluble in water, in which it forms a solution of density 1.83 g/cm³ at 20 °C. The β°-value for LiBr is extremely high, so that the density gradient formed at equilibrium is very shallow. This limits its use to very special purposes such as the separation of density-labelled proteins (Hu, Bock and Halvorson, 1962).

DATA FROM ISOPYCNIC GRADIENTS

Measurement of Density

The buoyant density of the particles in a zone at equilibrium in a density gradient is simply determined by measuring the density of the zone. This can be done directly with a pycnometer of capacity 0.1–0.3 ml. Alternatively, and much more conveniently, density can be calculated from refractive index readings, which not only can be taken much more rapidly but also require less solution (only 25 μl is needed with a good Abbe refractometer). Tables of refractive indices and densities of many salts solutions have been published (International Critical Tables and Wolf and Brown, 1968), from which graphs of refractive index *versus* density can be drawn. The exact relationship between the refractive indices and densities of solutions of a variety of density-gradient solutes have been studied, and polynomials derived which enable very precise densities to be calculated (see, for example, Ifft and Vinograd, 1966; Ifft, Martin and Kinzie, 1970). However, such precision is applicable only to data from the analytical ultracentrifuge. Over the range of densities commonly used, the relationship between refractive index and density is almost linear, and is described by an equation of the general form

$$\rho = a\eta - b \tag{6.29}$$

A convenient set of equations which are sufficiently accurate for data from experiments in preparative ultracentrifuges is given in *Table 6.3*. Others have been published by Vinograd and Hearst (1962).

These equations apply to solutions of the salts in distilled water, and do not take into consideration other components (buffers, salts, EDTA, etc.) normally present in gradients. Although the latter are usually present in such low concentrations that they do not contribute significantly to the density, they make a significant contribution to the refractive index and a correction for this must be included. This is easily done, since

$$\eta_{\text{corrected}} = \eta_{\text{observed}} - (\eta_{\text{buffer}} - \eta_{\text{water}}) \tag{6.30}$$

The buffer (and other salts) will also form equilibrium gradients but, since they usually have very high β°-values, no correction for this is normally required. There are exceptions of course; some materials (for example, urea, dimethyl sulphoxide and formamide) even form steep reverse gradients. In such cases, accurate estimates of density can only be obtained by pycnometry.

Measurement and Recovery of Particles

With very few exceptions, buoyant density-gradient solutes do not interfere with the measurement of the concentration of macromolecules or particles in fractions cut from a salt gradient. The exceptions most commonly met are noted on pp. 192–195. Similarly, there is rarely any difficulty in recovering the components of a fractionated mixture. Particles can easily be recovered by centrifugation after dilution of the gradient fraction, or by, for example, dialysis

Table 6.3 COEFFICIENTS OF EQN (6.29) ($\rho = a\eta - b$) RELATING DENSITY TO REFRACTIVE INDEX* FOR SOME AQUEOUS SALT SOLUTIONS

Salt	Temperature (°C) for		Coefficients		Valid range for coefficients†		Source‡
	η	ρ	a	b	Density	Refractive index	
CsCl	20	20	9.968 8	12.292	1.00–1.29	1.333–1.360	1
	20	20	10.927 6	13.593	1.22–1.90	1.355–1.418	1
	25	25	10.240 2	12.648	1.00–1.38	1.333–1.370	2
	25	25	10.860 1	13.497	1.25–1.90	1.357–1.418	2
Cs_2SO_4	25	25	12.120	15.166	1.10–1.40	1.342–1.366	3
	25	25	13.698 6	17.323	1.40–1.80	1.366–1.396	3, 4
NaBr	20	20	5.876 1	6.839	1.00–1.42	1.333–1.405	1
	25	25	5.888 0	6.852	1.00–1.50	1.333–1.418	5
NaI	20	20	5.333 0	6.118	1.10–1.80	1.333–1.485	6
	25	25	5.328 3	6.103	1.00–1.29	1.333–1.388	5
KBr	20	20	6.506 5	7.683	1.00–1.37	1.333–1.392	1
	25	25	6.478 6	7.643	1.00–1.37	1.333–1.390	3
KI	20	20	5.731 7	6.645	1.00–1.40	1.333–1.404	1
	25	25	5.658 1	6.543	1.00–1.22	1.333–1.372	5
	25	25	5.835 6	6.786	1.10–1.70	1.350–1.454	7
LiBr	25	25	4.916 9	5.555	1.00–1.30	1.333–1.394	5

* Taking the refractive index of water at 20 °C to be 1.333 0.
† Within the ranges stated, the error in the calculated density is mostly less than ±0.00 2 g/cm³, but can be up to ±0.005 g/cm³ at the extremes of the ranges. In many cases the coefficients are also valid for densities outside these ranges, but with increasing errors.
‡ The coefficients were obtained:
  1. By calculation from data by Wolf and Brown (1968).
  2. From Bruner and Vinograd (1965).
  3. From Vinograd and Hearst (1962).
  4. By calculation from data in Szybalski (1968).
  5. By calculation from data in the International Critical Tables.
  6. By calculation from measurements in the author's laboratory.
  7. From Wolf (1975).

or gel exclusion chromatography. Nucleic acids and proteins can be recovered by precipitation with acid, or with ethanol after appropriate dilution of the salt; they can also be recovered by dialysis or gel exclusion chromatography.

Assessment of Data

One of the most valuable features of isopycnic density-gradient centrifugation is that analytical data for the components of a mixture are obtained at the same time as they are separated. Two parameters can be measured, the buoyant density and the bandwidth. The former is related to the partial specific volume of the material in the zone and its solvation (see Chapters 2 and 8), and from it much can be deduced about the material, including information about (*a*) base composition, and primary and secondary structure (nucleic acids); (*b*) amino acid composition and structure (proteins); (*c*) relative proportions of nucleic acid and protein (nucleoprotein complexes such as chromatin and ribonucleoprotein particles); and (*d*) relative amounts of lipid and protein (lipoproteins and membranes). The bandwidth depends on the homogeneity and molecular or particle weight of the material in the zone (see p. 185 and Chapters 2 and 8), and information about either of these can be obtained if the other is known.

These relationships have been determined largely by the use of the analytical ultracentrifuge and it is important to realize that they can be applied to data obtained from experiments done in preparative ultracentrifuges. However, it is equally important to realize that data obtained from the latter are not so precise as those from the analytical ultracentrifuge. One reason for this is the lower precision of the temperature control in a preparative ultracentrifuge; another is that absolutely pure salts may not have been used because of their cost. The major errors arise, however, because the measurements are made after the rotor has been stopped and the gradient fractionated.

First, as noted previously (p. 175), a certain amount of redistribution of the salt forming the gradient occurs during deceleration of the rotor. This occurs at the top and bottom of the gradient, which is the reason for adjusting the initial density of the mixture to ensure that any particles for which an accurate estimate of buoyant density is required will band in the middle portion of the tube. Secondly, substantial errors are also introduced if the gradients are disturbed while awaiting fractionation, for example, by changes in temperature or exposure to vibration. Thirdly, viscous or heavily loaded bands may drag along the tube during fractionation and thus be displaced from their original positions in the density gradient; an example of this is shown in *Figure 6.2(b)*. Fourthly, there may be a considerable lapse of time between the gradient being fractionated and the densities of the fractions being measured, during which time there could be a significant loss of water by evaporation from each fraction, particularly if the fractions are small and they have been kept at room temperature. Fifthly, unless it is under proper control, the temperature of the refractometer may vary from day to day, or even from beginning to end of a series of measurements.

Most of these errors can be minimized, if not eliminated, by proper and careful control of conditions and, thus, quite precise measurements can be made. The precision of buoyant-density determinations can be substantially increased by ensuring that the gradient is carrying only the minimum quantity of material. Very often the greatest precision is required from experiments with DNA, in which case it is essential to avoid the formation of viscous bands, by reducing the molecular weight of the DNA to $1-5 \times 10^6$ daltons. Data precise enough for most purposes is obtained by including a marker DNA of known buoyant density, close to that of the 'unknown' DNA, in the same gradient and correcting the observed buoyant density by adding or subtracting the difference between the observed and known buoyant densities of the marker DNA. The greatest precision is achieved by including two marker DNAs whose buoyant densities straddle that of the 'unknown' DNA. However, even when the most stringent precautions are taken it is unwise to assume that the measurements are absolutely accurate, particularly if the data are used to estimate a parameter like the G + C content of a DNA. For native DNA, the latter can be calculated from its buoyant density in neutral CsCl (Schildkraut, Marmur and Doty, 1962):

$$\text{Mole fraction G + C} = \frac{\rho - 1.660}{0.098} \tag{6.31}$$

Thus, the G + C content corresponding to a density of 1.700 g/cm^3 is 40.8%, whereas that corresponding to 1.702 g/cm^3 is 42.9%; that is, a very small error in the estimate of buoyant density gives rise to a significant degree of uncertainty in the estimate of G + C content. Estimates of molecular weight from bandwidths are even more prone to error because they are very sensitive to the unavoidable disturbances which occur during deceleration of the rotor and unloading of the gradient, even under the best conditions.

These comments concern the difficulties and some of the possible sources of error encountered when the buoyant density of a particle or macromolecule is measured by centrifugation in a preparative ultracentrifuge. The interpretation of the data in terms of the structure or chemical composition of the material is subject to the same rules as that of the data from experiments with the analytical ultracentrifuge. Consequently, the same caution is required when the data are evaluated in regard to the possibility that some particles and macromolecules may behave anomalously because, for example, a base in a nucleic acid, or an amino acid in a protein, is chemically modified (see pp. 199–209 and Szybalski and Szybalski, 1971).

SEPARATIONS BY ISOPYCNIC BANDING

Since it is impossible to attempt, even briefly, to review this whole field, only a few examples will be given of the separations which have been made by isopycnic centrifugation in preparative ultracentrifuges. These particular examples have been chosen because they illustrate some of the points about the technique which have been discussed in previous sections, and they go some way towards indicating its overall scope. The bibliographies of the papers cited contain references to many other pertinent papers. In addition, reference should be made to some of the excellent reviews of various aspects of the subject which have been published. These include: Vinograd and Hearst (1962) and Fritsch (1975) – a general introduction to the theory and practice of isopycnic separations; Szybalski (1968) and Szybalski and Szybalski (1971) – a discussion of separations, especially of DNA, in CsCl and Cs_2SO_4 gradients; Flamm, Birnstiel and Walker (1972) – a description of the fractionation of DNA in CsCl gradients; Parish (1972) – a discussion of the fractionation of nucleic acids; and Schumaker and Rees (1972) – a description of density-gradient centrifugation for purifying and fractionating plant viruses. In addition, Colowick and Kaplan (1973) give a number of articles of general relevance.

Separation of DNA, RNA and DNA–RNA Hybrid

The distributions of native and denatured DNA, RNA and DNA–RNA hybrid after sedimentation to equilibrium in density gradients formed *in situ* from CsCl, Cs_2SO_4 and NaI are shown in *Figure 6.4*, and the buoyant densities of these macromolecules are listed in *Table 6.4*. These data illustrate a number of the points discussed in previous sections.

First, the slope of the equilibrium gradient formed decreases in the order Cs_2SO_4, CsCl, NaI, that is, inversely with the β°-values for these salts (see *Table 6.1*). Secondly, the buoyant densities of the macromolecules depend on the salt of which the gradient is composed. Thirdly, in all three salts, the buoyant density of RNA is greater than that of denatured DNA which, in turn, is greater than that of native DNA. However, the resolution of the three species of macromolecule is dependent on the salt constituting the gradient. Fourthly, in CsCl

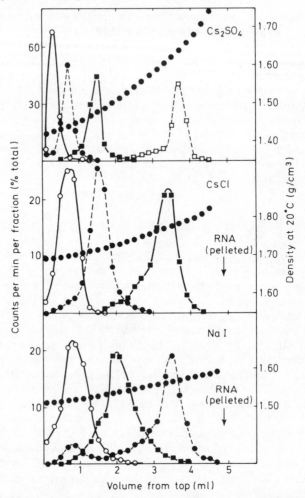

Figure 6.4 Isopycnic banding of native DNA, denatured DNA, DNA—RNA hybrid and RNA from mouse in density gradients formed from Cs_2SO_4, CsCl and NaI. The $[^3H]$-labelled nucleic acids were mixed with Cs_2SO_4 (initial density 1.54 g/cm^3), CsCl (initial density 1.75 g/cm^3 or NaI (initial density 1.55 g/cm^3), and 4.6 ml portions (5—10 μg of nucleic acid) were centrifuged at 25 °C for 68 h at 45 000 rev/min in the MSE 10 × 10 Ti fixed-angle rotor as described on pp. 170—172; o—o, native DNA; •—•, denatured DNA; ■—■, DNA—RNA hybrid; □----□, RNA. See Birnie (1972) for details of the preparation of the nucleic acid samples

and Cs_2SO_4, the buoyant density of DNA—RNA hybrid is, as expected, intermediate between those of native or denatured DNA and RNA. In contrast, in NaI DNA—RNA hybrid is less dense than denatured DNA, and bands at an intermediate position between native and denatured DNA (Birnie, 1972). It would appear that the buoyant density of a nucleic acid in NaI is more dependent on its secondary than its primary structure. The mechanism of this is unknown, although it can be surmised that it involves, to some extent at least, differences in the extents to which the molecules are hydrated in the different solvents (Birnie, MacPhail and Rickwood, 1974). Fifthly, the zones formed at equilibrium are relatively wide, due partly to heterogeneity in the molecular weights,

Table 6.4 BUOYANT DENSITIES OF SOME MACROMOLECULES AND PARTICLES

		Buoyant density (g/cm³) in			
		CsCl	Cs_2SO_4	NaI	KI
Native DNA:					
Cl. perfringens	(31% G + C)	1.692 [1]	1.421 [1]	1.520 [2]	1.488 [2]
Mouse	(40% G + C)	1.700 [1]	1.423 [1]	1.522 [3]	
E. coli	(50% G + C)	1.710 [1]	1.426 [1]	1.535 [2]	1.500 [2]
M. lysodeikticus	(71% G + C)	1.731 [1]	1.435 [1]	1.551 [2]	1.512 [2]
Denatured DNA:					
Mouse	(40% G + C)	1.715 [3]	1.445 [4]	1.574 [3]	
E. coli	(50% G + C)	1.725 [1]	1.450 [1]	1.580 [4]	
RNA:					
Mouse nuclear		>1.9	1.64 [4]	1.62–1.65 [3]	
Mouse ribosomal		>1.9		1.67 [2]	1.62 [2]
E. coli ribosomal		>1.9	1.663* [1]		
Yeast ribosomal		>1.9			1.62 [2]
Globin messenger		>1.9		1.68 [4]	
Reovirus (double-stranded)		>1.9	1.610 [1]	1.57 [2]	1.57 [2]
DNA–RNA hybrid:					
Mouse (synthetic)		1.775 [3]	1.485 [4]	1.540 [3]	
Protein:					
Bovine serum albumin		1.30 [5]			
'Average' mixture		1.33 ± 0.005 [6]			
Nucleoproteins:					
Chromatin (HCHO-fixed)		1.384 [7]			
Polysomes (HCHO-fixed)		1.51–1.53 [8]			
Polysomes (with 50 mM Mg²⁺)			1.46–1.48[9]		
Viruses:					
Bittner virus		1.18 [10]			
Polyoma virus		1.29, 1.32 [11]			
Phage λ		1.51 [12]			

*Precipitated.
Sources: [1] Szybalski (1968); [2] De Kloet and Andrean (1971); [3] Birnie (1972);
[4] Birnie (unpublished); [5] Vinograd and Hearst (1962); [6] Hu *et al.* (1962);
[7] Hancock (1970); [8] Miller (1972); [9] Greenberg (1977); [10] Manning
et al. (1970); [11] Crawford *et al.* (1962); [12] Kellenberger *et al.* (1961).

but mainly to heterogeneity in the base compositions of the nucleic acid species.
The resolving power of each gradient is, therefore, considerably poorer than
might be forecast by simply considering the mean buoyant densities of the
macromolecules (*Table 6.4*). Much narrower zones are formed with homogeneous
species of molecules (see *Figure 6.5*).

The very large difference between the buoyant densities of RNA and DNA,
and between those of DNA and protein, has led to the development of buoyant
density-gradient methods for the large-scale isolation of DNA (Flamm, Birnstiel
and Walker, 1972; Wolf, 1975; Butterworth, 1976) and RNA (Glisin, Crkvenjakov
and Byus, 1974; Wolf, 1975; Affara and Young, 1976; Young, Birnie and Paul,
1976) from tissues, cells and cell organelles (see pp. 209–213).

Fractionation of DNA

DNA can be fractionated by buoyant density according to a number of parameters. First, single- and double-stranded DNAs differ in the degree to which
they are hydrated (Birnie, MacPhail and Rickwood, 1974) and, consequently,

they differ in buoyant density (*Table 6.4*). However, in CsCl and Cs_2SO_4 the difference is small and, because of the heterogeneity in the base-composition of eukaryotic DNAs, the resolution of denatured and native mammalian DNA is generally too poor to allow buoyant density-gradient centrifugation in either of these salts to be used to separate a mixture of single- and double-stranded DNAs (*Figure 6.4*). The resolution can be enhanced by including actinomycin D, for example, in the gradient since this antibiotic forms complexes with native DNA but not denatured DNA. These complexes are stable at high ionic strengths, and have lower buoyant densities than the uncomplexed DNA in CsCl and Cs_2SO_4 (Kersten, Kersten and Szybalski, 1966). Unfortunately, with eukaryotic DNAs the decrease in buoyant density of the native DNA is not uniform since the antibiotic has greater affinity for GC-rich regions in the DNA, and this can blur the separation between native and denatured DNA. On the other hand, single-stranded DNA binds Ag^+ and Hg^{2+} ions much more avidly than does native DNA, and centrifugation in Cs_2SO_4 containing either Ag^+ or Hg^{2+} ions has been used to separate native and denatured DNA (Summers and Szybalski, 1967).

The difference between the buoyant densities of denatured and native DNA is much larger in NaI (*Table 6.4*) and this, coupled with the greater resolving power of a NaI gradient, allows a mixture of single- and double-stranded DNAs to be separated completely in NaI gradients (*Figure 6.4*). Centrifugation in KI gradients also separates native and denatured DNAs, but with poorer resolution (De Kloet and Andrean, 1971). The separation obtainable in NaI gradients is so large that centrifugation to equilibrium in NaI has been proposed as a method of estimating the relative proportions of single- and double-stranded regions in re-annealed DNA duplexes and thus determining the quality of such hybrids (Hell, MacPhail and Birnie, 1974). However, Clegg and Borst (1974) found that heat-denatured mitochondrial DNA from *Tetrahymena* re-anneals while being centrifuged to equilibrium in a NaI gradient (72 h at 25 °C). The extent to which DNA will re-anneal in NaI depends on its base-sequence complexity and G + C content, and it can be calculated that only relatively simple, AT-rich DNAs will re-anneal rapidly enough in NaI for this effect to be detected. Indeed, Clegg and Borst (1974) showed that phage T4 DNA annealed to only a small extent during centrifugation to equilibrium in NaI; mammalian DNA does not re-anneal under these conditions, except for the simple-sequence, AT-rich mouse satellite DNA (see *Figure 6.4*). Re-annealing of simple DNA sequences in NaI gradients can be suppressed by centrifuging at 5 °C instead of 25 °C.

One problem which arises with alkali iodide gradients is the difficulty of detecting zones of nucleic acids by their absorption at 260 nm because I^- absorbs strongly in the ultraviolet. Wolf (1975) circumvented this by measuring absorption at 285 nm. Anet and Strayer (1969b) found that complexing DNA with ethidium bromide has little effect on the buoyant density of DNA in NaI, and the fluorescence of the complexes of the dye with DNA and RNA has been used to detect, and measure the concentration of, DNA and RNA in NaI and KI (De Kloet and Andrean, 1971).

Secondly, DNA can be fractionated on the basis of its tertiary structure (helicity) by virtue of the fact that linear and relaxed circular DNA duplexes bind more ethidium bromide than do supercoiled circular DNAs; consequently, the decrease in buoyant density in CsCl or Cs_2SO_4 which follows the binding of the dye is less for supercoiled duplexes than for linear or relaxed circular

DNAs (Radloff, Bauer and Vinograd, 1967; Bauer and Vinograd, 1968, 1970a, 1970b). The difference between the buoyant densities of the complexes (about 0.04 g/cm^3 in CsCl) is sufficient to allow supercoiled circular DNAs to be separated from other DNAs by buoyant density-gradient centrifugation. One such separation is illustrated by *Figure 6.5*, which shows the separation of the DNA of an *E. coli* plasmid from its host-cell DNA. This is the method of choice for isolating bacterial plasmid DNAs on a relatively large scale (see p. 210), and is also applicable to the isolation of other supercoiled circular DNA duplexes such as mitochondrial, chloroplast and some viral DNAs. Since ethidium bromide has little effect on the buoyant density of DNA in NaI or KI (Anet and Strayer, 1969b) centrifugation in gradients of these salts cannot be used to isolate super-coiled circular DNAs. The technique of centrifugation in CsCl with ethidium

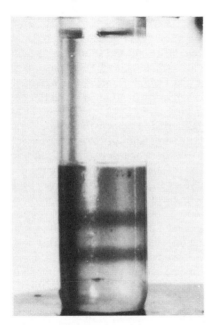

Figure 6.5 Separation of E. coli *DNA and plasmid (col E1 kan) DNA by isopycnic banding in* CsCl *with ethidium bromide. The DNA was isolated and centrifuged as described on pp. 210–211*

bromide has recently been refined to such an extent that it can be used to estimate the number of superhelical turns per base-pair in a supercoiled DNA by measurement of the change in buoyant density induced by the formation of the DNA–dye complex under standard conditions (Burke and Bauer, 1977).

Thirdly, the buoyant density of a DNA in most ionic media is dependent on its base composition (see *Table 6.4*). In particular, within wide limits (20–80%) the G + C content of native DNA bears a linear relationship to its buoyant density in neutral CsCl gradients (Schildkraut, Marmur and Doty, 1962), and similar correlations between base composition and buoyant density in NaI and KI have been found (De Kloet and Andrean, 1971). However, very AT-rich and GC-rich DNAs, and some synthetic copolymer duplexes, display anomalous buoyant densities in CsCl (see Szybalski and Szybalski, 1971). The G + C content of a DNA also influences its buoyant density in Cs_2SO_4, but the effect is less than in CsCl, and the relationship between the two is not linear. Modifications of the bases in DNA (for example, methylation, hydroxymethylation and glucosylation) also cause changes in buoyant density (Szybalski and Szybalski, 1971); and these are sometimes large enough to be detected in Cs_2SO_4 and/or CsCl gradients in preparative ultracentrifuges. One type of modification to DNA, the introduction of a density label, has been used extensively to study the replication of DNA *in vivo* and *in vitro* (Meselson and Stahl, 1958; see also Flamm, Birnstiel and Walker, 1972). Relatively large increases in buoyant density can be induced by replacing ^{14}N with ^{15}N (0.016 g/cm^3), ^{12}C with ^{13}C (0.036 g/cm^3) and carbon-bound H with deuterium (0.035 g/cm^3), while replacement of all thymine in a 50% G + C DNA with 5-bromouracil increases the density of the DNA by 0.08 g/cm^3 (Hu, Bock and Halvorson, 1962).

The dependence of buoyant density of a DNA on its base composition not only enables DNAs from different organisms to be separated (*Figure 6.2*), but also allows eukaryotic DNA to be subfractionated according to differences in base composition between one region and another of the eukaryotic genome. In this way, mouse satellite (simple-sequence) DNA has been isolated in CsCl gradients (see *Figure 6.2* and Flamm, Birnstiel and Walker, 1972), as have the ribosomal genes from *Xenopus laevis*, although the latter constitute only 0.3% of the total genomic complement of DNA (see Flamm, Birnstiel and Walker, 1972 This approach has also been used to demonstrate that, in mouse myeloma cells, GC-rich sequences replicate early, and AT-rich ones late, in S-phase (Flamm, Birnstiel and Walker, 1972). Fractionations like this can be enhanced and, indeed, new satellite DNAs can be revealed, by the inclusion in a density gradient of compounds or ions which have a particular affinity for regions in the DNA with some characteristic structure which distinguishes them from other regions in the same genome.

For example, netropsin binds specifically to regions of a DNA which contain A-T base-pairs, and thereby alters their buoyant density in CsCl (Wartell, Larson and Wells, 1974; Guttann, Votavova and Pivek, 1976). Inclusion of Hg^{2+} or Ag^+ ions in a Cs_2SO_4 gradient causes marked changes in the buoyant density of a DNA (Nandi, Wang and Davidson, 1965; Wang *et al.*, 1965; Jensen and Davidson, 1966), and results in the separation of a series of 'satellite' DNAs. The fractionation obtained depends on the DNA, and on the particular concentration of ions in, and pH of, the density gradient (Filipski, Thiery and Bernardi, 1973; Rinehart and Schmid, 1976). The reaction of Hg^{2+} and Ag^+ with DNA, and the effect of the binding of these ions on buoyant density, is complex. Although fractionations basically depend on the affinity of Hg^{2+} ions for A-T base-pairs and that of Ag^+ for G-C base-pairs (Lieberman *et al.*, 1976; Rinehart and Schmid, 1976), the precise sequence of the nucleotides appears to modify the interaction of these ions with DNA (Filipski, Thiery and Bernardi, 1973).

Finally, it should be mentioned that the complementary strands of some DNAs can be separated by buoyant density-gradient centrifugation in alkaline CsCl. This fractionation is also dependent on base composition, but this time on the G + T content of the individual strands (see Flamm, Birnstiel and Walker, 1972).

Fractionation of RNA

Much less has been done to explore the fractionation of RNA by buoyant density-gradient centrifugation for the simple reason that it has been difficult to find a gradient solute suitable for this purpose. CsCl is not soluble enough to provide a solution denser than RNA except at 40–50 °C, a temperature which is inconvenient from the point of view both of the RNA and of the centrifuge. Concentrated solutions of caesium formate and caesium acetate are dense enough, but their use is hampered by their high viscosity. Cs_2SO_4 would be very suitable for the isopycnic banding of RNA were it not for the marked propensity of most RNAs to aggregate and precipitate in solutions of this salt. Lozeron and Szybalski (1966) reasoned that precipitation of RNA might be due to the formation of non-specific hydrogen-bonded aggregates and, therefore, that it could be prevented by the inclusion of HCHO in the gradient to inhibit the formation of hydrogen bonds. They found that most RNAs could be banded isopycnically in Cs_2SO_4 gradients containing 1% HCHO, but that the buoyant density of the RNA was greater than predicted and most species of RNA formed bands near the bottom of the gradient.

Lozeron and Szybalski (1966) formulated a second solvent, containing 1% HCHO and equal volumes of saturated CsCl and Cs_2SO_4, with which RNA could be banded in the middle of the tube. However, the presence of HCHO is not always acceptable, for example, when infectious viral RNA, or RNA for molecular hybridization studies, is required. In these circumstances, gradients formed by centrifugation from a mixture of water, saturated Cs_2SO_4 and saturated CsCl (1 : 1 : 8 by vol.) permit the isopycnic banding of unaggregated RNA, but again near the bottom of the tube (Lozeron and Szybalski, 1966). Other mixtures of CsCl and Cs_2SO_4 have been found to be satisfactory for some RNA species. For example, nuclear RNA and RNA transcribed *in vitro* from chromatin bands near the middle of a gradient formed from water, saturated CsCl and saturated Cs_2SO_4 (1 : 2 : 2 by vol.); moreover, in gradients of this composition there is good resolution between unlabelled and density-labelled RNAs (*Figure 6.6a*). Unfortunately, however, these gradients have a low capacity for RNA; above about 10 μg/ gradient the RNA aggregates and precipitates, so precluding the use of these gradients for preparative-scale fractionation of RNA.

Williams and Vinograd (1971) have explored the effect of pH and of the destacking agent dimethyl sulphoxide (DMSO) on the behaviour of RNA and DNA in Cs_2SO_4 buoyant density gradients. The solubility of some RNA species is increased at low pH (4–5), but inclusion of DMSO (10% by vol.) was generally more successful at preventing aggregation and precipitation of RNA. Gradients formed by centrifugation from aqueous Cs_2SO_4 solutions containing 10% DMSO are capable of resolving double-stranded RNA, single-stranded RNA and double-stranded DNA (Williams and Vinograd, 1971). However, the discrimination between normal and density-labelled RNA is very poor in Cs_2SO_4–DMSO gradients (*Figure 6.6b*), whereas incorporation of DMSO into the mixed CsCl–Cs_2SO_4 gradient system (*Figure 6.6a*) decreases the separation between these RNAs (*Figure 6.6c*), although it does increase the capacity of the gradients for RNA.

The solubility of RNA in KI (De Kloet and Andrean, 1971) or in NaI from which all heavy metal ions have been removed (Birnie, 1972) does not appear to be a problem. RNA bands isopycnically in gradients of either salt and, moreover, displays considerable heterogeneity in buoyant density in KI (De Kloet and Andrean, 1971) and, particularly, in NaI (Birnie, 1972; Hell, MacPhail and Birnie, 1974). Andrean and De Kloet (1973) have explored the use of gradients formed from mixtures of KI and NaI and have concluded that it is possible to

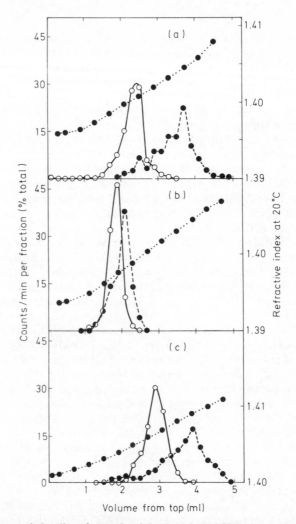

Figure 6.6 Isopycnic banding of normal and density-labelled RNA. (a) A mixture of 2 ml
of saturated CsCl, 2 ml *of saturated* Cs$_2$SO$_4$ *and* 1 ml *of buffer* (100 mM tris-HCl, 10 mM
EDTA, pH 7.5) *containing the RNA sample (about 5 μg) was centrifuged at* 41 000 rev/min
for 68 h *at* 25 °C *in the MSE* 10 × 10 Ti *fixed-angle rotor. (b) A mixture of* 3.15 ml *of*
saturated Cs$_2$SO$_4$, 1.35 ml *of buffer containing the RNA sample (about 5 μg) and* 0.50 ml
of dimethyl sulphoxide (final concentration, 10% v/v) *was centrifuged at* 45 000 rev/min
for 18 h *then* 35 000 rev/min *for* 48 h *at* 25 °C *in the MSE* 10 × 10 Ti *fixed-angle rotor.*
(c) A mixture of 1.9 ml *of saturated* CsCl, 1.7 ml *of saturated* Cs$_2$SO$_4$, 1.15 ml *of buffer*
containing the RNA sample (about 5 μg) and 0.25 ml *of dimethyl sulphoxide (final concen-*
tration, 5% v/v) *was centrifuged as described for (b). (Diane Donaldson and G.D. Birnie,*
unpublished)

The normal RNA was either isolated from lysed nuclei of mouse Friend cells which had
grown in [³H]uridine for 24 h or synthesized *in vitro* by transcription of a mouse Friend-
cell chromatin template essentially as described by Alonso *et al.* (1976), except that the
radioactive triphosphate was [³H]CTP (the buoyant density-gradient profiles of these
RNAs were indistinguishable); the density-labelled RNA was synthesized *in vitro* by transcription of Friend-cell chromatin, with 5-bromo-UTP replacing UTP in the transcription
mixture. (Note: In these complex three- and four-component systems, density is not
related directly to refractive index, and can only be determined by pycnometry (see p. 196).
This was not done because the interest in these experiments was in the resolution between
the RNAs; consequently, the gradients are recorded only as refractive indices at 20 °C.)

fractionate RNA partly on the basis of base composition and partly on secondary structure. It would appear that gradients of this kind might provide a basis other than size or electrophoretic mobility for fractionating RNAs, particularly mRNAs and nuclear RNAs (Hell, MacPhail and Birnie, 1974).

Isolation and Fractionation of Proteins

Proteins in general have low buoyant densities in ionic buoyant density-gradient media. Not a great deal of use of the technique has been made for the fractionation of proteins (but see Fritsch, 1975), despite the natural variation in the amino acid composition of proteins giving rise to differences in densities of up to 0.1 g/cm³ (Hu, Bock and Halvorson, 1962). Recently, buoyant density-gradient methods for the isolation of proteins, in particular chromosomal proteins, have been developed (see p. 213).

Lipoproteins can readily be fractionated by buoyant density-gradient sedimentation (Hinton *et al.*, 1974; Wilcox and Heimberg, 1970), whereas flotation of lipoproteins through a gradient of NaBr is a useful method for examining various protein components of pathological sera (see p. 214). Hu, Bock and Halvorson (1962) investigated isopycnic sedimentation as a means of differentiating between newly-synthesized and pre-existing proteins; they found that excellent resolution could be obtained in RbCl gradients between natural β-galactosidase and β-galactosidase which had been density-labelled with ^{13}C, or ^{13}C and ^{15}N, or ^{2}H and ^{15}N. A recent study of a similar fractionation led Boudet, Humphrey and Davies (1975) to conclude that KBr gradients gave better resolution than CsCl gradients, as would be expected from a consideration of the β°-values of the two salts (see *Table 6.1*).

Fractionation of Nucleoproteins

At the concentration of salt required to band nucleoprotein particles isopycnically, the ionic strength of the solution is such that most nucleoprotein complexes are disrupted into their constituent nucleic acids and proteins. Hancock (1970) circumvented this problem for chromatin by first treating it with HCHO. In a CsCl gradient the fixed chromatin formed a broad symmetrical band, indicating that the ratio of protein to DNA was not uniform throughout the chromatin. The reaction with HCHO is irreversible, and constituents of the chromatin fractions could not be examined further. Complexes of protein with nucleic acids are not all equally sensitive to dissociation by high ionic strengths, and this fact has been exploited recently for two interesting, and potentially very useful, purposes. First, Naito and Ishihama (1975) found that the transcription complex formed with template DNA, the growing RNA chain and functioning RNA polymerase is stable in high ionic strength media, and have exploited this to separate transcription complexes from uncomplexed polymerase and complexes of DNA with RNA polymerase, by buoyant density-gradient centrifugation in CsCl and Cs_2SO_4. Secondly, the integrity of messenger ribonucleoprotein particles is maintained in concentrated Cs_2SO_4 under conditions in which the bulk of the protein is stripped from ribosomal subunits so that buoyant density-gradient centrifugation of polysomes in Cs_2SO_4 gives a clean separation on a preparative scale of the mRNA-containing ribonucleoproteins from the other components of the polysomes (Liautard and Liautard, 1977; Greenberg, 1977). A similar approach had been used earlier as an analytical technique (Miller, 1972). Some nuclear ribonucleoprotein particles are also resistant to dissociation with Cs_2SO_4, and Wilt, Anderson and Ekenberg (1973) used this fact to isolate these particles from sea-urchin nuclei by isopycnic banding in Cs_2SO_4. Under these conditions not all chromatin is dissociated, and some deoxyribonucleo-protein co-bands with the nuclear particles (Wilt, Anderson and Ekenberg, 1973). Genta, Kaufman and Kaufman (1975) have suggested this as an approach to the isolation of *in vivo* DNA replication complexes.

Although some nucleoproteins are resistant to dissociation in Cs_2SO_4, it should be noted that they can be completely dissociated by the higher ionic strength of concentrated CsCl solutions, particularly in the presence of detergents (see p. 211).

Isolation and Fractionation of Viruses

One of the earliest applications of buoyant density-gradient sedimentation was in the study of bacteriophages (see Vinograd and Hearst, 1962; Davidson and Szybalski, 1971). The technique has proved very useful for the isolation and purification of, for example, ϕX-174 (Sinsheimer, 1959) and phage λ (Kaiser and Hogness, 1960), and it has also been used to fractionate phage λ for genetic studies (Kellenberger, Zichichi and Weigle, 1961). Some plant viruses (for example tobacco mosaic virus) are also stable in solutions of high ionic strength, and thus buoyant density-gradient centrifugation in CsCl and other salts has been used for their purification, fractionation and characterization (see Schumaker and Rees, 1972). Poliovirus bands isopycnically in CsCl without significant loss of infectivity, and Levintow and Darnell (1960) made use of this to isolate and purify large quantities of the virus for analysis and characterization of its coat proteins. Crawford, Crawford and Watson (1962) separated polyoma virus into two fractions, one consisting of non-infectious ('empty') particles, the other, infectious (DNA-containing) particles, by buoyant density-gradient centrifuga-tion in CsCl.

More recently, Manning *et al.* (1970) have studied the behaviour of the mouse mammary-tumour agent (Bittner virus) in four buoyant density-gradient media. They found that the virus banded as a single zone in gradients of RbCl and CsCl, at about the same buoyant density as Rous sarcoma virus and Rauscher leukaemia virus; in contrast, in potassium tartrate and sucrose gradients Bittner virus was resolved into two distinct zones, the lighter of which had the appearance of incomplete particles. Equine herpes virus (type I) also can be purified to a high degree in CsCl gradients (Lawrence, 1976). It was found that more than 90% of the virus infectivity could be recovered when the virus was purified by flotation through a CsCl gradient, but less than 10% was recovered if the virus sample had been layered on top of the gradient. The loss of infectivity under the latter conditions may be due to physical disruption of the virus because of the rapidity with which it sediments from the top of the gradient (Lawrence, 1976).

Not all viruses are stable in high ionic strength solutions. Some are relatively stable in potassium tartrate and potassium citrate, which has led to the use of gradients of these salts for isopycnic banding of some viruses (McCrea, Epstein and Barry, 1961), whereas others seem to be most stable in solutions of very low ionic strength (see Chapter 7).

LARGE-SCALE PREPARATIVE PROCEDURES

Although all the separations done by isopycnic centrifugation are preparative separations (or, at least, potentially so), it is only relatively recently that the methods have been developed specifically for large-scale preparations of some materials. Outlines of six of these are given here; for the finer details of each, reference should be made to the original publications.

Isolation of DNA

Total or Nuclear DNA

The advantages of using isopycnic centrifugation to prepare DNA have been discussed in detail by Flamm, Birnstiel and Walker (1972) and Butterworth (1976). In essence, the method consists of disrupting tissues, cells or isolated nuclei in detergent and dense CsCl solution, and centrifuging the mixture to equilibrium under conditions in which the RNA pellets at the bottom of the tube, the DNA forms a band in the gradient and the protein and carbohydrate form a pellicle at the top. The method is very tolerant of variations in procedure, provided that the basic requirements are satisfied. The following method, adapted from Flamm, Birnstiel and Walker (1972) and Butterworth (1976), is generally applicable.

Step 1. The DNA-containing material is suspended in at least 10 vol. of dilute buffer (for example, 50 mM tris-HCl, 10 mM EDTA, pH 8), containing 1–2% (w/v) sodium dodecyl sulphate, and homogenized with a tight-fitting Teflon pestle in a glass Potter-Elvehjem homogenizer vessel. Although whole cells and tissues can be treated in this way, if large amounts of material are to be processed it is usually best to remove a large proportion of the protein first, either by isolating a crude preparation of nuclei – for example, by the sucrose–citric acid procedure described by Birnie (1977) – or by a preliminary extraction with phenol–chloroform.

Step 2. If the solution so formed is very viscous (because of a high concentration of DNA), the molecular weight of the DNA should be reduced by shearing, either by extruding the solution through a narrow-gauge hypodermic needle or by brief ultrasonication.

Step 3. Sufficient solid CsCl is added to the solution to give a density of 1.71 g/cm^3 (1.28 g/ml of homogenate), and the whole mixed thoroughly (vortex mixer) until the CsCl has dissolved. A copious white precipitate (protein and caesium dodecyl sulphate) forms and begins to float to the top of the tube. The density of the solution is checked by refractive index and adjusted if necessary to 1.71 g/cm^3. Some care should be taken over this since, if the initial density is too high, the band of DNA will overlap with the proteins at the top of the gradient, whereas if it is too low the DNA will pellet along with the RNA. The concentration of DNA in the solution should not exceed 0.5 mg/ml.

Step 4. The mixture is put into the centrifuge tubes of a fixed-angle rotor. Not more than one-third to one-half of the capacity of each tube should be taken up with the mixture, the rest of the tube being occupied by liquid paraffin (see p. 171). The tubes are then centrifuged at 20 °C for about 70 h at a speed which depends on the rotor being used (for example, 40 000–45 000 rev/min for small- and intermediate-volume titanium rotors; see p. 177).

Step 5. After centrifugation, the DNA is collected either by fractionating the whole gradient by upwards displacement or, more conveniently, by extracting the zone containing the DNA by the method illustrated in *Figure 6.3(c).* In both cases, it is advisable first to remove the plug of insoluble material which has formed at the top of the gradient.

Step 6. The DNA is recovered by mixing the solution with 3 vol. of water and 10 vol. of ethanol (to give 67% v/v ethanol); after at least 4 h at -20 °C the precipitated DNA is collected by centrifugation at 10 000g_{av} for 10 min, washed with 95% ethanol and dried under a stream of nitrogen.

DNA so prepared is the Cs^+ salt; it can be transformed to the Na^+ salt by dissolving it in dilute buffer and passing the solution through a short column of the Na^+ form of Dowex-50 ion-exchange resin. It commonly contains less than 1% by weight of protein and 2% by weight of RNA, which is better than the DNA prepared by most other procedures. If a higher degree of purity is required, the DNA can be incubated first with ribonuclease then with proteinase K in the presence of 1% sodium dodecyl sulphate, and reprecipitated after extraction with phenol–chloroform (Butterworth, 1976).

Plasmid DNA

The method exploits the difference between the effects of intercalated ethidium bromide on the buoyant densities of relaxed linear or circular duplex DNAs and supercoiled circular DNAs (see p. 202). The method described was developed for the isolation of plasmid DNA (specifically col El kan) from *E. coli* (J. Davison, personal communication) from other procedures like those described by Vinograd and his collaborators (see p. 203). This, and other similar methods, are equally applicable to the isolation of other supercoiled circular DNAs.

Step 1. The culture of *E. coli* (1600 ml; E_{650}, about 0.6) is killed by being shaken with chloroform (about 0.2% v/v) for 10 min; any undissolved chloroform is removed.

Step 2. The cells are collected by centrifugation at 8000g_{av} for 15 min at 4 °C, resuspended in 1200 ml of 50 mM tris-HCl, pH 8.0, and resedimented at 8000g_{av} for 15 min at 4 °C.

Step 3. The pellet of cells is resuspended in 130 ml of 25% (w/w) sucrose in 50 mM tris-HCl (pH 8.0) at 4 °C, and 13 ml of a lysozyme solution (10 mg/ml in 50 mM tris-HCl, pH 8.0) is added. After 5 min, 26 ml of 0.5 M EDTA, pH 8.2, is mixed with the suspension, and the whole is incubated at 30 °C for 15 min.

Step 4. The protoplasts are lysed by the addition of 120 ml of 3.6% (v/v) Triton X-100 (final concentration, 1.5% v/v) and incubation at 20 °C for 10 min.

Step 5. The mixture is centrifuged at 65 000g_{av} for 1 h at 4 °C (MSE 8 × 35 fixed-angle rotor at 30 000 rev/min).

Step 6. The non-viscous part of the supernatant fluid is removed by decantation and to the residue (about 200 ml) is added one-ninth vol. of 5 M NaCl and solid polyethylene glycol (mol. wt 6000, supplied by British Drug Houses, Ltd) to a final concentration of 10% (w/v).

Step 7. The mixture is stirred at 20 °C until the polyethylene glycol has dissolved, and the mixture is left overnight at 4 °C.

Step 8. The mixture is centrifuged at $8000g_{av}$ for 15 min at 4 °C and the pellet is redissolved in 10.8 ml of 0.1 X SSC (15 mM NaCl, 1.5 mM sodium citrate, pH 7.0); to this solution is added 11.7 g of CsCl and 1.2 ml of ethidium bromide solution (2 mg/ml).

Step 9. The CsCl is dissolved by vigorous mixing, and the refractive index of the solution is adjusted to 1.388 0–1.389 0 by addition of water or CsCl as appropriate.

Step 10. The solution (15 ml) is divided equally between two polycarbonate tubes of the MSE 10 X 10 Ti fixed-angle rotor which are topped-up with paraffin and centrifuged at 40 000 rev/min for 64 h at 20 °C. Polycarbonate tubes are used to permit the separated bands to be seen.

Step 11. The upper band of *E. coli* DNA is removed, and the lower band of plasmid DNA is recovered as described on p. 192 (*Figure 6.3c*).

Step 12. The dye is removed by repeated extraction with isoamyl or isopropyl alcohol (presaturated with CsCl), or by passage through Dowex 50 ion-exchange resin (Radloff, Bauer and Vinograd, 1967), and the plasmid DNA is dialysed extensively against 0.1 X SSC (3 days at 4 °C with several changes of buffer).

This method yields large quantities (up to 500 µg per gradient) of biologically active supercoiled plasmid DNA; contamination with non-plasmid DNA is very low (see p. 192).

Isolation of RNA

Nuclear RNA

The method is similar in principle to the first one. It was first described by Glisin, Crkvenjakov and Byus (1974); the following modified version was described by Affara and Young (1976), who have discussed the method and its advantages in detail.

Step 1. Pure nuclei are prepared by a convenient method which eliminates contamination with cytoplasm (see Roodyn, 1972; Birnie, 1977).

Step 2. The nuclei are suspended in about 20 vol. of 0.1 M tris-HCl, pH 8.0, containing proteinase K (0.5 mg/ml). Sodium lauroyl sarcosinate is added to a final concentration of 4% (w/v) and the mixture is homogenized until all aggregates of DNA are dispersed (5–6 up-and-down strokes of a hand-driven loose-fitting Teflon ball pestle in a Potter-Elvehjem homogenizer should suffice). The mixture is incubated at 37 °C for 45–60 min.

Step 3. Solid CsCl is added to the mixture (1 g per ml of homogenate) and the whole is mixed until the CsCl has dissolved. The concentration of DNA + RNA in this solution should not exceed 50–100 µg/ml, otherwise some of the DNA will pellet along with the RNA.

Step 4. The mixture is carefully layered on top of a solution of CsCl (in 0.1 M EDTA, pH 7.4), the density of which has been adjusted carefully to 1.71 g/cm³. The volume of the CsCl–homogenate mixture and that of the dense CsCl solution depends on the rotor being used; for example, 5 ml of homogenate is layered on top of 1.2 ml of dense CsCl in the tubes of MSE 3 × 6.5 Ti swing-out rotor, and 10 ml of homogenate is layered on top of 2.2 ml of dense CsCl in the tubes of the MSE 6 × 14 Ti swing-out rotor. It should be noted that swing-out rotors must be used for this method, since the RNA does not pellet cleanly to the bottom of a tube in a fixed-angle rotor.

Step 5. The tubes are centrifuged at 25 °C for 24 h at 39 000 rev/min (3 × 6.5 ml rotor) or for 48 h at 29 000 rev/min (6 × 14 ml rotor).

Step 6. The plug of protein at the top of the tube is removed, and the whole of the liquid removed by aspiration, care being taken to avoid disturbing the pellet of RNA.

Step 7. The pellet is dissolved in distilled water and mixed with one-tenth vol. of 20% (w/v) potassium acetate and 2.5 vol. of 95% ethanol. After a minimum of 4 h at −20 °C the precipitated RNA is collected by centrifugation ($10\,000g_{av}$ for 15 min), washed with ethanol and dried in a stream of nitrogen.

RNA prepared in this way contains no detectable protein, and less than 1% by weight of DNA; by gel electrophoresis it is found to be essentially undegraded (Glisin, Crkvenjakov and Byus, 1974; Affara and Young, 1976). The yield of RNA is almost quantitative (at least 90%), and quite large amounts can be prepared without overloading the gradient; for example, the nuclear RNA from 7×10^7 mouse cells can be isolated in one tube of the 3 × 6.5 Ti rotor.

Polysomal RNA

A simplified version of the previous method, described by Young, Birnie and Paul (1976) and Affara and Young (1976), can be used to prepare RNA from polysomes, or other RNA-containing materials which do not contain DNA. The major difference is the absence of the prolonged centrifugation of the CsCl extract, which can be omitted because there is no need to pellet the RNA to separate it from DNA or proteins.

Step 1. A pellet of polysomes is dissolved in 0.15 M NaCl, 10 mM EDTA, 50 mM HEPES–NaOH, pH 7.5 (0.5–1 ml/mg of polysomes) which has been sterilized by treatment with diethyl pyrocarbonate followed by autoclaving.

Step 2. One-tenth vol. of a 10% (w/v) solution of sodium dodecyl sulphate is added, followed by solid CsCl (1.4 g per ml of polysome suspension) which has previously been treated with chelating resin to remove heavy metal ions (see p. 194). The whole is mixed vigorously (vortex mixer) until all the CsCl has dissolved. Diethyl pyrocarbonate (5 μl/mg of protein or 2.5 μl/mg of polysomes) is added and the solution is mixed again (vortex mixer). The copious white precipitate which forms begins to float to the surface.

Step 3. The mixture is centrifuged in a swing-out rotor at $10\,000g_{av}$ for 30 min at 15–20 °C, during which time the precipitate (caesium dodecyl sulphate plus denatured protein) forms a densely packed pellicle.

Step 4. A long needle is inserted through the pellicle to the bottom of the tube and the CsCl solution is withdrawn into a syringe. It is not always possible to avoid withdrawing a few particles of the precipitate along with the solution, but these are readily removed by ejecting the solution through a Swinnex filter containing a disc of Whatman GF/C glass-fibre paper. Removing the whole precipitate by filtration is not recommended since a significant proportion of the RNA becomes absorbed to the heavy filter cake.

Step 5. The RNA is recovered by adding 3 vol. of distilled water and 10 vol. of ethanol (to give 67% v/v ethanol); after at least 4 h at −20 °C the precipitated RNA is collected by centrifugation at 10 000g_{av} for 15 min, washed with ethanol and dried under a stream of nitrogen.

The method can be adapted to the isolation of RNA from cytoplasmic extracts or cytosol preparations simply by adding the detergent and solid CsCl directly to the RNA-containing solution (step 2), although the capacity is somewhat reduced because of the high proportion of protein in such extracts. It has also proved very successful for the recovery of RNA (or DNA) from polyacrylamide gel slices (I. Pragnell, personal communication). The gel slices are simply macerated in water or a suitable buffer, and solid CsCl added to the mixture; the gel fragments float to the top of the solution during the centrifugation.

RNA prepared in this way contains no detectable protein; by gel electrophoretic analysis the RNA is undegraded (Young, Birnie and Paul, 1976; Affara and Young, 1976). The yield of RNA is at least 90% of theoretical, and large amounts can be prepared very rapidly (steps 1–4 take 1 h).

Isolation of Protein

Chromosomal Proteins

Until recently it has proved difficult to make preparations of chromatin proteins, in particular the non-histone proteins, which are undegraded and also totally devoid of the endogenous RNA, which is a constituent of chromatin as it is usually prepared (see MacGillivray, 1976). Chromatin is readily dissociated into its constituent nucleic acids and proteins by exposure to media of high ionic strength, and these components of chromatin can be separated by isopycnic centrifugation (Monahan and Hall, 1974; Shaw, Blanco and Mueller, 1975). The following method was developed in this laboratory by A. Alonso, A.J. MacGillivray and D. Rickwood, and has been used to isolate proteins from chromatin (A. Alonso, personal communication) and to purify non-histone proteins isolated by other means (MacGillivray, 1976).

Step 1. Isolated chromatin is homogenized in 2 M NaCl containing 5 M urea, 1 mM dithioerythritol, 0.1 mM EDTA and 10 mM tris-HCl, pH 7.5, and centrifuged at 12 000g_{av} for 10 min. The pellet of material is rehomogenized in the same buffer, and the suspension centrifuged at 12 000g_{av} for 10 min. The combined supernatant fractions are diluted until the concentration of chromatin proteins is 1 mg/ml (or less) and mixed with an equal volume of 55% (w/w) aqueous CsCl (5.5 M) containing 0.1 mM phenylmethylsulphonyl fluoride (included to inhibit proteases).

Fractions of non-histone proteins isolated by, for example, hydroxylapatite chromatography, are adjusted to contain not more than 1 mg of protein per ml, dialysed against 8 M urea, 0.2 M tris-HCl, pH 8, and mixed with an equal volume of 60% (w/w) aqueous CsCl (6.4 M).

Step 2. Five millilitre portions of the solution are put into the tubes of an MSE 10 × 10 Ti fixed-angle rotor, and the tubes are topped-up with paraffin oil and centrifuged at 50 000 rev/min for 68 h at 4 °C (for dissociated chromatin) or 40 000 rev/min for 44 h at 4 °C (for previously isolated non-histone proteins).

Larger capacity rotors may be used with appropriate adjustments to the speed and time of centrifugation. The volume of the aqueous solution in each tube should not be more than one-third to one-half of its maximum capacity.

Step 3. The gradients are unloaded by upwards displacement. All the protein is contained in the top 1.5 ml of a 5 ml gradient, and only this volume need be collected.

Step 4. The proteins are recovered by dialysis or gel exclusion chromatography, or by some other means appropriate to the use for which they are required.

The yield of proteins isolated in this way can be virtually theoretical. The preparations contain no detectable DNA or RNA, and they appear to be undegraded as judged by two-dimensional gel electrophoresis (MacGillivray, 1976). They retain their biological activity as judged by chromatin reconstruction experiments

Plasma Lipoproteins

The classical approach to the isolation of plasma lipoproteins is flotation through a gradient from a serum sample whose density has been increased by addition of salt. The use of this method to prepare and fractionate lipoproteins on a very large scale in zonal rotors is discussed in Chapter 4. The following method using a swing-out rotor was developed by Hinton *et al.* (1974).

Step 1. Linear density gradients, ranging in density from 1.003 g/cm^3 to 1.2 g/cm^3 and of total volume 20 ml, are prepared in the tubes of an MSE 3 $\times$ 25 swing-out rotor, using a gradient engine and a low- and a high-density stock solution. The former solution contains 11.4 g of NaCl and 0.1 g of EDTA per litre, the latter 263 g of NaBr, 11.4 g of NaCl and 0.1 g of EDTA per litre; both are adjusted to pH 7.8 with NaOH.

Step 2. The serum sample is adjusted to a density of 1.22 g/cm^3 by the addition of 0.32 g of NaBr per ml of serum; this is done immediately before the sample is fractionated.

Step 3. The sample (0.5–1.0 ml) is layered under the gradient, and underlayered with a dense solution containing 582 g of NaBr/ml (2.0–1.5 ml). This is most easily done by injection through a long needle from a syringe, but great care must be taken to avoid introducing air bubbles which would disrupt the gradient.

Step 4. The tubes are centrifuged at 30 000 rev/min for 2 h at 4 °C. It is important that the rotor is accelerated slowly from rest to 5000 rev/min to avoid disturbing the layers. Acceleration from rest to 1000 rev/min should take about 3 min, and from 1000 to 5000 rev/min a further 2 min; thereafter, the rotor is allowed to accelerate normally.

Step 5. After centrifugation the gradients are fractionated by upwards displacement. The very low-density lipoprotein is found at the top of the gradient, low-density lipoprotein between 10 ml and 15 ml from the top, and the high-density lipoprotein at the original sample position along with other serum proteins.

The main advantage of this method is its rapidity and simplicity as compared with other methods; from a 1 ml sample of serum sufficient of the protein fractions are obtained for analysis of the component lipids and their fatty acids (Hinton *et al.*, 1974). Larger quantities of these materials can be obtained with large-volume swing-out rotors, and much larger ones by using zonal rotors (see p. 96).

ACKNOWLEDGEMENTS

The Beatson Institute is supported by grants from MRC and CRC. I am grateful to the authors and publishers for permission to reproduce *Figures 6.1, 6.2* and *6.3(a)*, to Mr John Ellis for *Figure 6.5* and to Miss Diane Donaldson, Dr Angel Alonso, Dr Lesley Coggins and Dr Ian Pragnell for permission to quote unpublished data. I am also grateful to my co-editor for his patient encouragement, and I am indebted to Mrs June Peffer for transcribing a rather difficult manuscript.

REFERENCES

AFFARA, N.A. and YOUNG, B.D. (1976). MSE Application Information A12/6/76. Crawley, Sussex, UK; MSE Instruments Ltd

ALONSO, A., BIRNIE, G.D., KLEIMAN, L., MacGILLIVRAY, A.J. and PAUL, J. (1976). *Biochim. biophys. Acta,* **454**, 469

ANDREAN, B.A.G. and DE KLOET, S.R. (1973). *Archs Biochem. Biophys.,* **156**, 373

ANET, R. and STRAYER, D.R. (1969a). *Biochem. biophys. Res. Commun.,* **34**, 328

ANET, R. and STRAYER, D.R. (1969b). *Biochem. biophys. Res. Commun.,* **37**, 52

BALDWIN, R.L. and SHOOTER, E.M. (1963). In *Ultracentrifugal Analysis in Theory and Experiment,* p. 143. Ed. J.W. Williams. New York and London; Academic Press

BAUER, W. and VINOGRAD, J. (1968). *J. molec. Biol.,* **33**, 141

BAUER, W. and VINOGRAD, J. (1970a). *J. molec. Biol.,* **47**, 419

BAUER, W. and VINOGRAD, J. (1970b). *J. molec. Biol.,* **54**, 281

BIRNIE, G.D. (1972). *FEBS Letts,* **27**, 19

BIRNIE, G.D. (1977). In *Methods in Cell Biology,* vol. 17, p. 13. Eds G. Stein, J. Stein and L. Kleinsmith. New York; Academic Press

BIRNIE, G.D., MacPHAIL, E. and RICKWOOD, D. (1974). *Nucleic Acids Res.,* **1**, 919

BOUDET, A., HUMPHREY, T.J. and DAVIES, D.D. (1975). *Biochem. J.,* **152**, 409

BRUNER, R. and VINOGRAD, J. (1965). *Biochim. biophys. Acta,* **108**, 18

BURKE, R.L. and BAUER, W. (1977). *J. biol. Chem.,* **252**, 291

BUTTERWORTH, P.H.W. (1976). In *Subnuclear Components: Preparation and Fractionation,* p. 295. Ed. G.D. Birnie. London; Butterworths

CLEGG, R.A. and BORST, P. (1974). *Mol. biol. Rep.,* **1**, 477

COLOWICK, S.P. and KAPLAN, N.O. (Eds) (1973). *Methods in Enzymology,* vol. 27. New York; Academic Press

CRAWFORD, L.V., CRAWFORD, E.M. and WATSON, D.H. (1962). *Virology,* **18**, 170

DAVIDSON, N. and SZYBALSKI, W. (1971). In *The Bacteriophage Lambda,* p. 45. Ed. A.D. Hershey. New York; Cold Spring Harbor Laboratory

DE KLOET, S.R. and ANDREAN, B.A.G. (1971). *Biochim. biophys. Acta,* **247**, 519

FILIPSKI, J., THIERY, J.-P. and BERNARDI, G. (1973). *J. molec. Biol.,* **80**, 177

FLAMM, W.G., BIRNSTIEL, M.L. and WALKER, P.M.B. (1972). In *Subcellular Components: Preparation and Fractionation,* 2nd edn, p. 279. Ed. G.D. Birnie. London; Butterworths

FRITSCH, A. (1975). *Preparative Density Gradient Centrifugations.* Geneva; Beckman Instruments International S.A.

GENTA, V.M., KAUFMAN, D.G. and KAUFMAN, W.K. (1975). *Analyt. Biochem.,* **67,** 279

GLISIN, V., CRKVENJAKOV, R. and BYUS, C. (1974). *Biochemistry,* **13,** 2633

GREENBERG, J.R. (1977). *J. molec. Biol.,* **108,** 403

GUTTANN, T., VOTAVOVA, H. and PIVEC, L. (1976). *Nucleic Acids Res.,* **3,** 835

HANCOCK, R. (1970). *J. molec. Biol.,* **48,** 357

HELL, A., MacPHAIL, E. and BIRNIE, G.D. (1974). In *Methodological Developments in Biochemistry,* vol. 4, p. 111. Ed. E. Reid. London; Longmans

HINTON, R.H., AL-TAMER, Y., MALLINSON, A. and MARKS, V. (1974). *Clin. chim. Acta,* **53,** 355

HU, A.S.L., BOCK, R.M. and HALVORSON, H.O. (1962). *Analyt. Biochem.,* **4,** 489

IFFT, J.B., MARTIN, W.R. and KINZIE, K. (1970). *Biopolymers,* **9,** 597

IFFT, J.B. and VINOGRAD, J. (1966). *J. phys. Chem.,* **70,** 2814

IFFT, J.B., VOET, D.H. and VINOGRAD, J. (1961). *J. phys. Chem.,* **65,** 1138

JENSEN, R.H. and DAVIDSON, N. (1966). *Biopolymers,* **4,** 17

KAISER, A.D. and HOGNESS, D.S. (1960). *J. molec. Biol.,* **2,** 392

KELLENBERGER, G., ZICHICHI, M.L. and WEIGLE, J. (1961). *J. molec. Biol.,* **3,** 399

KERSTEN, W., KERSTEN, H. and SZYBALSKI, W. (1966). *Biochemistry,* **5,** 326

LAWRENCE, W.C. (1976). *J. gen. Virol.,* **31,** 81

LEVINTOW, L. and DARNELL, J.E., Jr. (1960). *J. biol. Chem.,* **235,** 70

LIAUTARD, J.-P. and LIAUTARD, J. (1977). *Biochim. biophys. Acta,* **474,** 588

LIEBERMAN, M.W., HARVAN, D.J., AMACHER, D.E. and PATTERSON, J.B. (1976). *Biochim. biophys. Acta,* **425,** 265

LOZERON, H.A. and SZYBALSKI, W. (1966). *Biochem. biophys. Res. Commun.,* **23,** 612

LUDLUM, D.B. and WARNER, R.C. (1965). *J. biol. Chem.,* **240,** 2961

McCREA, J.F., EPSTEIN, R.S. and BARRY, W.H. (1961). *Nature, Lond.,* **189,** 220

McEWEN, C.R. (1967). *Analyt. Biochem.,* **19,** 23

MacGILLIVRAY, A.J. (1976). In *Subnuclear Components: Preparation and Fractionation,* p. 209. Ed. G.D. Birnie. London; Butterworths

MANNING, J.S., HACKETT, A.J., CARDIFF, R.D., MEL, H.C. and BLAIR, P.B. (1970). *Virology,* **40,** 912

MESELSON, M. and STAHL, F.W. (1958). *Proc. natn. Acad. Sci. U.S.A.,* **44,** 671

MESELSON, M., STAHL, F.W. and VINOGRAD, J. (1957). *Proc. natn. Acad. Sci. U.S.A.,* **43,** 581

MILLER, A.O.A. (1972). *Archs Biochem. Biophys.,* **150,** 282

MONAHAN, J.J. and HALL, R.H. (1974). *Analyt. Biochem.,* **62,** 217

NAITO, S. and ISHIHAMA, A. (1975). *Biochim. biophys. Acta,* **402,** 88

NANDI, U.S., WANG, J.C. and DAVIDSON, N. (1965). *Biochemistry,* **4,** 1687

PARISH, J.H. (1972). *Principles and Practice of Experiments with Nucleic Acids.* London; Longmans

RADLOFF, R., BAUER, W. and VINOGRAD, J. (1967). *Proc. natn. Acad. Sci. U.S.A.,* **57,** 1514

RINEHART, F.P. and SCHMID, C.W. (1976). *Biochim. biophys. Acta,* **425,** 451

ROODYN, D.B. (1972). In *Subcellular Components: Preparation and Fractionation,* 2nd edn, p. 15. Ed. G.D. Birnie. London; Butterworths

SCHILDKRAUT, C.L., MARMUR, J. and DOTY, P. (1962). *J. molec. Biol.,* **4,** 430

SCHUMAKER, V. and REES, A. (1972). In *Principles and Techniques in Plant Virology,* p. 336. Eds C.I. Kado and H.O. Agrawal. New York; Van Nostrand Reinhold

SHAW, J.L., BLANCO, J. and MUELLER, G.C. (1975). *Analyt. Biochem.*, **65**, 125

SINSHEIMER, R.L. (1959). *J. molec. Biol.*, **1**, 37

SUMMERS, W.C. and SZYBALSKI, W. (1967). *J. molec. Biol.*, **26**, 107

SVEDBERG, T. and PEDERSON, K.O. (1940). *The Ultracentrifuge.* Oxford; Clarendon Press

SZYBALSKI, W. (1968). In *Methods in Enzymology*, vol. 12B, p. 330. Eds S.P. Colowick and N.O. Kaplan. New York; Academic Press. Reprinted in *Fractions*, No. 1, p. 1. Palo Alto, Calif.; Beckman Instruments Inc.

SZYBALSKI, W. and SZYBALSKI, E.H. (1971). In *Procedures in Nucleic Acid Research*, vol. 2, p. 311. Eds G.L. Cantoni and D.R. Davies. New York; Harper and Row

VAN HOLDE, K.E. and BALDWIN, R.L. (1958). *J. phys. Chem.*, **62**, 734

VINOGRAD, J. and HEARST, J.E. (1962). *Prog. Chem. Organic Natural Products*, **20**, 371

WANG, J.C., NANDI, U.S., HOGNESS, D.S. and DAVIDSON, N. (1965). *Biochemistry*, **4**, 1697

WARTELL, R.M., LARSON, J.E. and WELLS, R.D. (1974). *J. biol. Chem.*, **249**, 6719

WILCOX, H.G. and HEIMBERG, M. (1970). *J. Lipid Res.*, **11**, 7

WILLIAMS, A.E. and VINOGRAD, J. (1971). *Biochim. biophys. Acta*, **228**, 423

WILT, F.H., ANDERSON, M. and EKENBERG, E. (1973). *Biochemistry*, **12**, 959

WOLF, A.V. and BROWN, M.G. (1968). In *Handbook of Chemistry and Physics*, 48th edn, p. D144. Ed. R.C. Weast. Cleveland, Ohio; Chemical Rubber Company

WOLF, H. (1975). *Analyt. Biochem.*, **68**, 505

YOUNG, B.D. (1976). MSE Application Information 13/6/76. Crawley, Sussex, UK; MSE Instruments Ltd

YOUNG, B.D., BIRNIE, G.D. and PAUL, J. (1976). *Biochemistry*, **15**, 2823

7 Isopycnic Centrifugation in Non-Ionic Media

D. RICKWOOD

Department of Biology, The University of Essex, Colchester, Essex

Isopycnic centrifugation, which is used to fractionate biological materials according to their buoyant densities, has proved to be an extremely versatile technique in both molecular and cell biology. Since its introduction for the separation of DNA by Meselson, Stahl and Vinograd (1957), this technique has been widely used for the fractionation of nucleic acids. However, its use has been severely limited by the types of density-gradient solute available. For example, most cellular and subcellular structures are sensitive to both the ionic content and osmotic strength of the surrounding medium. In addition, the solutions of carbohydrate gradient media are extremely viscous at high concentrations and, unless modified by the introduction of heavy atoms, have only a very limited density range. However, the recent upsurge in interest in this field has led to the introduction of other types of non-ionic density-gradient media with markedly different properties.

GENERAL TECHNIQUES

Preparation and Loading of Gradients

Gradients can be formed prior to centrifugation, or in some cases they can be formed *in situ* during the centrifugation run. The choice as to whether the gradients should be preformed depends to a large extent on the nature of the gradient material; for example, sucrose solutions do not readily form gradients when centrifuged. In addition, if the particles to be fractionated are large enough to sediment to their isopycnic positions much more rapidly than the gradients are formed *in situ*, the use of preformed gradients will enable the centrifugation time to be very considerably reduced. This consideration is particularly important when the particles being centrifuged are labile.

In the case of low-molecular-weight density-gradient media (mol. wt ≤ 1000), linear gradients can be prepared by allowing step gradients to diffuse over a period of 12–24 h (Hell, 1972; Birnie *et al.*, 1973a). The time necessary to obtain linear gradients is dependent both on the molecular weight and diffusion coefficient of the gradient solute and the temperature and concentration range of the gradient. Linear and exponential gradients can be prepared using simple gradient-making machines (see Noll, 1969), whereas non-linear gradients with defined shapes can be generated by using the more complex commercial devices which can be programmed as required. The various methods of preparing density gradients are discussed fully in Chapter 3.

Gradients formed *in situ,* 'self-forming' gradients, are produced as a result of the centrifugal field sedimenting the gradient solute molecules against the diffusion gradient. The rate of formation, shape and stability of gradients are determined by a number of physical parameters including molecular weight and diffusion coefficient of the solute, densities of the gradient solute and solvent, the viscosity and temperature of the solution, and the centrifugal force applied (for a discussion of these factors, and their relative importance, see Chapter 6). This restricts the shape, range and, hence, the resolution of such gradients. However, in some cases the shape of the gradient can be manipulated either by the addition of other substances to the medium (Pertoft, 1966; Olsson, 1969) or by utilizing the high viscosity of some gradients (Hell, Rickwood and Birnie, 1974; Birnie and Rickwood, 1976). One problem encountered with particulate media (for example, colloidal silica and glycogen) is that during centrifugation significant amounts of the gradient material may be pelleted, which subsequently makes gradients difficult to fractionate. A detailed description of the formation of gradients for each medium is given in the appropriate section.

Another important factor to be taken into consideration when deciding on the conditions for fractionation is how the sample is to be loaded on to the gradient. For rate-zonal separations the sample *must* be loaded on to the top in a small volume; unfortunately, the use of this technique can cause difficulties. Problems commonly encountered are the loss or inactivation of biological material during concentration and, particularly in the case of cells and viruses, the formation of artefacts arising from the aggregation of particles. Also, top-loaded samples reach their isopycnic positions in an exponential manner, so when separating low-molecular-weight material, for example proteins, it is important to show that the position of the material on an isopycnic gradient reflects its true density. An alternative procedure, used with both preformed and self-forming gradients, is to layer the sample into the bottom of the gradient and allow the material to float up to its isopycnic position. This technique is particularly useful where the material of interest is lighter than the bulk of the material, and it has been used for the separation of cells (Freshney, 1976; Seglen, 1976), membranes (Hinton, 1972) and nucleoproteins (Rickwood, 1976b). The most common procedure is to mix the sample throughout the gradient, in which case by using concentrated stock solutions of the density-gradient medium it is possible to use relatively large sample volumes. Moreover, the use of this technique reduces the possibility that artefactual peaks may arise as a result of material remaining in the loading zone.

Centrifugation of Gradients

Preformed gradients are generally centrifuged in swing-out rotors, whereas self-forming gradients can be centrifuged in either swing-out or fixed-angle rotors; however, as originally described by Flamm, Bond and Burr (1966), and as discussed in Chapter 6, the resolution of gradients is better in fixed-angle rotors. Gradients reorientate during centrifugation in fixed-angle rotors and in order to prevent the cap disrupting the gradient the tubes are only partially filled, for example 5–7 ml of solution in a 10 ml tube, depending on the angle of the tube in the rotor. However, many types of centrifuge tube can only be centrifuged at high speed when they are full, so the gradient solutions are generally overlayered with light paraffin oil prior to centrifugation. Care must also be taken in choosing which type of centrifuge tube is used for gradients. When metrizamide gradients are centrifuged in fixed-angle rotors in polycarbonate or polypropylene tubes the relatively large density differences across the gradients cause these fairly rigid tubes to crack and in some cases collapse; this does not occur if the more flexible polyallomer tubes are used.

The centrifugation conditions necessary for isopycnic banding depend on a number of factors, including density and size of the particle and the viscosity of the gradient. Frequently, the exact time and speed of centrifugation are chosen in a fairly arbitrary manner to suit social rather than scientific considerations. However, as a guide, throughout this chapter the time of centrifugation and average relative centrifugal force (*g*) applied to the particle is given for each type of separation. The exact centrifugation conditions employed can be obtained from the original reference.

Fractionation of Gradients

The chief consideration when deciding how gradients should be fractionated is how to minimize mixing of the gradients and, thus, loss of resolution. The major sources of difficulty are disturbances arising from (*a*) manipulations of the tubes and gradients, and (*b*) the presence of insoluble material within the gradient, or as a skin on top of the gradient, or as a pellet at the bottom of the tube. Of these, perhaps the most common problem is the formation of pellets either of the gradient material or of some part of the sample. The various methods of fractionating gradients, and the relative importance of these difficulties, particularly in regard to losses in resolution, are discussed in Chapter 6.

Extremely viscous gradients cannot be satisfactorily fractionated by any of the standard techniques and, in such cases, the gradient is best fractionated by slicing the tube itself (De Duve, Berthet and Beaufay, 1959; Hossainy, Zweidler and Bloch, 1973). Also, it is extremely difficult to fractionate gradients containing particulate material and, although gradients can be fractionated using the techniques already described, the sticky nature of many types of biological aggregate means that they tend to stick to both the centrifuge tube and the gradient unloading apparatus. These gradients can be fractionated by slicing the tubes, or a simpler but cruder method is to collect the light-scattering bands from the gradient using a pasteur pipette with the tip bent at right-angles to the stem. Another somewhat novel approach which can be used to fractionate gradients is to include acrylamide, riboflavin and a cross-linker in the gradients. After centrifugation, the gradients are photopolymerized and sliced (Cole, 1971). The cross-linker can be either N,N'-methylenebisacrylamide or a labile cross-linker such as diallyltartardiamide or ethylene diacrylate, the latter being particularly useful if recovery of the material trapped within the gel matrix is required.

Analysis of Gradients

It is extremely important that the density of each fraction of a gradient is measured accurately. The obvious way to do this is to weigh known volumes accurately using a pycnometer; however, such techniques are extremely laborious when the analysis involves several hundred samples. Another technique is to drop a small sample of each fraction into a calibrated gradient of water-saturated non-aqueous solvents (for example, benzene mixed with either bromobenzene or chloroform) and read off the density at which the drop reaches equilibrium (Linderström-Lang, 1937; Miller and Gasek, 1960). Frequently, the density of a fraction can be determined from the refractive index of the solution since there is generally a linear relationship between the refractive index, concentration and density of solutions. Many fractions can be analysed in this way in a relatively short time. Wherever possible, the equation relating the refractive index of a solution to its density has been included in the following sections. However, these parameters are temperature dependent, and the temperature is not always quoted. When it has been, it is included in the equation.

DENSITY-GRADIENT SOLUTES

The molecular species used for non-ionic density-gradient media are extremely varied in nature. *Figure 7.1* summarizes the physicochemical properties of some of these media, and compares them with those of CsCl. As can be judged from these data, some media are only suitable for particular types of separation, and a summary of the uses of each particular type of gradient media is given in *Table 7.1*. It is important when choosing the medium and centrifugation conditions for isopycnic separations to remember that often the speed and time of

Table 7.1 USES OF MAIN TYPES OF NON-IONIC DENSITY-GRADIENT MEDIA IN COMPARISON WITH CAESIUM CHLORIDE

Solute	Macro-molecules	Nucleo-proteins	Cell organelles	Cells	Viruses
Monosaccharides and disaccharides	−	some	+	−	+
Polysaccharides	−	−	+	+	+
Iodinated compounds	+	+	+	+	+
Colloidal silica (Ludox)	−	−	+	+	+
CsCl	+	some	−	−	+

centrifugation directly affect the integrity of the fractionated particles (see, for example, Mateyko and Kopac, 1963; Wattiaux, 1973; Collot, Wattiaux-De Coninck and Wattiaux, 1976). In the following sections the characteristics and uses of each class of density-gradient medium are described, together with details of the conditions used to obtain the required fractionation.

Monosaccharides and Disaccharides

One of the most commonly used compounds for density-gradient centrifugation is sucrose. It has become almost universally accepted as the medium of choice for most rate-zonal separations (see Chapters 3 and 4), although glucose and glycerol (an analogous compound) have also been used for both rate-zonal and

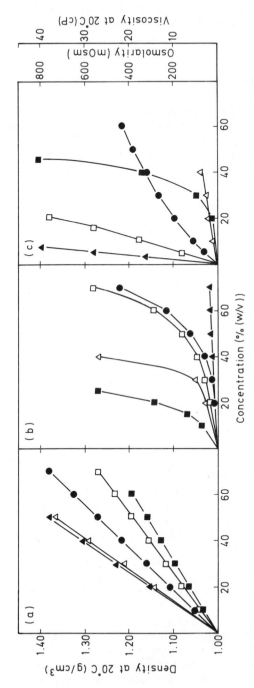

Figure 7.1 Physicochemical properties of non-ionic gradient media: the variations of (a) density, (b) viscosity and (c) osmolarity with the concentration of solutions of sucrose □——□; Ficoll ■——■; metrizamide ●——●; Ludox △——△ and CsCl ▲——▲. The data are taken from Rickwood and Birnie (1975) and Table 7.8.

isopycnic separations. The characteristics of sucrose and other similar gradient media have been well documented; in addition, the low cost of these compounds has resulted in their use as buoyant density-gradient media for a limited number of applications.

Structure and Properties

The structures of glucose and sucrose, examples of monosaccharide and disaccharide sugars (*Figure 7.2*) illustrate the basic structural features of this class of compounds. The presence of C–C bonds linked to hydroxyl groups makes the basic sugar molecule both stable and hydrophilic. On the other hand, the glucosidic links of disaccharides, especially that of sucrose, are liable to acid hydrolysis

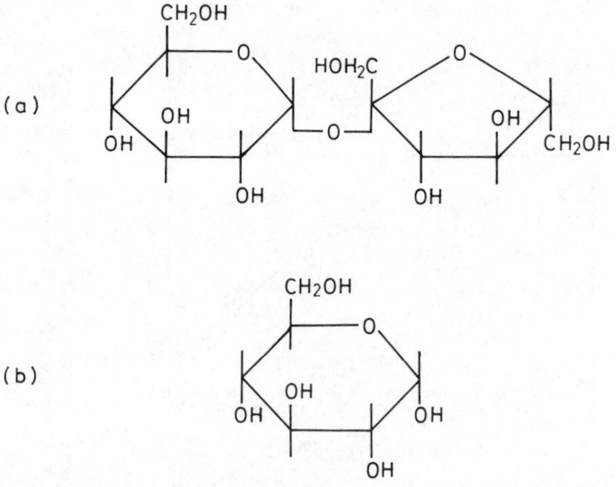

Figure 7.2 Structure of (a) sucrose and (b) glucose

at pH values less than 3 and at elevated temperatures. Moreover some sugars, but not sucrose, contain potential aldehydic groups which can also undergo modification under certain conditions. Fortunately, the conditions under which these compounds are modified very rarely occur in normal gradient separations. However, if the use of non-standard conditions is contemplated, it might prove useful to consult one of the standard works on carbohydrate chemistry (for example, Pigman, 1957).

One problem which does arise in sterilizing media, particularly concentrated sugar solutions, is the occurrence of 'caramelization' during autoclaving. An alternative, safer, method is to shake the solution with diethyl pyrocarbonate (approximately 0.1% w/v) at room temperature. It is important to hydrolyse the excess diethyl pyrocarbonate (for example, by heating the solution at 60 °C for 1–2 h), otherwise it may react with biological material in the sample.

Although the solubilities of sucrose and other saccharides in aqueous media are extremely high, concentrated solutions become extremely viscous, especially at 5 °C, and can even form a 'glass'. The high viscosity of such solutions means that it takes particles a very long time to reach their isopycnic positions; moreover, most biological materials become dehydrated in such solutions. Another restriction on the use of these gradient solutes is that for densities in excess of 1.03 g/cm^3 the hypertonic solutions required can adversely affect the integrity of the more osmotically sensitive particles such as cells, lysosomes and mito-chondria.

Formation and Fractionation of Gradients

Solutions of most of these compounds do not readily form gradients when cen-trifuged in either swing-out or fixed-angle rotors. Therefore, gradients must be preformed either by using a gradient-making machine (Britten and Roberts, 1960) or by allowing a step gradient to diffuse (see, for example, Birnie *et al.*, 1973a). However, when preparing gradients from viscous solutions care must be taken to form gradients slowly when using the former method and to allow adequate time for diffusion when using the latter.

After centrifugation, the gradients can be unloaded by any of the standard techniques (see p. 188). However, when dealing with very viscous gradients, piercing a large enough hole in the bottom of the tube to drip out the contents may lead to air bubbles passing up through the gradient. Thus, with the more viscous gradients it is generally best to unload the gradients by upwards displace-ment and for this purpose it is easiest to use an inert medium (for example, Fluorochemical FC43) rather than a dense, very viscous, sucrose solution.

The density of each fraction is most easily determined by measuring its refractive index. In the case of sucrose, the density of each fraction can be calculated from the equation

$$\rho_{0°} = 2.732\ 9\ \eta_{20°} - 2.642\ 5$$

Applications

Sugar molecules are highly hydrated in aqueous media; thus, although solutions with densities as high as 1.39 g/cm^3 can be prepared (Raynaud and Ohlenbusch, 1972), in such solutions macromolecules, particularly nucleic acids, are partially dehydrated and band at densities higher than are obtainable with these compounds. For sucrose solutions, the most commonly used density range is 1.00–1.20 g/cm^3 which is sufficient to band membranes, membrane-associated organelles and viruses (*Table 7.2*). This range includes the density of most cells, although the high osmolarity of sucrose solutions makes this medium unsuitable for separating cells, when their viability after separation is important. Detailed descriptions of the methods for the isolation of membranes and cell organelles using sucrose gradients have been reviewed recently (Hinton, 1972; Reid, 1972). However, sucrose gradients are of limited use for separating mitochondria, lysosomes and other organelles isopycnically, since in hypertonic solutions they band at essen-tially the same densities (*Table 7.2*; see Reid, 1972), unless one or other is speci-fically modified (Reid, 1972). Isopycnic separations in sucrose gradients have become a standard procedure for obtaining many species of purified virus from a variety of cell types (*Table 7.2*). However, the suitability of sucrose gradients for virus purification is directly related to the type of host cell and, to a much greater extent, to the species of virus, since a certain amount of cellular material usually either co-bands or aggregates with the virus. General reviews on the comparative uses of sucrose and ionic media for the preparation of viruses by isopycnic centrifugation have been published (Brakke, 1967; Anderson and Cline, 1967).

Sucrose solutions of relatively high density can be prepared, but the viscosity of such solutions is extremely high. To decrease their viscosity, deuterium oxide (D_2O) with a density of 1.105 g/cm^3 can be used as a solvent, thus enabling informosome-like particles to be banded at 1.29 g/cm^3 (Kempf et al., 1972).

Table 7.2 BUOYANT DENSITIES OF BIOLOGICAL PARTICLES IN SUCROSE SOLUTIONS

Particles	Medium	Centrifugation conditions	Density range g/cm^3	Ref.
Liver plasma membranes	sucrose	100 000g for 1.5 h	1.13–1.18	Hinton (1972)
Lysosomes	sucrose		1.21–1.22	
Mitochondria	sucrose	59 000g for 4 h	1.19	Wattiaux-De Coninck and Wattiaux (1971)
Microbodies (peroxisomes)	sucrose		1.23	
Plant microbodies	sucrose		1.25	
Thylakoids	sucrose	65 000g for 4 h	1.17	Theimer et al. (1975)
Chloroplasts	sucrose		1.22	
Chloroplasts	sucrose	81 000g for 3 h	1.21	Rocha and Ting (1970)
Chromatin	sucrose–glucose	50 000g for 40 h	1.357	Raynaud and Ohlenbusch (1972)
Informosomes	sucrose–D_2O	180 000g for 20 h	1.29	Kempf et al. (1972)
Mouse sarcoma virus	sucrose	65 000g for 1 h	1.16	Rickwood and Birnie (1975)
Mouse mammary tumour virus	sucrose sucrose–D_2O	240 000g for 1 h	1.17 1.20	Manning et al. (1970)
Canine distemper virus	sucrose	88 000g for 16 h	1.20	Armitage et al. (1975)

However, the exchange of water and D_2O molecules between the particles and the surrounding medium can make them denser, thus partly negating the advantage of using D_2O as a solvent (Beaufay et al., 1964). Sucrose alone does not form dense enough solutions to band chromatin, and instead a composite solution of sucrose and glucose has been used (*Table 7.2*; Raynaud and Ohlenbusch, 1972). However, these gradients are so viscous that only very high-molecular-weight chromatin can band isopycnically at a density of 1.357 g/cm^3. This density is close to that calculated (1.35 g/cm^3) for isolated rat-liver nuclei (Johnston et al., 1968), although true isopycnic banding was not achieved in these experiments.

Reactivity with Biological Materials

There are two types of interaction with biological materials. First, monosaccharides and disaccharides inhibit the activity of many enzymes (Hinton, Burge and Hartman, 1969; Hartman et al., 1974). However, this inhibition is linear with solute concentration and is readily reversible on dilution, although the mechanism remains unknown. Secondly, sucrose solutions, probably as a result of their high osmolarity, have been reported to damage the respiratory control of mito-chondria (Zimmer, Keith and Packer, 1972) and disrupt other vesicular struc-tures (Kurokawa, Sakamoto and Kato, 1965). Another source of interference is degradation of the gradient materials by the enzymes present in the sample. However, this can generally be avoided by careful choice of the gradient medium.

Factors Affecting the Analysis of Gradients

As stated previously, saccharidic solutes inhibit the activity of enzymes, and some enzymes are more sensitive than others. Therefore, when studying the dis-tribution of marker enzymes across a gradient the sucrose concentrations in each assay should be adjusted so that they are similar. Sucrose also interferes with the estimation of glycogen, the orcinol estimation of RNA, the diphenylamine estimation of DNA, the micro-biuret estimation of protein and decreases the sensitivity of the Lowry protein assay procedure (Hinton, Burge and Hartman, 1969; Hartman *et al.*, 1974). However, monosaccharides and disaccharides can easily be removed by, for example, acid-precipitation of the sample, or pelleting the material after dilution, or dialysis.

When measuring the radioactivity of gradient fractions, the addition of water-soluble scintillant can give rise to precipitation of the sugar solute. However, such problems can be avoided by diluting samples or by careful choice of the scintillant (see p. 54). Even in the absence of precipitation, the sucrose in the samples can quench β-emitters, particularly the weak β-emission of ^{3}H-labelled compounds (Dobrota and Hinton, 1973).

Polysaccharides

One of the main disadvantages of using monosaccharides and disaccharides as density-gradient media is the high osmolarity of concentrated solutions. There-fore, in order to overcome this particular problem, polysaccharide density-gradient media have been used for the rate-zonal and isopycnic separations of osmotically sensitive particles, such as mitochondria and cells. For isopycnic separations, several different types of polymer have been used, including glyco-gen (Beaufay *et al.*, 1964), a synthetic polymer of glucose (Oroszlan *et al.*, 1964), and dextrans (Mach and Lacko, 1968). However, the most commonly used medium is a polymer of sucrose called Ficoll (Pharmacia Fine Chemicals AB, Uppsala, Sweden) and, therefore, most of the discussion will be centred on this material, the physicochemical properties of which are similar to most other polysaccharides. In some cases, high-molecular-weight dextrans and Ficoll have been added during density-gradient fractionations, not as a density-gradient solute as such, but rather because of some specific property, for example the induction of cell agglutination; such cases will not be considered here.

Structure and Properties

Although some polysaccharides can be prepared from biological sources, Ficoll is prepared by the random chemical copolymerization of epichlorohydrin and sucrose; it has a bead form but its molecular structure is somewhat ill-defined. The structures of some dextrans have been characterized and the structure of that marketed by Pharmacia Fine Chemicals AB is shown in *Figure 7.3* (Lindberg and Svensson, 1968); this dextran has a random-coil configuration. Two types of Ficoll are produced with weight-average molecular weights of 70 000 and 400 000, whereas dextrans are available with varying molecular weights. As expected, the

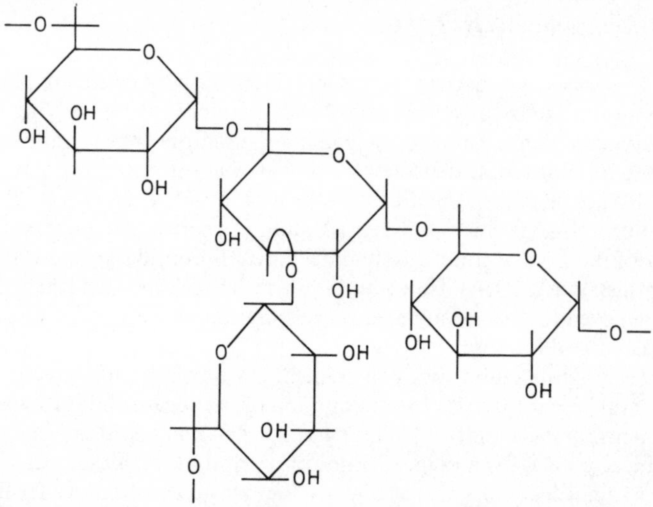

Figure 7.3 Partial structure of dextran from Leuconostoc mesenteroides B 512 *as obtainable from Pharmacia Fine Chemicals AB, Uppsala, Sweden*

stability and chemical reactivity of such gradient solutes are related to those of the monomeric material and the covalent linkages present. However, the high molecular weight of these solutes makes them relatively inert osmotically, at least, at concentrations below 30% (w/v) (density 1.115 g/cm^3). On the other hand, the viscosity of such solutions becomes very high at concentrations greater than 20% (w/v) (*Figure 7.1*), thus drastically increasing the time required for particles to reach their isopycnic positions.

Formation and Fractionation of Gradients

Polysaccharide materials diffuse only slowly and it can take several days for step gradients to become linear (Lyons and Moore, 1965). Hence, linear gradients are best prepared using a standard gradient maker, whereas discontinuous gradients can be made by overlayering solutions of the required concentration. Gradients can be unloaded in a similar way to sucrose gradients. After unloading the gradients, the density of each fraction can be determined from its refractive index and the cells or cell organelles in each fraction estimated by direct counting, spectrophotometric analysis, or radioactive labelling. Fractionated cells or organelles can be recovered from gradient fractions by centrifuging the diluted fractions and collecting the pelleted material. The concentration of Ficoll and dextran in gradient fractions can be calculated from the equation

$$\rho_{20^\circ} = 2.381\,\eta_{20^\circ} - 2.175$$

Applications

Ficoll gradients have been used for both isopycnic and rate-zonal separations. Many types of cell, including blood cells (Boyd *et al.*, 1967), fibroblasts (Boone, Harell and Bond, 1968), tumour cells (Sykes *et al.*, 1975), rat-liver cells (Castagna and Chauveau, 1969) and mouse-liver cells (Pretlow and Williams, 1973) have been fractionated by isopycnic centrifugation in Ficoll gradients (*Table 7.3*).

Table 7.3 BUOYANT DENSITIES OF CELL ORGANELLES, CELLS AND VIRUSES IN GRADIENTS OF POLYSACCHARIDES

Particles	Medium*	Centrifugation conditions	Density range g/cm^3	Ref.
Membranes	F	100 000g for 16 h	1.05	Wallach and Kamat (1964)
Chromatophores	F	195 000g for 36 h	1.07	Ketchum and Holt (1970)
Brain vesicles	F	21 000g for 15 min	—	Kurokawa *et al.* (1965)
Mitochondria	F, G, PEG	128 000g for 2.5 h	1.136, 1.127, 1.128	Beaufay *et al.* (1964)
Microsomes	D	64 000g for 1.5 h	—	Mach and Lacko (1968)
Erythrocytes	D	750g for 10 min	—	
Hepatocytes	F	63 000g for 2 h	1.10–1.15	Castagna and Chauveau (1969)
	F	950g for 1.5 h	1.09–1.14	Pretlow and Williams (1973)
Fibroblasts	F	8000g for 60 min	1.05	Sykes *et al.* (1975)
Ehrlich ascites cells	F	1400g for 45 min	1.069	Warters and Hofer (1974)
Mammary tumour virus	F	59 000g for 60 min	1.14	Lyons and Moore (1965)

*D, Dextran; F, Ficoll; G, glycogen; PEG, polyethylene glycol

However, Pretlow and Williams (1973) found that with mouse-liver cells rate-zonal separations in Ficoll were superior to those obtained by isopycnic banding. Many types of osmotically sensitive organelles from animal (De Duve, 1964), plant (Honda, Hongladarom and Laties, 1966) and fungal (Ketchum and Holt, 1970) cells have also been separated using polysaccharidic density-gradient media (*Table 7.3*). Separations in these media reduce the possibility of damaging organelles and, in addition, their buoyant density is lower in polysaccharidic media than in hypertonic sucrose solutions, partly because they are more highly hydrated. However, the degree of resolution of organelles depends on the composition of the medium (Beaufay *et al.*, 1964).

Reactivity with Biological Materials

The reactivity is similar to that of oligosaccharide material in that glycosidic bonds are liable to enzymic hydrolysis by the sample particles. In addition, some cells may also phagocytose the polysaccharide particles. These substances can also cause the aggregation of specific cell types (Bøyum, 1968).

Factors Affecting the Analysis of Gradients

Polysaccharides, like monosaccharides and disaccharides, interfere with the chemical estimations of glycogen, nucleic acids and protein. However, dextrans and Ficoll inhibit enzymes to a lesser extent than equivalent concentrations of sucrose (Hartman *et al.*, 1974). An additional problem with polysaccharides is that they cannot be removed by dialysis or ultrafiltration, so particulate material is generally recovered by diluting and centrifuging the fractions.

Iodinated Compounds

The need for density-gradient media which combine the stability and inertness of the saccharidic media with the flexibility of the large density range of the alkali metal halide solutions has led to an upsurge in interest in iodinated density-gradient media. Such compounds were originally developed as X-ray contrast media and their use for centrifugal separations is only relatively recent. Although, with the exception of metrizamide, all of these media are to some extent ionic, the ionic strength of gradients prepared from such compounds is low and thus they are included in this chapter. A detailed review of these media and their applications for biological separations has been published recently (Rickwood, 1976a).

Structure and Properties

The structures, trivial names and chemical names of some of the more commonly used compounds are given in *Table 7.4*; a more comprehensive list has been published elsewhere (Hinton and Mullock, 1976). With one exception, they consist of a tri-iodobenzene ring substituted with hydrophilic groups to improve their solubility in aqueous solutions. However, in the case of those compounds which contain a free carboxyl group it is the salt rather than the free acid which is water soluble, and the characteristics of their solutions are dependent also on the counter ion which may be either a metal ion or glucosamine derivative. The presence of glucosamine greatly increases the viscosity of the solutions. In the case of metrizamide the carboxyl group is blocked by the introduction of a peptide bond through the amino group of 2-deoxyglucosamine which thus makes the compound completely non-ionic. Concentrated solutions of metrizamide, like those of the glucosamine derivative salts, are also quite viscous.

One of the main problems of these gradient solutes is the stability of the iodine atoms, since the ultraviolet absorption of the iodine atoms makes these compounds sensitive to ultraviolet light. Heating solutions of these compounds can also lead to the release of iodine. Any iodine released from these compounds can, however, be neutralized by the addition of either a reducing agent such as sodium bisulphite (Buckingham and Gros, 1975) or sodium thiosulphate, although this does increase the ionic strength of the solution. Metrizamide is also sensitive to environments which can result in the breakage of the peptide linkage and which can modify the 2-deoxyglucosamine moiety. However, it has been found that metrizamide is stable for long periods at room temperature over the range pH 2.5–12.5. Although at alkaline pH the solutions become yellow over a period of days,

Table 7.4 THE STRUCTURE AND NOMENCLATURE OF IODINATED COMPOUNDS

Structure	Mol. wt	Counter ions	Trivial names	Chemical name
	402	$(HOC_2H_4)_2NH_2^+$	Diodone Diatrast Umbradil Vasiodone	3,5-di-iodo-4-pyridone-N-acetate
	614	Na^+ MGN^{+*} Na^+	Iothalamate Conray Angio-Contrix 48	5-acetamido-2,4,6-tri-iodo-N-methyl-iso-phthalamic acid
	614	MGN^+ Na^+ Na^+, MGN^+	⎧ Diatrizoates ⎫ ⎨ Cardiografin ⎬ Gastrografin ⎩ Renografin ⎭ Hypaque Urografin	3,5-diacetamido-2,4,6-tri-iodobenzoic acid
	628	Na^+ Na^+	Isopaque Ronpacon Triosil	3-acetamido-5(N-methyl-acetamido)-2,4,6-tri-iodobenzoic acid
	789	none	metrizamide	2-(3-acetamido-5-N-methylacetamido-2,4,6-tri-iodo(benzamido)-2-deoxy-D-glucose

$*MGN^+ = N$-methylglucamine.

this is not due to the release of iodine but rather it may be the effect of alkali on the 2-deoxyglucosamine which appears to be reversed on neutralization (G. Russev and D. Rickwood, unpublished).

With the exception of metrizamide and the glucosamine salts, these compounds form dense non-viscous solutions in aqueous media. The viscosities of metrizamide and glucosamine salts are markedly higher than those of other iodinated compounds, but not unacceptably so up to densities greater than 1.3 g/cm³, that is, over the range of densities required for most separations. When higher densities are required the viscosity of solutions can be reduced markedly by using D_2O (density 1.105 g/cm³) as solvent (Huttermann and Guntermann, 1975) although, as in the case of sucrose, some particles band at slightly higher densities in D_2O

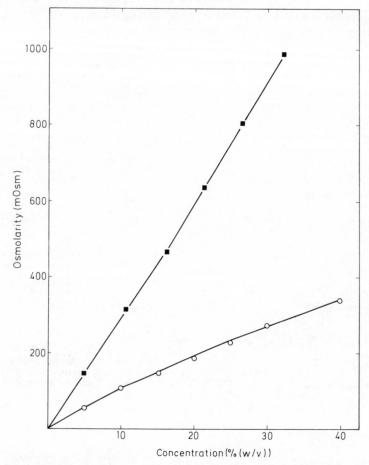

Figure 7.4 The relationship between concentration and the osmolarity of metrizamide
○——○ *and sodium metrizoate* ■——■

(Buckingham and Gros, 1976) as a result of D_2O substituting for H_2O in the
hydration shell of the particle. When used for separating osmotically sensitive
organelles an additional useful attribute of these media is the almost linear
relationship between concentration and osmolarity. However, as shown by
Holtermann (1973), at all densities the osmolarity of metrizamide is much less
than that of its ionic derivative sodium metrizoate (*Figure 7.4*).

Formation and Fractionation of Gradients

Gradients are formed when solutions of iodinated density-gradient media are
centrifuged at high speed (Rickwood, Hell and Birnie, 1973; Hell, Rickwood
and Birnie, 1974; Serwer, 1975), although in the case of metrizamide, for short
centrifugation times, shallow continuous gradients are only formed in fixed-
angle rotors, not in swing-out rotors (Hell, Rickwood and Birnie, 1974; Huttermann and Guntermann, 1975; Birnie and Rickwood, 1976). The rate of formation

of metrizamide gradients is generally somewhat slower than that of CsCl gradients, and depends on the viscosity of the solutions and on the centrifugal force applied. Although the shape of the gradient at equilibrium is not optimal for isopycnic separations, because the gradient forms only slowly it is possible to manipulate its shape to optimize the resolution over any particular range of densities (Birnie and Rickwood, 1976). Gradients can also be preformed by any of the standard techniques, although, in the case of metrizamide gradients formed by the diffusion of step gradients, the rate of diffusion is slower than that of sucrose. Centrifugation of preformed gradients in swing-out rotors does not markedly affect the shape or density range of the gradients (Hell, Rickwood and Birnie, 1974; Huttermann and Guntermann, 1975).

Although gradients can be unloaded by any of the standard procedures, usually upwards displacement with fluorochemical FC43 yields the best results. The density of each fraction can be determined either with a pycnometer or, more conveniently, from the refractive index:

Iothalamate

$$\rho_{25^\circ} = 3.904\,\eta_{25^\circ} - 4.201 \qquad \text{(Serwer, 1975)}$$

Renografin

$$\rho_{4^\circ} = 3.541\,9\,\eta_{24^\circ} - 3.719\,8 \quad \text{(Meistrich and Trostle, 1975)}$$

Metrizoate

$$\rho_{5^\circ} = 3.839\,\eta_{20^\circ} - 4.117$$

Metrizamide

$$\rho_{5^\circ} = 3.453\,\eta_{20^\circ} - 3.601 \qquad \text{(Birnie et al., 1974)}$$

$$\rho_{20^\circ} = 3.350\,\eta_{20^\circ} - 3.462 \qquad \text{(Birnie et al., 1973b)}$$

Metrizamide is not toxic to cells; hence, cells in gradient fractions can be simply plated out after fractionation (Freshney, 1976). Cell organelles and viruses can be recovered from gradients directly by pelleting the diluted fractions. Alternatively, in the case of macromolecules and macromolecular complexes which are difficult to pellet, the iodinated media can be removed by ultrafiltration or dialysis (Rickwood, Birnie and MacGillivray, 1975), although metrizamide is too large to pass easily through standard dialysis tubing. One advantage of using metrizamide is that, unlike some of its ionic counterparts, it is soluble in both ethanolic and acidic solutions (Mullock and Hinton, 1973); hence, nucleic acids and, in some cases, proteins can be isolated from gradient fractions by simple precipitation procedures.

Applications

Iodinated density-gradient media are the most versatile of the non-ionic media because they can be used over a wide density range and are relatively inert and non-viscous.

Macromolecules. Nucleic acids band at exceptionally low densities in these media (*Table 7.5*). Thus, in metrizamide DNA bands at 1.12 g/cm^3, approaching the density of fully hydrated DNA (Birnie, Rickwood and Hell, 1973b). In ionic iodinated media, DNA bands at a slightly greater density than in metrizamide (Chan and Scheffler, 1974; Doenecke and McCarthy, 1975; Serwer, 1975), probably as a result of the ionic nature of these compounds reducing the water activity of the gradients. It is possible to vary the ionic environment of metriza-mide at will, and thus it has been shown that the buoyant density of DNA in metrizamide is dependent on the ions present and the pH of the solutions (Birnie, MacPhail and Rickwood, 1974; Rickwood, 1976b).

As in CsCl solutions, denatured DNA bands at a higher density than native DNA, although at alkaline pH the density difference is much less than at neutral pH (Rickwood, 1976b). However, in contrast to the results obtained with CsCl, the G + C composition of the DNA only has a small effect on its buoyant density

Table 7.5 BUOYANT DENSITIES OF MACROMOLECULES IN IODINATED DENSITY-GRADIENT MEDIA

Macromolecule		Medium	Centrifugation conditions	Buoyant density g/cm^3	Ref.
Native DNA:					
E. coli*	Na$^+$	metrizamide	60 000g for 44 h	1.120	Birnie *et al.* (1973b)
mouse†	Na$^+$	metrizamide	60 000g for 44 h	1.118	Birnie *et al.* (1973b)
mouse†		Conray	70 000g for 96 h	1.17	Doenecke and McCarthy (1975)
mouse†		Renografin	60 000g for 46 h	1.14	Chan and Scheffler (1974)
mouse†	Cs$^+$	metrizamide	60 000g for 68 h	1.179	Birnie *et al.* (1974)
bacteriophage T7	Na$^+$	Iothalamate	84 000g for 49 h	1.131	Serwer (1975)
Denatured DNA:					
E. coli*	Na$^+$	metrizamide	60 000g for 44 h	1.150 }	Birnie *et al.*
mouse†	Na$^+$	metrizamide	60 000g for 44 h	1.147 }	(1973b)
RNA:					
ribosomal‡	Na$^+$	metrizamide	60 000g for 44 h	1.170	Birnie *et al.* (1973b)
		Conray	70 000g for 96 h	1.22	Doenecke and McCarthy (1975)
heterogeneous nuclear†	Na$^+$	metrizamide	60 000g for 44 h	1.168	Birnie *et al.* (1973b)
Proteins:					
		Conray	70 000g for 96 h	1.30(1.27)	Doenecke and McCarthy (1975)
catalase, urease		metrizamide	163 000g for 68 h	1.27(1.42) }	
a-casein		metrizamide	163 000g for 92 h	1.24(1.42) }	Birnie *et al.*
Cohn fraction V serum albumin		metrizamide	163 000g for 92 h	1.22(1.42) }	(1973b)
Carbohydrates:					
blue dextran		metrizamide	60 000g for 44 h	1.195	Birnie *et al.* (1973b)

*50% G + C; †40% G + C; ‡60% G + C.

(Birnie, Rickwood and Hell, 1973b; Serwer, 1975; Rickwood, 1976b). Similarly, nuclear RNA (40% G + C) and ribosomal RNA (60% G + C) both band at the same density in metrizamide (Birnie, Rickwood and Hell, 1973b). Iodinated density-gradient media, as judged by the results obtained with metrizamide, neither interact with DNA (Birnie, Rickwood and Hell, 1973b) nor alter its buoyant density, even after prolonged exposure. In addition, the buoyant density of DNA in metrizamide is independent of temperature over the range 1–30 °C (Rickwood, 1976b), in contrast to the results obtained with CsCl (Vinograd, Greenwald and Hearst, 1965).

There is some evidence to show that iodinated aromatic compounds do interact with proteins. In the case of some ionic iodinated compounds the binding can be quite tight (Lundh, 1973), so that all of the protein bands at densities greater than 1.30 g/cm^3 (D. Rickwood, unpublished). Apparently, these complexes are not formed in Urografin (Gschwender and Popescu, 1976). Metrizamide interacts only weakly with protein ($K_{diss} > 10^{-2}$ M) (Rickwood *et al.*, 1974a; Huttermann and Wendlberger-Schieweg, 1976). The formation of dense metrizamide–protein complexes is completely reversible and can be eliminated by banding proteins in lower concentrations of metrizamide using D_2O as the gradient solvent and by layering the sample on top of the gradient (Huttermann and Guntermann, 1975). The buoyant density of hydrated non-complexed proteins in metrizamide is about 1.27 g/cm^3, whereas the presence of prosthetic groups conjugated to the protein may reduce the density to as low as 1.22 g/cm^3, depending on the density and the quantity of the conjugated material bound to the protein (Birnie, Rickwood and Hell, 1973b; Rickwood and Birnie, 1975). The apparent absence of interaction in gradients of metrizamide in D_2O makes this a good medium for studying directly the *de novo* synthesis of proteins using ^{15}N-labelled amino acids; the resolution obtained is better than in gradients of CsCl or RbCl (Huttermann and Guntermann, 1975; Huttermann and Wendlberger, 1976). The real advantage of this technique is that the synthesis of proteins can be studied without introducing ambiguities resulting from the use of metabolic inhibitors.

Nucleoproteins. Nucleoprotein complexes are extremely sensitive to the ionic environment; hence, intact chromatin and ribonucleoproteins can only be banded in ionic media like CsCl if they are first fixed with formaldehyde (see, for example, Hancock, 1970). Ionic iodinated density-gradient media have been used to band unfixed chromatin (Chan and Scheffler, 1974; Doenecke and McCarthy, 1975). However, chromatin is extremely sensitive to ionic strength and in some cases the concentration of ions in these gradients is high enough to affect the solubility and, more important, the composition of the chromatin (Wray, 1976). Hence, the non-ionic nature of metrizamide makes it an ideal medium for banding nucleoproteins. Metrizamide does not interact with nucleoproteins nor dissociate proteins from nucleoprotein complexes (Birnie *et al.*, 1973b; Mullock and Hinton, 1973; Buckingham and Gros, 1975, 1976).

As shown in *Figure 7.5*, the buoyant density of nucleoprotein complexes in metrizamide is directly related to the relative amounts of protein and nucleic acid in the complex (Rickwood and Birnie, 1975). However, all nucleoprotein complexes band at a higher density than expected, possibly as a result of conformational effects which reduce the hydration of the complex (Birnie *et al.*, 1973b; Rickwood and Birnie, 1975). Both high-molecular-weight interphase chromatin (Rickwood *et al.*, 1974b) and metaphase chromosomes (Wray and

Stefos, 1974) form single bands in metrizamide (*Table 7.6*). However, chromatin forms two bands after shearing (*Table 7.6*), reflecting the presence of protein-rich regions distributed throughout the chromatin (Rickwood *et al.*, 1974b). The presence of divalent cations only marginally affects the buoyant density of chromatin in metrizamide (Wray and Stefos, 1974; D. Rickwood, A. Hell and G.D. Birnie, unpublished). On the other hand, the presence of divalent cations

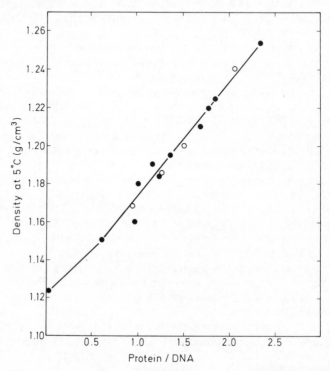

Figure 7.5 The relationship between the composition of deoxyribonucleoprotein complexes and their buoyant densities in metrizamide dissolved in 0.14 M NaCl, 0.1 mM EDTA, 10 mM tris-HCl, *pH 7.5. The data were obtained from chromatin* ○ *and nucleohistone complexes* ●, *as given in Rickwood* et al. *(1975)*

in the gradient during the fractionation of ribonucleoprotein particles gives a separation which is based primarily on the different amounts of metal ion bound to the particles, rather than on the ratio of protein to RNA in the particle (Buckingham and Gros, 1975, 1976). Thus, as shown in *Table 7.6*, polysomes, ribosomes and their subunits band at very high densities in the presence of magnesium ions, whereas the ribonucleoproteins containing the mRNA band close to their expected density. In contrast, Houssais (1975) has found that nuclear ribonucleoproteins separate into two populations, one banding at 1.18 g/cm^3 and the other at 1.31 g/cm^3, the latter of which presumably binds magnesium ions.

Another application of metrizamide gradients is the study of the interactions between macromolecules, particularly between DNA and proteins. Thus, it has been possible to study the binding of both the histones (Rickwood, Birnie and MacGillivray, 1975) and the chromatin non-histone proteins (Rickwood, 1976b;

Table 7.6 BUOYANT DENSITIES OF SUBCELLULAR COMPONENTS IN IODINATED DENSITY-GRADIENT MEDIA

Particles		Medium	Centrifugation conditions	Density g/cm^3	Ref.
50 S ribosomal subunit	Na$^+$			1.214	
30 S ribosomal subunit	Na$^+$			1.223	
60 S ribosomal subunit	Mg^{2+}	metrizamide	150 000g for 68 h	1.315	Buckingham and Gros (1975)
40 S ribosomal subunit	Mg^{2+}			1.230	
80 S ribosome	Mg^{2+}			1.305	
Myosin mRNP	Mg^{2+}			1.205	
Nuclear RNP	Mg^{2+}	metrizamide	83 000g for 65 h	1.31/1.18	Houssais (1975)
Chromatin (high mol. wt)		metrizamide	60 000g for 44 h	1.20	Rickwood et al. (1975)
		Conray	70 000g for 96 h	1.24	Doenecke and McCarthy (1975)
		Renografin	60 000g for 46 h	1.24/1.20	Chan and Scheffler (1974)
Sheared chromatin		metrizamide	60 000g for 44 h	1.185/1.24	
Metaphase chromosomes		metrizamide	16 000g for 2 h	1.24	Wray and Stefos (1974)
Microsomal membranes		metrizamide	–	1.14–1.26	Aas (1973)
Lyosomes Mitochondria		metrizamide	–	1.15–1.20 1.20–1.25	Aas (1973)
Lysosomes Mitochondria Peroxisomes		metrizamide	60 000g for 3 h	1.140 1.145 1.230	Collot et al. (1976)
Lysosomes Mitochondria		metrizamide	105 000g for 3 h	1.15 1.18	Van Den Hove-Vandenbroucke and De Nayer (1976)
Brain nuclei: neuronal oligodendroglial		metrizamide	55 000g for 3 h	1.275–1.283 1.262–1.268	Mathias and Wynter (1973)

Rickwood and MacGillivray, 1977) to DNA. In the latter case, using the technique of layering the sample into the bottom of the gradient it is possible to obtain nucleoprotein complexes free of contaminating non-complexed DNA and protein. The unique feature of this technique is that it is not only possible to study the proteins bound to the DNA over a very wide range of solvent environments, but it is also possible to identify and study the DNA sequences to which the proteins are bound.

Subcellular organelles. Membranes, being lipoprotein complexes, band at densities as low as 1.13–1.14 g/cm^3 in metrizamide gradients. *Table 7.6* shows that crude microsomal membranes band over a wide range of densities (1.14–1.26 g/cm^3), probably reflecting the heterogeneity of the complexes, particularly with respect to the amount of ribonucleoprotein material associated with the membranes (Aas, 1973). Lysosomes and mitochondria are both membrane-enveloped structures and, as shown in *Table 7.2*, in hypertonic sucrose solutions they both band at similar densities. An early but brief report from Aas (1973) claimed that, as determined by marker enzymes, lysosomes and mitochondria banded in metrizamide at 1.18 g/cm^3 and 1.22 g/cm^3, respectively (*Table 7.6*), thus enabling these two organelles to be completely separated. However, a recent detailed study (Collot, Wattiaux-De Coninck and Wattiaux, 1976) indicated that in metrizamide rat-liver mitochondria and lysosomes banded at essentially the same density ($1.140-1.145$ g/cm^3), although peroxisomes were well separated from other organelles, banding at 1.230 g/cm^3. On the other hand, Van Den Hove-Vandenbroucke and De Nayer (1976), working with pig thyroid tissue, were able to separate lysosomes (banding at 1.15 g/cm^3) and mitochondria (banding at 1.18 g/cm^3) by centrifugation through a pad of 20% (w/v) metrizamide. They also found that damaged mitochondria formed two bands at 1.14 g/cm^3 and 1.16 g/cm^3. Therefore, at present it appears that the situation with regard to the separation of these subcellular organelles in metrizamide gradients is not yet fully resolved.

Table 7.7 BUOYANT DENSITIES OF CELLS AND VIRUSES IN IODINATED DENSITY GRADIENT MEDIA

Particles*	Medium	Centrifugation conditions	Density g/cm^3	Ref.
Rat liver:				
parenchymal / non-parenchymal	metrizamide	3 200 g for 20 min	{1.12 / 1.08}	Munthe-Kaas and Seglen (1974)
Mouse testis:				
spermatogonia / spermatocytes	Renografin	13 000g for 20 min	(1.101 / 1.078 / 1.090)	Meistrich and Trostle (1975)
Human blood:				
monocytes			(1.128	
lymphocytes			1.129	
erythrocytes	Renografin	13 000g for 30 min	1.151	Grdina et al. (1973)
neutrophils			1.161	
eosinophils			1.169	
basophils			1.173)	
Chinese hamster ovary cells	Renografin	13 000g for 30 min	1.102 / 1.148	Grdina et al. (1974b)
Sporozoites	Renografin/ BSA	13 000g for 30 min	1.1	Krettli et al. (1973)
LCP	Urografin	300 000g for 12 h	1.14	Gschwender et al. (1975)
MSV	metrizamide	60 000g for 24 h	1.13	
VSV	metrizamide	195 000g for 6 h	1.11	Birnie and
VSV core	metrizamide		1.13	Rickwood (1976)
RSV	metrizamide	189 000g for 6 h	1.15 to 1.17	Wunner, Buller and Pringle (1976)

Abbreviations:
*LCP, lymphocytic choriomeningitis virus; MSV, mouse sarcoma virus; VSV, vesicular stomatitis virus; RSV, respiratory syncytial virus; BSA, bovine serum albumin.

Nuclei have also been banded in metrizamide, such separations being done in the presence of magnesium ions to preserve their integrity (Mathias and Wynter, 1973). In this ionic environment the buoyant density of nuclei appears to be influenced not only by the chromatin but also by the amount of ribonucleoprotein material within the nuclear membrane. Thus, as shown in *Table 7.6*, transcriptionally active nuclei which contained a large amount of ribonucleoprotein material were found to band at a higher density than more inactive nuclei (Mathias and Wynter, 1973).

Cells. There are some drawbacks in using Ficoll gradients for the isopycnic separations of cells (see p. 229) and iodinated media have been used to overcome them (*Table 7.7*). The most commonly used medium, Renografin, has been used to fractionate cells from human blood (Grdina *et al.*, 1973; De Simone, Kleve and Schaeffer, 1974), fibrosarcoma (Grdina *et al.*, 1974a), mouse testis (Meistrich and Trostle, 1975) and cells with different growth rates (Grdina *et al.*, 1974b), as well as for preparing infective malarial sporozoites (Krettli, Chen and Nussenzweig, 1973). In addition, the use of sodium metrizoate, combined with Ficoll to aggregate erythrocytes, has been widely adopted for the isolation of lymphocytes (Harris and Ukacjiofo, 1969), and of leucocytes by unit gravity separations (Bøyum, 1968).

However, a major problem in using ionic iodinated density-gradient media is that solutions with densities greater than 1.066 g/cm³ are hypertonic; hence, cells banding in the dense parts of the gradient may be damaged by the high osmolarity of the surrounding medium. An additional problem is that hypertonic solutions give rise to changes in cell volume, which in turn affect the buoyant density of the cells (Grdina *et al.*, 1973; Meistrich and Trostle, 1975). It is possible to overcome the problems of high osmolarity by using metrizamide, since for any given density the osmolarity of metrizamide solutions is only about one-quarter of that of sodium metrizoate and, therefore, by combining metrizamide with sucrose or salts, cells and organelles can be separated using gradients of densities up to 1.19 g/cm³ without resorting to the use of hypertonic solutions. Initial studies by Munthe-Kaas and Seglen (1974) have shown that rat-liver cells can be fractionated in metrizamide gradients and the resolution obtained is superior to that obtained using Ficoll gradients. One rather disturbing feature of sodium diatrizoate (and possibly some other iodinated media) is that it markedly increases the intracellular level of cAMP (Patrick, Rengachary and Melnkovych, 1975).

Viruses. Frequently, and almost routinely in some cases, viruses are concentrated and at least partially purified by velocity sedimentation or isopycnic centrifugation in density gradients. The choice of density medium in each case is largely dependent upon the species of virus and the consequence of its interaction with medium. Virus size and morphology, in particular the degree of pleomorphism, are critical factors which affect viruses in different media. Conditions of high osmotic pressure or high ionic strength produced by concentrated solutions of sucrose or CsCl, for example, can adversely affect those viruses which are extremely pleomorphic on the one hand, or those which are salt-sensitive on the other, resulting in altered virus structure and loss of infectivity. Whereas a number of different media (including sucrose and CsCl) have been used successfully for virus preparations, the recent introduction of iodinated density-gradient media for biological separations provides extended facility for viruses as well. Relatively little is known, however, about the suitability of iodinated media for the separation of viruses. The initial work by Gschwender, Brummund and Lehmann-Grube (1975) using Urographin, the results quoted in a recent review on metrizamide (Rickwood and Birnie, 1975) and the work of Wunner, Buller and Pringle (1976) suggest that iodinated density media will prove extremely useful for separating viruses and viral components. On the other hand, there is some evidence that some species of virus are particularly sensitive to iodinated media (Vanden Berghe, Van Der Groen and Pattyn, 1976).

Reactivity with Biological Materials

Some of the ionic iodinated compounds bind to proteins (Lundh, 1973; Rickwood and Birnie, 1975) and a similar, but much weaker, binding has been observed for metrizamide (Rickwood *et al.*, 1974a; Huttermann and Wendlberger-Schieweg, 1976). However, in the latter case the binding is apparently eliminated by using D_2O as solvent, so reducing the concentration of metrizamide (see p. 235). In addition, metrizamide like sucrose inhibits the activity of some enzymes (Mathias and Wynter, 1973; Rickwood *et al.*, 1974b; Hinton, Mullock and Gilhuus-Moe, 1974). In some cases sucrose inhibits more than metrizamide and *vice versa*; however, this inhibition is lost after dilution or complete removal of the metrizamide. No further studies of the biological reactivity of these compounds have been reported.

Factors Affecting the Analysis of Gradients

The one common feature of all these gradient media is that they all absorb strongly below 300 nm. Therefore, direct optical analysis of the distribution of nucleic acids and proteins in such gradients is not possible. In the case of metrizamide and glucosamine salts of ionic iodinated compounds, the presence of the sugar moiety seriously interferes with the determination of RNA by the orcinol reaction, the determination of DNA with diphenylamine and the phenol-sulphuric acid estimation of glycogen (Hinton, Mullock and Gilhuus-Moe, 1974). In addition, metrizamide also interferes with both the Lowry estimation of protein (Hinton, Mullock and Gilhuus-Moe, 1974) and the micro-biuret method (D. Rickwood and A. Hell, unpublished). One advantage of metrizamide is that, in contrast to Urografin and Conray, it is soluble in both acid and ethanolic solutions, hence frequently samples can be precipitated and the metrizamide removed by washing prior to analysis. In addition, the acid-solubility of metrizamide makes it possible to estimate the protein content of each fraction using the method of Schaffner and Weissmann (1973). An additional problem in the case of analytical ultracentrifugation is that the change of refractive index throughout metrizamide gradients is so large that some makes of instrument are not suitable for work which involves the formation of steep gradients. The distribution of radioactive material across gradients can be determined by measuring the radioactivity of each fraction, although Bray's scintillator is not suitable for such analyses (see p. 54).

Colloidal Silica

Structure and Properties

The product used almost exclusively in all the work to date has been colloidal silicon dioxide in the form of 'Ludox', which is manufactured by E.I. du Pont de Nemours Co., Wilmington, USA. The structures of different forms are shown in *Figure 7.6* and some of their physicochemical properties are given in *Table 7.8*. Some forms differ only in counter ions or particle size; however, Ludox AM contains aluminium ions substituted in the surface of the particles

Table 7.8 PROPERTIES OF LUDOX COLLOIDAL SILICA

					Grades			
	HS-40%	HS-30%	Percoll**	SM	TM	AS§	AM†	130M
Stabilizing counter-ion	sodium	sodium	sodium	sodium	sodium	ammonium	sodium	chloride
Particle charge	negative	negative	negative	negative	negative	negative	negative	positive
Particle size, nm	13–14	13–14	14–15	7–8	21–24	13–14	13–14	13∓15
Silica (as SiO_2), wt %	40	30	22	30	49.5	30	30	30‡
pH (25 °C)	9.7	9.8	8.8	9.9	8.9	9.6	9.0	4.4
Viscosity (25 °C), cp*	17.5	4.5	10	5.5	50	20	17	5.15
Specific gravity (25 °C)	1.30	1.21	1.13	1.21	1.39	1.21	1.21	1.23
Conductance (20 °C) mhos	4730	–	–	4730	–	2630	2270	–
Osmolarity (mOsm)	80	58	20	–	–	–	30	–

*Measured by Ostwald-Fenske pipette No. 100 or 200 depending on viscosity range.
†Surface modified with aluminate ions.
‡Ludox 130M is 26% silica, 4% alumina (Al_2O_3).
§Ludox AS-40% is also available containing 40% silica solids.
**A combination of Ludox HS with polyvinylpyrrolidone (Pharmacia Fine Chemicals AB).

which makes the colloid more stable, particularly to changes in pH. With the exception of the SM grade, the particles are about 15 nm in diameter and give a slightly opalescent solution. Suspensions (30% w/v silicon dioxide) have a density of 1.21 g/cm^3; the viscosity of such solutions is relatively low and their osmolarity is very low (*Figure 7.1*).

The colloidal particles of silicon dioxide are dense, fairly inert and very stable. However, under certain conditions the colloidal suspension can break down. The suspensions are particularly sensitive to freezing and thawing unless stabilized

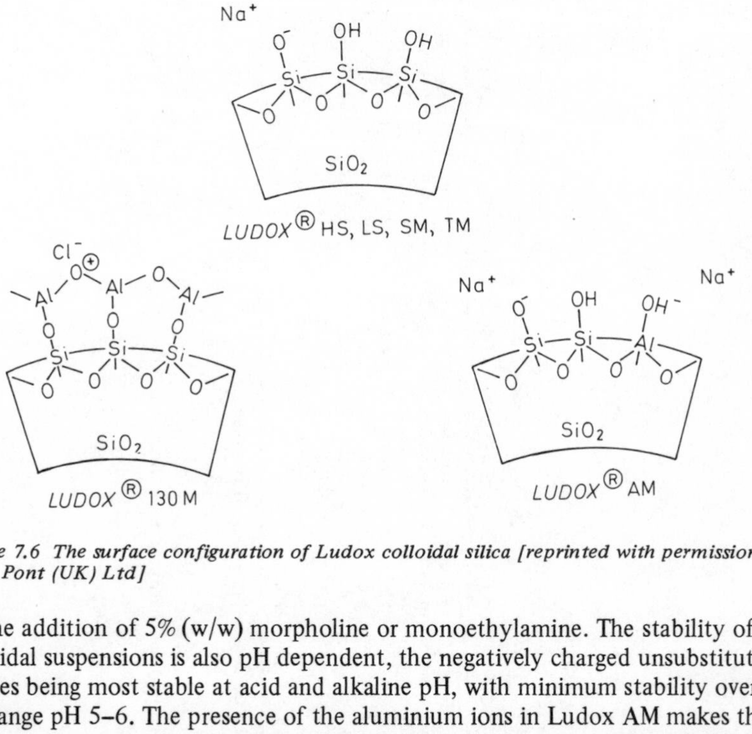

Figure 7.6 The surface configuration of Ludox colloidal silica [reprinted with permission of du Pont (UK) Ltd]

by the addition of 5% (w/w) morpholine or monoethylamine. The stability of the colloidal suspensions is also pH dependent, the negatively charged unsubstituted species being most stable at acid and alkaline pH, with minimum stability over the range pH 5–6. The presence of the aluminium ions in Ludox AM makes this colloidal suspension less sensitive to changes in pH; it is most stable in alkaline solutions. However, increasing the pH above 10.7 results in the solubilization of the silica to form alkali silicates. On the other hand, Ludox 130M is most stable over the range pH 4–5, but is degraded in more acid solutions. Finally, because these media are colloids, centrifugation results in pelleting of some of the material; they also exhibit absorption and light-scattering properties in the ultraviolet region of the spectrum (*Figure 7.7*).

Recently, a colloidal silica preparation has been specifically developed for centrifugation. The new gradient medium, Percoll, is available from Pharmacia Fine Chemicals AB. Percoll has a number of advantages over the colloidal silica preparation from which it is derived, since the particles are sized and polyvinylpyrrolidone is tightly bound to the silica particles. Coating the particles helps to remove some of the undesirable features of these suspensions in that not only does Percoll interact less with biological material but also the colloid is more stable; for example it is stable to freezing and thawing. However, it is more viscous than some other preparations (*Table 7.8*) and it is destabilized at acid pH and at high ionic strengths

Formation and Fractionation of Gradients

When adjusting the pH of Ludox it is important to stir the solution continuously to prevent gelation. Since these colloidal solutions diffuse very slowly, preformed gradients are best prepared using a gradient-making machine. Gradients of Ludox do self-form, giving densities up to 1.45 g/cm^3, when they are centrifuged at high speed (Pertoft *et al.*, 1967), and the shape of the gradient can be optimized by the addition of Ficoll (Pertoft, 1966; Olsson, 1969). After centrifugation, gradients can be unloaded by any of the standard techniques described on

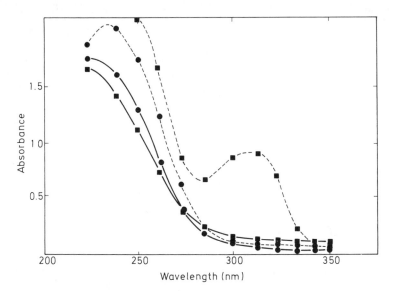

Figure 7.7 The absorption spectrum of 10% (w/v) Ludox grades SM and AM before ●- - -●, ■- - - ■ and after dialysis ●——●, ■——■, respectively

pp. 188–192. The initial density and the density of individual fractions can be determined from weighing samples of known volume or from the refractive index of the solutions using the equation

$$\rho = 8.204\eta - 9.936 \quad \text{(Loir and Lanneau, 1975)}$$

The colloidal silica can be removed from gradient fractions by precipitation with low concentrations of polyamines (Pertoft *et al.*, 1967).

Applications

Because of the tendency of the colloidal particles to pellet on prolonged centrifugation, this medium is not suitable for banding macromolecules or macromolecular complexes such as chromatin (A. Hell, personal communication). Ludox gradients have been used for the isolation of cell organelles (Olsson, 1969; Wolff and Pertoft, 1972a; Schmitt, Behnke and Herrmann, 1974; Morgenthaler, Marsden and Price, 1975), cells (Mateyko and Kopac, 1963; Pertoft, 1969;

Table 7.9 BUOYANT DENSITIES OF CELL ORGANELLES, CELLS AND VIRUSES IN LUDOX

Particles	Gradient medium	Centrifugation conditions	Density g/cm³	Ref.
Chloroplasts	Ludox HS	40 000g for 30 min	1.1	Schmitt *et al.* (1974)
Chloroplasts	Ludox AM–PEG*	8 000g for 10 min	1.12	Morgenthaler *et al.* (1975)
Leucocyte granules	Ludox HS–Dextran	26 000g for 40 min	1.23–1.27	Olsson (1969)
Liver cells	Ludox	–	1.05–1.06	Hayek and Tipton (1966)
Liver cells	Ludox	19 000g for 2 h	1.116–1.128	Mateyko and Kopac (1963)
Liver cells	Ludox HS–PEG	23 000g for 30 min	1.07	Pertoft (1969)
Spermatids	Ludox HS	4 000g for 23 min	1.05–1.15	Loir and Lanneau (1975)
Tobacco mosaic virus	Ludox SM and HS	100 000g for 60 min	1.065 ⎫	
Adenovirus		80 000g for 45 min	1.23 ⎬	Pertoft *et al.*
Poliovirus		80 000g for 60 min	1.19 ⎭	(1967)

*PEG, polyethylene glycol.

Pertoft and Laurent, 1969; Wolff and Pertoft, 1972b) and viruses (Pertoft *et al.*, 1967). The uses of Ludox are summarized in *Table 7.9*.

Interaction with Biological Materials

It has been reported that the colloidal silica particles interact with biological materials, in particular membrane material (Schmitt, Behnke and Herrmann, 1974). This interaction is minimized by the addition of high-molecular-weight polyethylene glycols (PEG) or polyvinylpyrrolidone (PVP) to the gradients (Pertoft, 1969; Wolff and Pertoft, 1972b; Morgenthaler, Marsden and Price, 1975). These materials appear to coat the silica particles, making them more inert towards cell membranes. However, the presence of PEG in the gradient can alter the density of cell organelles (Morgenthaler, Marsden and Price, 1975; Morgenthaler and Price, 1976); in addition, it is a powerful membrane fusogen (Bonnett and Erikson, 1974; Constabel and Kao, 1974). Whereas the inclusion of PEG reduces the inhibition of some enzymic activities it is even more important, at least for the isolation of intact chloroplasts, to purify the commercial Ludox AM with activated charcoal. Without this purification step these organelles are photosynthetically inactive after fractionation in Ludox gradients (Morgenthaler, Marsden and Price, 1975); however, the nature of this particular inhibition remains unknown. Prolonged dialysis also removes ultraviolet absorbing and toxic material from Ludox AM (*Figure 7.7*). Another problem is that cells may phagocytose the particles and hence change in density, although this can be minimized by carrying out the fractionation in the cold.

Factors Affecting the Analysis of Gradients

Perhaps the major difficulty in working with these gradients is the pelleting of colloidal silica that occurs during centrifugation. Thus such gradients can only be unloaded satisfactorily by upward displacement. Another problem is the absorption and light-scattering properties of these media which make it difficult to analyse gradients spectrophotometrically below 300 nm. Ludox does not inhibit hydrolases nor interfere with the Lowry protein estimation (Olsson, 1969). In cases where Ludox may interfere with assays it can be removed by precipitation with polyamines.

Other Gradient Media

The non-ionic density-gradient media described so far represent the most commonly used compounds for isopycnic centrifugation under non-ionic conditions. However, some other compounds have been used for a limited number of applications.

Chloral Hydrate

Chloral hydrate is a dense crystalline solid with the formula $CCl_3 CH(OH)_2$; it is extremely soluble in water, giving solutions with densities up to 1.6 g/cm^3 which are completely non-ionic. In addition, whereas the molar absorption of these solutions in the ultraviolet region of the spectrum is significant ($E_{260\,nm}^{50\%}$ = 0.65), it is relatively low compared with the iodinated compounds (Hartman et al., 1974). Thus, at first sight this compound would appear to be very suitable for isopycnic separations. However, chloral hydrate is an aldehyde and its solutions tend to become acidic on standing; in alkaline solutions ($> pH\ 8$) chloral hydrate decomposes into chloroform and formic acid (Hossainy, Zweidler and Bloch, 1973). One serious disadvantage of chloral hydrate is that concentrated solutions dissolve Perspex (Lucite, Plexiglas), precluding its use in any apparatus used with chloral hydrate solutions. In addition, concentrated solutions are extremely viscous and gradients only self-form extremely slowly. Preformed gradients can be prepared either by the diffusion of step gradients or by using a gradient machine. Gradients can be unloaded by any of the standard methods.

The initial density and density of individual fractions can be calculated from the equation

$$\rho_{4^\circ} = 3.676\ 5\eta_{4^\circ} - 3.906\ 6 \quad \text{(Hossainy, Zweidler and Bloch, 1973)}$$

The one application of these gradients that has been reported in detail is for the isopycnic centrifugation of chromatin (Hossainy, Zweidler and Bloch, 1973), in which case banding of the chromatin occurred after centrifugation at 180 000g for 44 h. The results from this study suggest that the water activity of chloral hydrate solutions is low so that nucleic acids band at densities similar to those in CsCl, and chromatin (with a protein to DNA ratio of 1.5) bands close to 1.4 g/cm^3.

In spite of its aldehydic nature there is no evidence that chloral hydrate covalently modifies proteins or nucleic acids. However, it does irreversibly inhibit RNA polymerase (Hossainy and Bloch, 1973) and denatures free DNA (Hossainy, Zweidler and Bloch, 1973), although DNA complexed with protein is not affected. It also inhibits protein synthesis and cell division (McMahon and Gopel, 1976). Because of the limited amount of work done with this gradient medium, it is not known whether chloral hydrate interacts with other biological materials besides protein and nucleic acids. Similarly, the factors affecting the analysis of gradients have yet to be studied.

Bovine Serum Albumin

It has been recognized for some time that protein solutions protect enzymic activities and, more particularly, help to preserve the integrity of cells. Therefore, it is not surprising that in the search for suitable density-gradient media some workers have used concentrated solutions of bovine serum albumin (Leif and Vinograd, 1964; Zucker and Cassen, 1969; Grdina *et al.*, 1974b). Bovine serum albumin fraction V is generally used; it is readily soluble in aqueous media but concentrated solutions absorb very strongly in the ultraviolet region of the spectrum ($E_{280 \text{ nm}}^{1\%} \simeq 7$) and are very viscous. On the other hand, the high molecular weight of the protein makes it relatively inert osmotically, in marked contrast with other media such as Renografin. Continuous gradients are best preformed using a gradient-making machine. After fractionation the density of the gradient fractions can be calculated using the equation

$$\rho = 1.412\, 9\eta_{24^\circ} - 0.881\, 4 \quad \text{(Grdina *et al.*, 1974b)}$$

Cells can be separated in 10–35% (w/v) gradients of bovine serum albumin in balanced salt solutions either by dispersing the cells throughout the gradient or by layering them on top or into the bottom of the gradient, and centrifuging at about 13 000*g* for 30 min. The distribution of cells can be determined by direct particle counting or by scintillation counting where the cells have been labelled with radioisotopes.

The most frequent problem encountered with these gradients is the great variability between different batches of bovine serum albumin, some of which cause cells to aggregate. In addition, albumin appears to promote phagocytosis by cells, resulting in the formation of cell vacuoles (R.I. Freshney, personal communication), although this can be minimized by fractionating the cells in the cold.

Non-aqueous Organic Media

The fractionation of cellular material in non-aqueous media is a well-established procedure. For example, the preparation of acetone powders and the non-aqueous isolation of nuclei (Siebert *et al.*, 1965) have shown that a number of enzymic activities are unaffected by this treatment. As a result of these observations, Stubblefield and Wray (1974) have devised a composite aqueous/non-aqueous gradient system for the fractionation of metaphase chromosomes. This medium has some advantages in that some of the problems of chromosome fractionation, such as their 'stickiness', is reduced. On the other hand, there is evidence that such procedures can strip proteins from the chromosomes (Stubblefield and Wray, 1974). It may be that such gradients could be adapted for other biological particles.

The use of organic solvents presents some problems quite apart from their effect on biological particles. For example, certain organic solvents cannot be used with some types of plastic centrifuge tubes (see Chapter 9), whereas others can react with aluminium or Perspex which precludes their use in a number of types of zonal rotors. In addition, the volatility of such gradient materials means that the composition of gradients in unstoppered tubes can change during centrifugation, although in the case of the gradients of Stubblefield and Wray (1974) this problem was minimized by the aqueous nature of the top of the gradients.

ACKNOWLEDGEMENTS

I should like to thank the many people who have advised me on the content and arrangement of this chapter, in particular Professor Wattiaux, Ian Freshney and Bill Wunner. In addition, I should like to thank the commercial firms for providing me with additional information.

REFERENCES

AAS, M. (1973). 9th Internat. Congr. Biochem., Stockholm, p. 31

ANDERSON, N.G. and CLINE, G.B. (1967). In *Methods in Virology*, vol. 2, p. 137. Eds K. Maramorosch and H. Koprowski. New York; Academic Press

ARMITAGE, A.M.T., CORNWALL, H.J.C., WRIGHT, N.G. and WEIR, A.R. (1975). *Archs Virol.*, **47**, 319

BEAUFAY, J., JACQUES, P., BAUDHUIN, P., SELLINGER, O.Z., BERTHET, J. and DE DUVE, C. (1964). *Biochem. J.*, **92**, 184

BIRNIE, G.D., HELL, A., SLIMMING, T.K. and PAUL, J. (1973a). In *Methodological Developments in Biochemistry*, vol. 3, p. 127. Ed. E. Reid. London; Longmans

BIRNIE, G.D., RICKWOOD, D. and HELL, A. (1973b). *Biochim. biophys. Acta*, **331**, 283

BIRNIE, G.D., MacPHAIL, E. and RICKWOOD, D. (1974). *Nucleic Acids Res.*, **1**, 919

BIRNIE, G.D. and RICKWOOD, D. (1976). In *Biological Separations in Iodinated Density-Gradient Media*, p. 193. Ed. D. Rickwood. London; Information Retrieval Ltd.

BONNETT, H.T. and ERIKSON, T. (1974). *Planta*, **120**, 71

BOONE, C.W., HARELL, C.S. and BOND, H.E. (1968). *J. Cell Biol.*, **36**, 369

BOYD, E.M., THOMAS, D.R., HARTON, B.F. and HUISMAN, T.H.J. (1967). *Clin. chim. Acta*, **16**, 333

BØYUM, A. (1968). *Scand. J. clin. Lab. Invest.*, **21**, Suppl. 97

BRAKKE, M.K. (1967). In *Methods in Virology*, vol. 2, p. 93. Eds K. Maramorosch and H. Koprowski. New York; Academic Press

BRITTEN, R.J. and ROBERTS, R.B. (1960). *Science, N.Y.*, **131**, 32

BUCKINGHAM, M.E. and GROS, F. (1975). *FEBS Lett.*, **53**, 355

BUCKINGHAM, M.E. and GROS, F. (1976). In *Biological Separations in Iodinated Density-Gradient Media*, p. 71. Ed. D. Rickwood. London; Information Retrieval Ltd

CASTAGNA, M. and CHAUVEAU, J. (1969). *Expl Cell Res.*, **57**, 211

CHAN, R.T.L. and SCHEFFLER, I.E. (1974). *J. Cell Biol.*, **61**, 780

COLE, T.A. (1971). *Analyt. Biochem.*, **41**, 276

COLLOT, M., WATTIAUX-DE CONINCK, S. and WATTIAUX, R. (1975). *Eur. J. Biochem.*, **51**, 603

COLLOT, M., WATTIAUX-DE CONINCK, S. and WATTIAUX, R. (1976). In *Biological Separations in Iodinated Density-Gradient Media*, p. 89. Ed. D. Rickwood. London; Information Retrieval Ltd

CONSTABEL, F. and KAO, K.N. (1974). *Can. J. Bot.*, **52**, 1603

DE DUVE, C. (1964). *Harvey Lect.*, **59**, 49

DE DUVE, C., BERTHET, J. and BEAUFAY, H. (1959). *Prog. Biophys. biophys. Chem.,* **9,** 325

DE SIMONE, J., KLEVE, L. and SCHAEFFER, J. (1974). *J. Lab. clin. Med.,* **84,** 517

DOBROTA, M. and HINTON, R.H. (1973). *Analyt. Biochem.,* **56,** 270

DOENECKE, D. and McCARTHY, B.J. (1975). *Biochemistry,* **14,** 1366

FLAMM, W.G., BOND, H.E. and BURR, H.E. (1966). *Biochim. biophys. Acta,* **129,** 310

FRESHNEY, R.I. (1976). In *Biological Separations in Iodinated Density-Gradient Media,* p. 123. Ed. D. Rickwood. London; Information Retrieval Ltd

GRDINA, D.J., MILAS, L., HEWITT, R.R. and WITHERS, H.R. (1973). *Expl Cell Res.,* **81,** 250

GRDINA, D.J., MILAS, L., MASON, K.A. and WITHERS, H.R. (1974a). *J. natn. Cancer Inst.,* **52,** 253

GRDINA, D.J., MEISTRICH, M.L. and WITHERS, H.R. (1974b). *Expl Cell Res.,* **85,** 15

GSCHWENDER, H.H., BRUMMUND, M. and LEHMANN-GRUBE, F. (1975). *J. Virol.,* **15,** 1317

GSCHWENDER, H.H. and POPESCU, M. (1976). In *Biological Separations in Iodinated Density-Gradient Media,* p. 145. Ed. D. Rickwood. London; Information Retrieval Ltd

HANCOCK, R. (1970). *J. molec. Biol.,* **48,** 357

HARRIS, R. and UKACJIOFO, E.V. (1969). *Lancet,* **327,** 7615

HARTMAN, G.C., BLACK, N., SINCLAIR, R. and HINTON, R.H. (1974). In *Methodological Developments in Biochemistry,* vol. 4, p. 93. Ed. E. Reid. London; Longman

HAYEK, D.H. and TIPTON, S.R. (1966). *J. Cell Biol.,* **29,** 405

HELL, A. (1972). MSE Application Information A6/6/72. Crawley, Sussex, UK; MSE Instruments Ltd

HELL, A., RICKWOOD, D. and BIRNIE, G.D. (1974). In *Methodological Developments in Biochemistry,* vol. 4, p. 117. Ed. E. Reid. London; Longmans

HINTON, R.H. (1972). In *Subcellular Components: Preparation and Fractionation,* 2nd edn, p. 119. Ed. G.D. Birnie. London; Butterworths

HINTON, R.H., BURGE, M.L.E. and HARTMAN, G.C. (1969). *Analyt. Biochem.,* **29,** 248

HINTON, R.H. and MULLOCK, B.M. (1976). In *Biological Separations in Iodinated Density-Gradient Media,* p. 1. Ed. D. Rickwood. London; Information Retrieval Ltd

HINTON, R.H., MULLOCK, B.M. and GILHUUS-MOE, C.C. (1974). In *Methodological Developments in Biochemistry,* vol. 4, p. 103. Ed. E. Reid. London; Longmans

HOLTERMANN, H. (1973). *Acta radiol. Suppl.,* **335,** 1

HONDA, S.I., HONGLADAROM, T. and LATIES, G.G. (1966). *J. exp. Bot.,* **17,** 460

HOSSAINY, E.M. and BLOCH, D.P. (1973). *J. Cell Biol.,* **59,** 149a

HOSSAINY, E.M., ZWEIDLER, A. and BLOCH, D.P. (1973). *J. molec. Biol.,* **74,** 283

HOUSSAIS, J.F. (1975). *FEBS Lett.,* **56,** 341

HUTTERMANN, A. and GUNTERMANN, U. (1975). *Analyt. Biochem.,* **64,** 360

HUTTERMANN, A. and WENDLBERGER, G. (1976). In *Biological Separations in Iodinated Density-Gradient Media,* p. 25. Ed. D. Rickwood. London; Information Retrieval Ltd

HUTTERMANN, A. and WENDLBERGER-SCHIEWEG, G. (1976). *Biochim. biophys. Acta,* **453,** 176

JOHNSTON, I.R., MATHIAS, A.P., PENNINGTON, F. and RIDGE, D. (1968). *Biochem. J.*, **109**, 127

KEMPF, J., EGLY, J.L., STRICKER, Ch., SCHMITT, M. and MANDEL, P. (1972). *FEBS Lett.*, **26**, 130

KETCHUM, P.A. and HOLT, S.C. (1970). *Biochim. biophys. Acta*, **196**, 141

KRETTLI, A., CHEN, D.H. and NUSSENZWEIG, R.S. (1973). *J. Protozool.*, **20**, 662

KUROKAWA, M., SAKAMOTO, T. and KATO, M. (1965). *Biochem. J.*, **97**, 833

LEIF, R.C. and VINOGRAD, J. (1964). *Proc. natn. Acad. Sci., U.S.A.*, **51**, 520

LINDBERG, S. and SVENSSON, S. (1968). *Acta chem. scand.*, **22**, 1907

LINDERSTRÖM-LANG, K. (1937). *Nature, Lond.*, **139**, 713

LOIR, M. and LANNEAU, M. (1975). *Expl Cell Res.*, **92**, 499

LUNDH, S. (1973). *Int. J. Pept. Prot. Res.*, **5**, 309

LYONS, M.J. and MOORE, D.H. (1965). *J. natn. Cancer Inst.*, **35**, 549

MACH, O. and LACKO, L. (1968). *Analyt. Biochem.*, **22**, 393

McMAHON, D. and GOPEL, G. (1976). *Mol. cell. Biochem.*, **10**, 27

MANNING, J.S., HACKETT, A.J., CARDIFF, R.D., MEL, H.C. and BLAIR, P.B. (1970). *Virology*, **40**, 912

MATEYKO, G.M. and KOPAC, M.J. (1963). *Ann. N.Y. Acad. Sci.*, **105**, 219

MATHIAS, A.P. and WYNTER, C.V.A. (1973). *FEBS Lett.*, **33**, 18

MEISTRICH, M.L. and TROSTLE, P.K. (1975). *Expl Cell Res.*, **92**, 231

MESELSON, M., STAHL, F.W. and VINOGRAD, J. (1957). *Proc. natn. Acad. Sci., U.S.A.*, **43**, 581

MILLER, G.L. and GASEK, J.MacG. (1960). *Analyt. Biochem.*, **1**, 78

MORGENTHALER, J.J., MARSDEN, M.P.F. and PRICE, C.A. (1975). *Archs Biochem. Biophys.*, **168**, 289

MORGENTHALER, J.J. and PRICE, C.A. (1976). *Biochem. J.*, **153**, 487

MULLOCK, B.M. and HINTON, R.H. (1973). *Trans. Biochem. Soc.*, **1**, 578

MUNTHE-KAAS, A.C. and SEGLEN, P.O. (1974). *FEBS Lett.*, **43**, 252

NOLL, H. (1969). In *Techniques in Protein Biosynthesis*, vol. 2, p. 101. Eds P.N. Campbell and J.R. Sargent. New York and London; Academic Press

OLSSON, I. (1969). *Expl Cell Res.*, **54**, 325

OROSZLAN, S.J., RIZUI, S., O'CONNOR, T.E. and MORA, P.T. (1964). *Nature, Lond.*, **202**, 780

PATRICK, J.C., RENGACHARY, S. and MELNKOVYCH, G. (1975). *In vitro*, **11**, 404

PERTOFT, H. (1966). *Biochim. biophys. Acta*, **126**, 594

PERTOFT, H. (1969). *Expl Cell Res.*, **57**, 338

PERTOFT, H. and LAURENT, T.C. (1969). In *Modern Separations of Macromolecules and Particles*, vol. 2, p. 71. New York; Wiley-Interscience

PERTOFT, H., PHILIPSON, L., OXELFELT, P. and HOGLUND, S. (1967). *Virology*, **33**, 185

PIGMAN, W. (1957). *The Carbohydrates*. New York; Academic Press

PRETLOW, T.G. and WILLIAMS, E.E. (1973). *Analyt. Biochem.*, **55**, 114

RAYNAUD, A. and OHLENBUSCH, H.H. (1972). *J. molec. Biol.*, **63**, 523

REID, E. (1972). In *Subcellular Components: Preparation and Fractionation*, 2nd edn, p. 93. Ed. G.D. Birnie. London; Butterworths

RICKWOOD, D. (Ed.) (1976a). *Biological Separations in Iodinated Density-Gradient Media*. London; Information Retrieval Ltd

RICKWOOD, D. (1976b). In *Biological Separations in Iodinated Density-Gradient Media*, p. 27. Ed. D. Rickwood. London; Information Retrieval Ltd

RICKWOOD, D. and BIRNIE, G.D. (1975). *FEBS Lett.*, **50**, 102

RICKWOOD, D., BIRNIE, G.D. and MacGILLIVRAY, A.J. (1975). *Nucleic Acids Res.*, **2**, 723

RICKWOOD, D., HELL, A. and BIRNIE, G.D. (1973). *FEBS Lett.*, **33**, 221

RICKWOOD, D., HELL, A., BIRNIE, G.D. and GILHUUS-MOE, C.C. (1974a). *Biochim. biophys. Acta*, **342**, 367

RICKWOOD, D., HELL, A., MALCOLM, S., BIRNIE, G.D., MacGILLIVRAY, A.J. and PAUL, J. (1974b). *Biochim. biophys. Acta*, **353**, 353

RICKWOOD, D. and MacGILLIVRAY, A.J. (1977). *Expl Cell Res.*, **104**, 287

ROCHA, V. and TING, I.P. (1970). *Archs Biochem. Biophys.*, **140**, 398

SCHAFFNER, W. and WEISSMANN, C. (1973). *Analyt. Biochem.*, **56**, 502

SCHMITT, J.M., BEHNKE, H.O. and HERRMANN, R. (1974). *Expl Cell Res.*, **85**, 63

SEGLEN, P.O. (1976). In *Biological Separations in Iodinated Density-Gradient Media*, p. 107. Ed. D. Rickwood. London; Information Retrieval Ltd

SERWER, P. (1975). *J. molec. Biol.*, **92**, 433

SIEBERT, G., HUMPHREY, G.C., THEMANN, H. and KERSTEN, W. (1965). *Hoppe-Seyler's Z. physiol. Chem.*, **340**, 51

STUBBLEFIELD, E. and WRAY, W. (1974). *Cold Spring Harbor Symp. quant. Biol.*, **38**, 835

SYKES, J.A., WHITECARVER, J., BRIGGS, L. and ANSON, L.H. (1975). *J. natn. Cancer Inst.*, **44**, 855

THEIMER, R.R., ANDING, G. and SCHMID-NEUHAUS, B. (1975). *FEBS Lett.*, **57**, 89

VANDEN BERGHE, D.A.R., VAN DER GROEN, G. and PATTYN, S.R. (1976). In *Biological Separations in Iodinated Density-Gradient Media*, p. 175. Ed. D. Rickwood. London; Information Retrieval Ltd

VAN DEN HOVE-VANDENBROUCKE, M.F. and DE NAYER, Ph. (1976). *Excerpta Medica*, Internat. Congr. Series 361, 7th Internat. Thyroid Congr., Boston, Abstract 82

VINOGRAD, J., GREENWALD, R. and HEARST, J.E. (1965). *Biopolymers*, **3**, 109

WALLACH, K.B. and KAMAT, D.F.H. (1964). *Proc. natn. Acad. Sci., U.S.A.*, **52**, 721

WARTERS, R.L. and HOFER, K.G. (1974). *Expl Cell Res.*, **87**, 143

WATTIAUX, R. (1973). *Archs int. Physiol. Biochim.*, **81**, 343

WATTIAUX-DE CONINCK, S. and WATTIAUX, R. (1971). *Eur. J. Biochem.*, **19**, 552

WOLFF, D.A. and PERTOFT, H. (1972a). *Biochim. biophys. Acta*, **286**, 197

WOLFF, D.A. and PERTOFT, H. (1972b). *J. Cell Biol.*, **55**, 579

WRAY, W. (1976). In *Biological Separations in Iodinated Density-Gradient Media*, p. 57. Ed. D. Rickwood. London; Information Retrieval Ltd

WRAY, W. and STEFOS, K. (1974). *J. Cell Biol.*, **63**, 380a

WUNNER, W.H., BULLER, R.M.L. and PRINGLE, C.R. (1976). In *Biological Separations in Iodinated Density-Gradient Media*, p. 159. Ed. D. Rickwood. London; Information Retrieval Ltd

ZIMMER, G., KEITH, A.D. and PACKER, L. (1972). *Archs Biochem. Biophys.*, **152**, 105

ZUCKER, R.M. and CASSEN, B. (1969). *Blood*, **34**, 591

8 Analytical Ultracentrifugation

ROBERT EASON
AILSA M. CAMPBELL
Department of Biochemistry, The University of Glasgow

HISTORY AND SCOPE

It is now fifty years since the first analytical ultracentrifuge was constructed by Svedberg and Rinde (1924) with the original intention of investigating the properties of colloidal solutions. Sedimentation studies began on proteins about two years later when they were found to be representative of 'monodisperse colloids' and consequently provided data which were more easily interpreted than those from polydisperse systems (Svedberg and Fahraeus, 1926; Svedberg and Chirnoaga, 1928). This was at a time when the concept of the existence of giant molecules such as DNA and proteins had not been suggested and all 'macromolecules' were thought to be clusters of small molecules of variable mass in common with the current usage of the term 'colloid'.

In the twenty years that followed, development of analytical ultracentrifugation reflected three different branches of research. First, the mechanics of the process were streamlined and the centrifuge which emerged from a series of prototypes from different laboratories was the Spinco Model E made by Beckman (Bauer and Pickels, 1936; Beams *et al.*, 1955). This electrically driven model has dominated the analytical ultracentrifuge field for twenty-five years with successive editions reflecting improvements in optics and electronics but little alteration in basic design. Other analytical ultracentrifuges are available (Bowen, 1970) but retain a relatively small proportion of the market. The second field of research was in optics, leading to the development of sensitive systems for the detection of refractive index changes. First, schlieren systems, then a variety of interferometric systems were developed. However, the interest in nucleic acids, many species of which show large concentration dependence in their sedimentation properties and which are difficult to obtain in large amounts, led to renewed interest in absorption systems (see pp. 252–255). Paradoxically, absorption systems had been the first ones employed by Svedberg, but in the early days ultraviolet photography was less well developed and led to less reliable results than refractive index techniques (Schachman, 1959). The third branch of research which was active was the theoretical branch. The early theory is well summarized in Svedberg's own book (Svedberg and Pederson, 1940), and this is an area of development which has continued to the present day with refinements for polydisperse systems, interacting systems and zonal systems still appearing, as it has become apparent that few macromolecular systems correspond to the ideal.

The major contribution made by the analytical ultracentrifuge up to this point in time is the determination of the molecular weights of large numbers of different macromolecules, especially proteins. The equilibrium technique has in particular been the major biochemical method for the determination of absolute molecular weights. The larger nucleoprotein and polynucleotide molecules which have fallen outside the range of this method have been extensively studied by sedimentation velocity analysis which, in common with other hydrodynamic techniques, has led to good estimates of size in a range where, until recently, few other biochemical techniques have been available.

Sedimentation velocity analysis is a transport process. In recent years other transport processes which share many common features have been developed. Although the basic transport driving forces may differ and the phenomenological properties of the solute may vary in the different systems, many of the basic flow equations developed for velocity centrifugation are very similar to those of new transport techniques for molecular-weight determination. Such methods are often used with empirical calibration by standards whose molecular weight is known, as in the case of gel electrophoresis of proteins in sodium dodecyl sulphate (Weber, Pringle and Osbern, 1972) and gel chromatography (Ackers, 1970; Winzor, 1970). Extension of the range of the latter technique in particular has made it comparable to velocity sedimentation in many respects. Both processes depend on the separation of large and small molecules by 'flow' down a column of solution, with the larger molecules migrating faster than the smaller. Gel chromatography is being increasingly extended to the study of interacting protein systems and may, in future, supersede the ultracentrifuge in this area.

There have been many attempts to use hydrodynamic methods to study macromolecular shape and these have perhaps been the least satisfactory area of development in ultracentrifugal work. The sedimentation data have been of little value by themselves unless combined with other hydrodynamic data (see pp. 279–280), and only crude shape factors such as the ellipsoidal axial ratio can be determined. This contrasts with the precision of X-ray crystallography in protein structure analysis. In the nucleic acid and nucleoprotein field, electron microscopy (Upholt, Gray and Vinograd, 1971), laser light scattering (Campbell and Jolly, 1973) and neutron scattering (Baldwin *et al.*, 1975) have contributed more in terms of pure shape analysis despite the drawbacks of each of these techniques in themselves, in particular the requirement for comparatively large amounts of material for scattering techniques.

Many reviews of the current use of ultracentrifugation have appeared (Creeth and Pain, 1967; Bowen, 1970; Coates, 1970; Brewer, Pesce and Spencer, 1974). In this chapter, the theory and uses of equilibrium analytical ultracentrifugation will be described, together with the theory and major uses of the velocity technique.

INSTRUMENTATION

Optical Systems

Analytical ultracentrifuges usually provide two types of optical system for recording sedimentation events. The schlieren system and the Rayleigh interferometric system detect changes in refractive index in the solution and, as an alternative, optical systems which measure light absorption can be used for appropriate macromolecular systems. Although the absorption system is more sensitive, a large emphasis in previous years has been placed on experimentation using the schlieren and Rayleigh systems. The main reason for this has been the difficulty in making quantitative measurements with the absorption photographic system. The schlieren and Rayleigh optical systems have been exhaustively described in detail in the literature, and equations in a form particularly relevant to their use have been fully described (Van Holde, 1967; Creeth and Pain, 1967; Richards, Teller and Schachman, 1968; Teller *et al.*, 1969; Bowen, 1970).

In recent years, the use of absorption optical systems has greatly increased due to the development of the automatic split-beam photoelectric scanning system (Schachman and Edelstein, 1966). A simplified diagram of the scanning system is shown in *Figure 8.1.* In ultracentrifuges with the earlier ultraviolet photographic system, the monochromator and photomultiplier are replaced with a

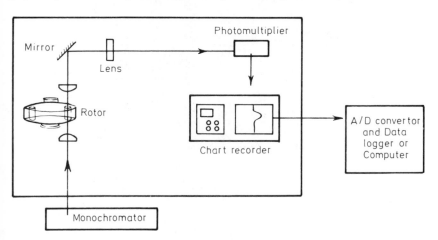

Figure 8.1 A representation of the photoelectric scanning system for the Beckman Model E analytical ultracentrifuge showing alternative forms of data output and recording

light source plus filter system and film holder. Such absorption systems are extremely sensitive. For example, using the scanner system, sedimentation studies have been carried out on aspartate transcarbamylase in solutions of 3 μg/ml at 230 nm (Schachman and Edelstein, 1966), whereas in band sedimentation analysis of DNA less than 0.1 μg is easily detected. In the course of this chapter, examples will be drawn from experiments in which either the ultraviolet photographic system or the photoelectric scanner was used.

The ultraviolet photographic system represents the more usual means for detecting absorption changes in ultracentrifuge experiments. It is particularly useful in sedimentation velocity analysis in which exposures are taken at intervals and the photograph is developed at the end of the experiment (*Figure 8.2*). Appropriate radial distances are measured from microdensitometer tracings of such photographs (see p. 274). Quantitative analysis of solute concentration in ultraviolet photographs has always presented difficult technical problems so that the introduction of the photoelectric scanner represents a most important development in ultracentrifugal analysis.

Figure 8.2 *Ultraviolet absorption photographs from sedimentation velocity experiments.* Top: *Moving boundary analysis of MS2 RNA centrifuged at 44 000 rev/min, exposures taken at 4 min intervals.* Bottom: *Band sedimentation analysis of φX 174 RF DNA centrifuged at 39 460 rev/min, exposures taken at 8 min intervals*

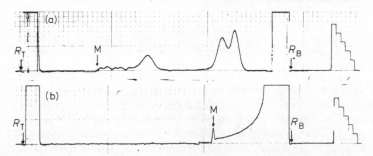

Figure 8.3 *Traces obtained from the photoelectric scanner. (a) Band sedimentation analysis in alkaline CsCl, density 1.5 g/cm³, of viral DNA extracted from cells infected with SV40. The speed was 40 000 rev/min, time at speed before scanning was 40 min, temperature was 21 °C and wavelength of scanning was 265 nm. Sedimentation is from left to right. Supercoiled molecules of wild type SV40 DNA and defective SV40 DNA have migrated well down the cell while single-stranded linear and circular molecules derived from nicked supercoils are near the top of the cell. (b) Short-column, low-speed sedimentation equilibrium analysis of serum albumin in 0.15 M NaCl, pH 7.0. (Such a system is usually considered to represent a two-component system to a good approximation; see pp. 257–261). The speed of centrifugation was 10 000 rev/min, temperature was 25 °C, initial concentration was 0.8 mg/ml and the cell was scanned at 280 nm. The arrows indicate the top and bottom reference edges, R_T and R_B, and the meniscus is indicated by the arrow marked M. The calibration stairsteps, in increments of 0.2 O.D. units, are on the right of each trace*

The scanner system was developed by Schachman and co-workers (Schachman and Edelstein, 1966) and an introduction to the operation of the system has recently appeared (Chervenka, 1971). The scanner offers direct visualization of results during the course of an experiment, an advantage shared with the refractive index systems but lacking in the older ultraviolet photographic method. It incorporates the advantages of double-beam spectrophotometry and permits data at different wavelengths to be recorded during an experiment. Thus, solute components with different optical properties can be detected in the same mixture (Steinberg and Schachman, 1966). Examples of scanner traces are shown in *Figure 8.3*. Very precise quantitative data from solutes at low concentration can be derived from scanner results. Having ensured that a stable baseline is present on the trace, optical density values are derived from chart height measurements by means of the calibration steps. For this conversion, least squares linear or quadratic curve-fitting is employed (Bevington, 1969). As is true for any optical recording device, electronic noise may distort the accuracy of the readings. It is possible to generate multiple traces, average the data manually and improve the signal-to-noise ratio by appropriate smoothing procedures (Savitzky and Golay, 1964; Bevington, 1969; Dyson and Isenberg, 1971). Clearly, however, such processing can be accomplished in a more satisfactory manner using a computer. A number of commercially available devices for this purpose include the Beckman Model 3801 or 3800 data systems which employ analogue–digital converters and either paper tape or magnetic tape output in addition to the chart recorder results. The advantages of using a computer are the expected ones of increased amounts of more accurate data and the ease of applying statistical methods to data processing.

Analytical Ultracentrifuge Cells

A wide variety of cells is available for the analytical ultracentrifuge. An exploded view of a typical cell assembly for the Beckman Model E ultracentrifuge is shown in *Figure 8.4*. The cell can be assembled and tightened with a torque wrench to give an airtight unit. The sample can be introduced through the holes in the cell housing and centrepiece, except for special types of experiment. In work with absorbance optical systems, single- and double-sector centrepieces can be used (*Figure 8.5a* and *8.5b*). In the case of ultraviolet photography, single-sector centrepieces are used. These may have either a 4° or 2° sector (the angle subtended at the centre of rotation) and the thickness may vary from 1.5 to 30 mm, so that the optical path can be chosen according to the absorbance of the sample to be analysed. Double-sector centrepieces are used in general with the photoelectric scanner. These incorporate a 2.5° sector and the thickness can vary from 12 to 30 mm. The centrepieces are usually made of Kel-F or Epon, although aluminium centrepieces are available.

An array of special centrepieces is available and some of these are shown in *Figure 8.5c, 8.5d* and *8.5e*. In the case of a type I band-forming centrepiece (*Figure 8.5c*; Vinograd, Radloff and Bruner, 1965), the sample is introduced into the well on the right of the centrepiece when the cell is partly assembled and the corresponding buffer is introduced on the left. After the cell is assembled, the appropriate solvent is inserted in the sectors so that when the centrifugal force is applied, the sample layers on the solvent as a narrow band. The levels on either side of the centrepiece are kept the same by means of the capillary at the bottom of the centrepiece. The synthetic boundary centrepiece (*Figure 8.5d*) is used for diffusion experiments or in studying moving boundaries of solutes having a low sedimentation coefficient. A small amount of solution is introduced on the right and a larger volume of buffer on the left. When the centrifugal force is applied, buffer will run through the lower capillary, displacing air through the upper

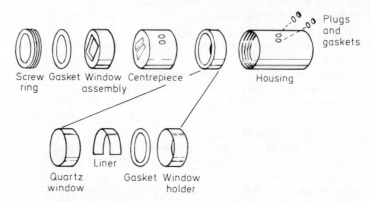

Figure 8.4 *Exploded view of a typical cell assembly for the analytical ultracentrifuge*

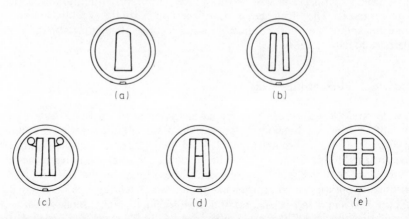

Figure 8.5 *Centrepieces for analytical ultracentrifuge cells: (a) single sector, 4°; (b) double sector, 2.5°; (c) double sector, band-forming type I; (d) double sector, synthetic boundary; (e) six-sector Yphantis type*

capillary, so as to form a sharp boundary between solution and buffer. The six-channel centrepiece (*Figure 8.5e*) is used for short-column equilibrium runs and three samples, with the corresponding buffers, can be analysed simultaneously. In these cases, the sample at highest concentration is placed in the innermost compartment in the partly assembled cell.

An essential element in most ultracentrifuge runs is the counterbalance (*Figure 8.6*). In addition to balancing the rotor for the weight of the cell, the counter-balance contains the reference edges which allow conversion of distances on the scanner trace or photograph to real distances in the ultracentrifuge cell. The reference holes are set to a precisely known distance and are far enough apart

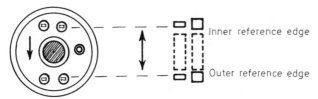

Inner reference edge

Outer reference edge

Figure 8.6 Diagram of counterbalance showing distance between inner and outer reference edges. The shaded area is a variable weight to adjust the counterbalance for different types of cell assembly. The smaller reference holes correspond to those which appear above the reference sector in the double-sector mode of operation of the photo-electric scanner

Figure 8.7 Rotors for the analytical ultracentrifuge. The six-hole rotor on the left allows the use of up to five double-sector cell assemblies and the counterbalance by means of the multiplexer accessory on the ultracentrifuge. With the other rotor, one cell assembly plus counterbalance can be used or two cells can be spun together and the holes in the rotor, seen at the top and bottom, employed as reference points

to encompass the image of the sectors as the rotor spins past the light source. When the photoelectric scanner is used, from one to five cells can be used with the appropriate rotor (*Figure 8.7*).

SEDIMENTATION EQUILIBRIUM ANALYSIS

Sedimentation Equilibrium in Two-component Systems

One of the simplest problems to which sedimentation equilibrium analysis can be applied is to determine the molecular weight of a pure solute species. When such experiments are performed and the photoelectric scanner is used to record the data, short-column, low-speed conditions are to be preferred (*Figure 8.3b*). Solution columns of from 1 to 3 mm are usually employed, since the time to reach equilibrium is proportional to the square of the depth of solution (Van Holde and Baldwin, 1958). Low speeds are used so that pressure-dependent effects are minimized compared with the high-speed (or meniscus depletion) method, often used in Rayleigh interference photographs (Yphantis, 1964; Van Holde, 1967).

In sedimentation equilibrium experiments, a system consisting of a protein with a net charge of zero (electroneutral protein) in dilute salt solution is usually considered to approximate to a two-component system in which, by convention, the solvent is component 1 and the macromolecular solute is component 2. It is, of course, perfectly clear that this constitutes the ternary system, water–protein–salt, but in practice it is found that to apply two-component theory in this instance introduces negligible errors. The advantage is that simple equations may be applied to yield answers with a high degree of accuracy. A full understanding of the significance of the parameters in the relevant equations is gained on the basis of thermodynamic derivation of the equations (Goldberg, 1953). The thermodynamic approach avoids the necessity of supplementary molecular assumptions and allows the effects of non-ideality, which operate in any real system, to be taken into account. Furthermore, in addition to the molecular-weight data, information on the thermodynamic properties of the system can be obtained (Williams *et al.*, 1958).

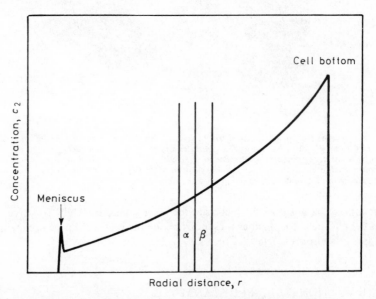

Figure 8.8 Diagram of a two-component system at sedimentation equilibrium. A protein in dilute salt is often taken to be a very good approximation of a system of this type. The concentration of solute increases in an exponential manner from the top to the bottom of the cell

In a two-component system at sedimentation equilibrium, the macromolecule is distributed in the centrifugal field as shown in *Figure 8.8*. If we consider the system to be composed of thin slices, such as α and β in *Figure 8.8*, then, at equilibrium, the total potential of each component, i, is a constant in all such slices.

The total potential, $\bar{\mu}_i$, is the sum of the chemical potential, μ_i, and the potential energy (per mole), so that in the two-component system:

$$\bar{\mu}_1 = \mu_1 - \tfrac{1}{2}M_1\omega^2 r^2 \quad \text{and} \quad \bar{\mu}_2 = \mu_2 - \tfrac{1}{2}M_2\omega^2 r^2 \tag{8.1}$$

where ω is the angular velocity in rad/s ($= \pi/30$ rev/min), r is the radial distance, in cm, from the centre of rotation and M is the molecular weight.

Consider how the chemical potential of the macromolecule varies as a function of radial position. Since the chemical potential is a function of concentration, temperature and pressure,

$$\frac{d\mu_2}{dr} = \left(\frac{\partial \mu_2}{\partial P}\right)\frac{dP}{dr} + \left(\frac{\partial \mu_2}{\partial c_2}\right)\frac{dc_2}{dr} \tag{8.2}$$

at constant temperature.

From eqn (8.1)

$$\frac{d\mu_2}{dr} = M_2\omega^2 r$$

and consideration of elementary thermodynamics (Williams *et al.*, 1958) indicates that

$$\left(\frac{\partial \mu_2}{\partial P}\right) = M_2\bar{v}_2$$

where $\bar{v}_2$ is the partial specific volume of the macromolecule and

$$\frac{dP}{dr} = \rho\omega^2 r$$

where ρ is the density of the solution. Substitution into eqn (8.2) gives

$$M_2\omega^2 r(1 - \bar{v}_2\rho) = \left(\frac{\partial \mu_2}{\partial c_2}\right)\frac{dc_2}{dr} \tag{8.3}$$

This is the fundamental equation for a two-component system at equilibrium in a centrifugal field. The equation can be converted to a practical form by writing the relation of chemical potential to concentration:

$$\mu_2 = \mu_2^\circ + RT \ln c_2 y_2$$

where μ_2° is the standard chemical potential, y_2 the activity coefficient, and R the gas constant; then

$$\frac{\partial \mu_2}{\partial c_2} = \frac{RT}{c_2}\left[1 + c_2 \frac{\partial \ln y_2}{\partial c_2}\right]$$

By assuming that $\ln y_2$ can be expressed in a power series of the form

$$\ln y_2 = BM_2 c_2 + CM_2 c_2^2 + \ldots$$

where B and C are virial coefficients, eqn (8.3) gives

$$\frac{M_2\omega^2 r(1-\bar{v}_2\rho)c_2}{RT} = [1 + BM_2c_2 + 2CM_2c_2^2 + \ldots]\frac{dc_2}{dr}$$

The essential working equation then becomes, to a good approximation:

$$\frac{d\ln c_2}{dr^2} = \frac{M_2(1-\bar{v}_2\rho)\omega^2}{2RT(1+BM_2c_2)} \tag{8.4}$$

In 'ideal' solutions, that is, at very low concentration, virial effects, that is non-ideal effects, vanish and BM_2 becomes zero.

In sedimentation equilibrium experiments on pure protein samples in which non-ideal effects operate, it can be seen from eqn (8.4) that an apparent molecular weight, M_2^{app}, can be found, where

$$\frac{1}{M_2^{app}} = \frac{1}{M_2}[1 + BM_2c_2]$$

Appropriate extrapolation to zero concentration of $1/M_2^{app}$ against c_2 will yield the value of the molecular weight. Many alternative extrapolation procedures have been used in this context, but new forms of the equations for sedimentation

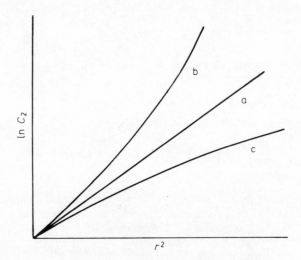

Figure 8.9 The possible forms of a graph of $\ln c_2$ *against* r^2 *for proteins at sedimentation equilibrium: a, linear; b, concave upwards; c, concave downwards*

equilibrium have been derived which allow evaluation of the 'ideal' molecular weight of a macromolecule from a single experiment (Rowe and Rowe, 1970).

Experimental values of c_2 and r can be obtained from the scanner trace (*Figure 8.3b*), and it is clear from eqn (8.4) that a graph of $\ln c_2$ against r^2 is required to calculate the molecular weight of the protein. The significance of such a graph should be considered very carefully (*Figure 8.9*). In an ideal solution, where virial effects vanish, the graph should be linear (line a). Such a result is often taken as proof of the homogeneity and purity of a protein preparation.

However, this procedure is unjustifiable and, although a linear graph is consistent with homogeneity of the macromolecular preparation, it does not constitute proof that this is the case. The reasons for this can be seen by considering curves b and c in *Figure 8.9*. When the macromolecular solution is heterogeneous either due to polydispersity or self-association, a curve such as b is observed. The effect of thermodynamic non-ideality because of concentration dependence is to produce a result of the type shown in curve c. In many cases, these two effects of heterogeneity and non-ideality operating together can produce a graph which is effectively linear and an erroneous conclusion as to the homogeneity of the protein preparation can be reached. If sedimentation equilibrium experiments are to be employed as a criterion of homogeneity instead of one of the many alternative techniques available in protein chemistry, a more reliable criterion is the independence of the value of the apparent molecular weight on radial position (Bowen, 1970; Williams, 1972).

Once the plot of $\ln c_2$ against r^2 has been evaluated, a value for $\bar{v}_2$ is needed to calculate the molecular weight of the protein. Most conceptual and practical difficulties have been removed from the task of obtaining a reliable value for $\bar{v}_2$ by measuring the density increment of a dialysed preparation of the sample (see p. 265). In the case of a two-component system, values based on amino acid composition or derived from sedimentation equilibrium experiments in H_2O and D_2O can be used with confidence (Edelstein and Schachman, 1967; Kupke, 1973). Dangers lie in the application of such values of $\bar{v}_2$ in situations where two-component theory is not appropriate and this problem is discussed in the section on multi-component systems (pp. 265–270).

Sedimentation Equilibrium Analysis of Associating Systems

It is often found in sedimentation equilibrium analysis of protein solutions, that $\ln c_2$ against r^2 graphs turns out to be concave upwards (*Figure 8.9*, line b). This is an indication of polydispersity and the question arises as to whether a heterogeneous mixture of non-reacting components or some form of self-associating system is present. To distinguish between these alternatives, a number of sedimentation equilibrium experiments at different initial concentrations are performed and $\ln c_2$ against r^2 data obtained. From each experiment, the $\ln c_2$ against r^2 data can be taken in sets of, say, five or more and local slopes derived by least-squares fitting. A corresponding value of the apparent molecular weight can then be calculated, eqn (8.4), and a set of such values determined for each experiment. By combining the sets of experimental values, a graph of apparent molecular weight over the entire range of concentration can be constructed. If the experiments give rise to divergent curves, then a heterogeneous mixture is present. If the data superimpose to form a continuous curve (*Figure 8.10*), then an associating system is present, and the composite graph can be constructed by interpolation or least-squares fitting. Such curves may be concave upwards or downwards, falling off at higher concentrations due to non-ideality.

In order to study an associating system, several important criteria must be observed. The protein solution must be dialysed exhaustively so that Cassassa and Eisenberg conventions apply (Cassassa and Eisenberg, 1964). This means that the macromolecular components are defined in such a manner that the equations for two-component theory can be used in analysing the associating

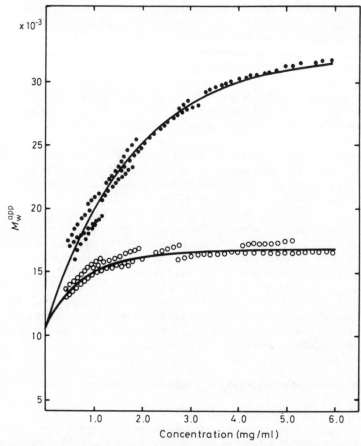

*Figure 8.10 Sedimentation equilibrium analysis of self-associating systems; apparent
weight-average molecular weight of H4 histone in 0.075 M and 0.15 M NaCl. Samples of
histone at different initial concentrations were dialysed against the appropriate salt
concentration at pH 7.0 then centrifuged at 26 000 rev/min for 40 h at 20 °C in an An–GTi
(six-place) rotor in a Beckman Model E analytical ultracentrifuge. Apparent weight-average
molecular weights (M_w^{app}) were determined from computer evaluation of $\ln c_2$ against r^2
data. The gradient at each point in the $\ln c_2$ against r^2 data was evaluated by fitting a
quadratic polynomial to successive and overlapping sets of five data points and differentiating the polynomial analytically. The molecular-weight data at 0.075 M NaCl (○) and at
0.15 M NaCl (●) are presented along with the best-fitting curves (solid lines) which
represent third degree polynomials which were constrained to pass through the monomer
molecular-weight value of 11 300 at zero concentration. (Reproduced with permission of*
Biochim. Biophys. Acta)

system. Dilution of the sample for runs over a wide range of concentration is
achieved using the dialysate. The short-column, low-speed approach is preferred.
since the theories for combined chemical and sedimentation equilibrium do not
account for pressure dependence of equilibrium constants. Careful analysis of
the associating system demands very accurate concentration data, otherwise
gross errors in apparent molecular weight and association constants will result
(Visser *et al.*, 1972).

The detailed theoretical analysis of associating systems has been the subject
of active study in recent years (Adams, 1967; Roark and Yphantis, 1969;

Williams, 1972), and several methods of data analysis have been developed (Chun *et al.*, 1972; Visser *et al.*, 1972; Tang and Adams, 1973; Tung and Steiner, 1974). Two general lines of approach can be recognized. One approach sets out to obtain a direct description of discrete molecular-weight distribution by expressing the observed concentration distribution in terms of a sum of exponential terms (Haschemeyer and Bowers, 1970). The exponential form of the concentration distribution expression for sedimentation equilibrium is obtained by integrating an ideal equation for i solute components in the form of eqn (8.4), between the meniscus at radius r_m and any radial position, r, thus

$$C_r = \sum_i C_{i,m} \exp[H_i(r^2 - r_m^2)]$$

where C_r, the total concentration at any radial position, is the sum of the concentration of the individual i solute components, and

$$H_i = \frac{M_i(1 - \bar{v}_i\rho)\omega^2}{2RT}$$

For an ideal self-associating system, the total concentration, C, can be expressed in terms of the monomer concentration, C_1, and the equilibrium constant for the ith association (Adams, 1967; Haschemeyer and Bowers, 1970), thus

$$C = \sum_i K_i C_1^i$$

At sedimentation equilibrium, the concentration distribution for the self-associating system is given by

$$C_r = \sum_i K_i C_{1,m}^i \exp[H_i(r^2 - r_m^2)]$$

The experimentally derived concentration distribution is then fitted to a sum of exponentials by linear programming methods to extract the unknown parameters.

Another type of approach, which has been widely adopted, is concerned with the evaluation of molecular-weight averages and related parameters, that is, the apparent weight-average and apparent number-average molecular weights and the apparent weight fraction monomer present, from the concentration distribution data (Adams, 1967). The apparent weight-average molecular weight at concentration c, M_{wc}^{app}, is obtained from $\ln c_2$ against r^2 data as described before, and is related to the ideal molecular weight by

$$\frac{M_1}{M_{wc}^{app}} = \frac{M_1}{M_{wc}} + B_1 M_1 c + B_2 M_1 c^2$$

The apparent number-average molecular weight, M_{nc}^{app}, is obtained from the weight-average molecular weight by

$$\frac{M_1}{M_{nc}^{app}} = \frac{1}{c} \int_0^c \frac{M_1}{M_{wc}^{app}} \, dc$$

These values are obtained by numerical integration. The apparent weight fraction monomer, f_1^{app}, is given by

$$\ln f_1^{\text{app}} = \int_0^c \left(\frac{M_1}{M_{\text{wc}}^{\text{app}}} - 1 \right) \frac{dc}{c}$$

which is again evaluated by numerical integration. Relationships between these parameters are then derived for various models of association. For example, a monomer–dimer–trimer model (Adams, 1967) predicts that

$$\frac{8M_1}{M_{\text{nc}}^{\text{app}}} - 6 = 3f_1^{\text{app}} \exp(-B_1 M_1 c) + 4B_1 M_1 c - \frac{1}{\left(\dfrac{M_1}{M_{\text{wc}}^{\text{app}}} - B_1 M_1 c \right)}$$

The non-ideality term, $B_1 M_1$, is found by successive approximation, varying the value then minimizing the residual (that is, the difference) between the left-hand side of the equation (experimental values) and the right-hand side (predicted for that particular value of $B_1 M_1$). Having thus obtained a value for $B_1 M_1$, the monomer concentration $c_1 = cf_1^{\text{app}} \exp(-B_1 M_1 c)$ can be calculated and the equilibrium constants evaluated from appropriate relations between the concentration of the various species and equilibrium constants (Adams, 1967). The use of only one virial coefficient in experiments covering a wide range of concentration has caused some discussion, since it is not clear at what point an additional virial coefficient will begin to make a significant contribution. There may remain some doubt therefore as to the completely accurate evaluation of a particular model.

A more recent approach which employs the molecular-weight averages concerns the use of graphical analysis (Chun *et al.*, 1972; Tang and Adams, 1973). In this method, virial coefficients are eliminated.

Since

$$\frac{M_1}{M_{\text{wc}}^{\text{app}}} = \frac{M_1}{M_{\text{wc}}} + B_1 M_1 c$$

and

$$\frac{M_1}{M_{\text{nc}}^{\text{app}}} = \frac{M_1}{M_{\text{nc}}} + \frac{B_1 M_1 c}{2}$$

then

$$\frac{2M_1}{M_{\text{nc}}^{\text{app}}} - \frac{M_1}{M_{\text{wc}}^{\text{app}}} = \frac{2M_1}{M_{\text{nc}}} - \frac{M_1}{M_{\text{wc}}} = \xi \qquad (8.5)$$

In the case, for example, of a monomer–$n$mer association (Tang and Adams, 1973),

$$\frac{1}{M_{\text{nc}}} = f_1 \left(1 - \frac{1}{n} \right) + \frac{1}{n} \quad \text{and} \quad \frac{1}{M_{\text{wc}}} = \frac{1}{n - f_1(n - 1)}$$

Substitution in eqn (8.5) and rearrangement gives

$$\frac{2f_1^2(n-1)^2}{n} - f_1(n-1)\left[2+\xi-\frac{2}{n}\right] - 1 + \xi n = 0$$

For each value of n, the equation is solved for f_1. The correct fit is obtained for that value of n which yields a linear graph of $(1-f_1)/f_1$ against $(cf_1)^{n-1}$, and the slope gives the equilibrium constant K_n, since $(1/f_1 - 1)$ equals $K_n(cf_1)^{n-1}$. A graph of M_1/M_{wc}^{app} [which equals $1/n - f_1(n-1)$] against c should give a straight line whose slope gives $B_1 M_1$, the non-ideality term.

Finally, a method is available using molecular-weight averages but which does not depend on the assumption of a specific model as in the previous cases (Tung and Steiner, 1974). The molecular-weight data is represented as a series expansion of the form

$$\frac{M_{wc}^{app}}{M_1 c_1 (1 - B_1 M_{wc}^{app})} = 1 + 4K_2 m_1 (\text{app}) \exp(-B_1 M_1 c)$$
$$+ 9K_3 m_1^2 (\text{app}) \exp(-2B_1 M_1 c) + \dots$$

where $m_1 (\text{app}) = $ apparent molarity of the monomer. Either least-squares polynomial fitting or linear programming can be used to locate the best values of B_1 and the association constants.

Methods for studying more complex types of association, such as $A + B \rightleftharpoons AB$, by sedimentation equilibrium analysis have been reviewed (Adams, 1969).

Sedimentation Equilibrium in Multi-component Systems

Solutions of biological macromolecules studied by sedimentation equilibrium analysis usually constitute a multi-component system comprising solvent (component 1), macromolecule (component 2), and a low-molecular-weight diffusible component (component 3), which may be ionic or non-ionic. In such a system, preferential interaction of the macromolecule with the other components will occur. Such preferential binding of low-molecular-weight components can make it difficult to define the macromolecular species precisely. Moreover, an important consequence of preferential interaction is that the apparent specific volume of the macromolecule can be modified in such a way that eqn (8.4) becomes inappropriate for use in sedimentation equilibrium experiments. The theoretical advances in the thermodynamics of multi-component systems by Cassassa and Eisenberg have greatly developed and clarified work in this area (Cassassa and Eisenberg, 1961, 1964; Eisenberg, 1962; Cassassa, 1969). It is now possible to determine the molecular weight of the macromolecule unambiguously without explicitly defining the interactions which occur in the multi-component system. Two well-known multi-component systems, proteins in denaturing conditions and DNA in concentrated solutions of CsCl, will be considered as examples.

The molecular weight of subunits from oligomeric proteins are commonly evaluated in either concentrated guanidine hydrochloride or in the presence of detergents (Huston *et al.*, 1972). However, despite the precision of sedimentation equilibrium analysis, there was for many years considerable uncertainty as to the molecular weight and hence the number of the subunits in such enzymes as aldolase (Reisler and Eisenberg, 1969; Lee and Timasheff, 1974). This uncertainty was due to effects of preferential interactions on apparent specific volumes and the misapplication of equations for two-component theory to a situation in which multi-component theory was appropriate. At sedimentation equilibrium in a three-component system, the distribution of an electroneutral macromolecular component is given by (Reisler and Eisenberg, 1969)

$$\frac{d \ln c_2}{dr^2} = \frac{\omega^2 M_2 \left(\frac{\partial \rho}{\partial c_2}\right)^0_\mu}{2RT} \tag{8.6}$$

In such systems, the macromolecule is dialysed exhaustively so that the subscript, μ, refers to constant chemical potential of the diffusible components and the superscript indicates the absence of virial effects at vanishing concentration of the macromolecule. The molecular weight obtained refers to the particular species defined in the term c_2. The preferential interactions are all reflected in the value of $(\partial \rho / \partial c_2)^0_\mu$ and, therefore, a precise definition of these can essentially be avoided (Noelken and Timasheff, 1967; Kirby-Hade and Tanford, 1967). This term is evaluated by measuring the difference in density between the protein solution and the dialysate using a precision densitometer (Stabinger, Leopold and Kratky, 1967; digital densitometer DMA-02, Anton Parr K.G., Graz, Austria). In the three-component system,

$$\left(\frac{\partial \rho}{\partial c_2}\right)^0_\mu \approx (1 - \phi'\rho)$$

where the partial specific volume of the protein becomes an apparent quantity, ϕ', whose value depends on the particular conditions and which cannot easily be predicted.

In solutions of concentrated guanidine hydrochloride, where reduced proteins exist as random coils (Tanford, Kawahara and Lapanje, 1967), it is clear that nonideality will be higher than for globular proteins. In such circumstances, heterogeneity can be easily masked, for example by obtaining linear plots of $\ln c_2$ against r^2. To help safeguard against incorrect interpretation in such systems, alternative forms of graphical analysis, such as $(dc/dr)/r$ against c, and $1/M^{app}$ against c, have been described (Munk and Cox, 1972). Exactly similar procedures are employed to analyse proteins in detergent solutions (Tanford *et al.*, 1974) where binding to protein is of a higher order of magnitude than in guanidine hydrochloride solutions. Other methods of analysing proteins in denaturing solvents, in particular gel chromatography (Fish, Reynolds and Tanford, 1970), can give accurate values of molecular weight.

Sedimentation equilibrium analysis of DNA in buoyant density gradients has been widely exploited in molecular biology. An important development in recent years has been the theoretical and practical advances which allow the method to be used for determining unambiguous molecular weights of homogeneous DNA preparations (Cohen and Eisenberg, 1968; Schmid and Hearst, 1969, 1971). Such absolute methods for molecular-weight determination are important in providing well-documented standards for the more commonly used comparative methods (Freifelder, 1970).

The classical work of Meselson *et al.* (1957) demonstrated that, when the three-component system H_2O–DNA–CsCl was centrifuged to equilibrium, the CsCl formed a well-defined density gradient and the DNA formed a narrow band in the gradient. The gradient formed by the CsCl alone is predicted from two-component theory (Ifft *et al.*, 1961) thus

$$\frac{d\rho}{dr} = \frac{d\rho}{d \ln a_3} \frac{(1 - \bar{v}_3 \rho)M_3 \omega^2 r}{RT}$$

where a_3 is the activity of CsCl [cf. eqn (8.4)]. At 1 atm,

$$\frac{d\rho^0}{dr} = \frac{\omega^2 r}{\beta^0}$$

where

$$\beta^0 = \frac{d \ln a_3}{d\rho^0} \frac{RT}{(1 - \bar{v}_3 \rho)M_3}$$

The superscripts indicate 1 atm and β^0 is the coefficient of the density gradient (the composition density gradient). The resolution between macromolecular species in the density gradient is proportional to the value of β^0 for the particular low-molecular-weight component used. Values of β^0 for a number of aqueous binary solvents have been published (Ifft *et al.*, 1970; see also *Table 6.1*).

The original three-component analysis by Hearst and Vinograd (1961a, 1961b) derived the relationship

$$M_2(1 + \Gamma') \left[1 - \frac{\bar{v}_2 + \Gamma' \bar{v}_1}{1 + \Gamma'} \rho \right] \omega^2 r = \left(\frac{\partial \mu_2}{\partial m_2} \right)_{\mu_1} \frac{dm_2}{dr} \tag{8.7}$$

where m_2 is a measure of the concentration of DNA, and Γ' is the 'solvation' parameter on a weight basis, and

$$\Gamma' = \left(\frac{M_1}{M_2} \right) \left(\frac{dm_1}{dm_2} \right)_{\mu_1}$$

[cf. eqn (8.3)].

At the position of the band maximum in the DNA distribution, dm_2/dr is 0. Substitution into eqn (8.7) then gives the definition of the buoyancy condition

$$\frac{1}{\rho^0} = \frac{\bar{v}_2 + \Gamma' \bar{v}_1}{1 + \Gamma'} = \bar{v}_{s,0}$$

where ρ^0 is the buoyant density and $\bar{v}_{s,0}$ is the partial specific volume of the 'solvated' macromolecule at band centre, radius r_0. The molecular weight of the 'solvated' macromolecule, $M_{s,0}$ at band centre, is given by

$$M_{s,0} = M_2(1 + \Gamma')$$

By using linear expansions about the band centre,

$$\rho = \rho^0 + \left(\frac{d\rho}{dr}\right)\delta$$

$$\bar{v}_2 = \bar{v}_{s,0} + \left(\frac{d\bar{v}_2}{dr}\right)\delta$$

$$M_2 = M_{s,0} + \left(\frac{dm_2}{dr}\right)\delta, \text{ and}$$

$$\delta = r - r_0$$

Substitution into eqn (8.7) gives

$$-M_{s,0}\left(\frac{\partial\rho}{\partial r}\right)_{\text{eff}} \bar{v}_{s,0} r_0 \omega^2 \delta d\delta = \left(\frac{\partial\mu_2}{\partial m_2}\right)_{\mu_1} dm_2 \qquad (8.8)$$

The effective density gradient

$$\left(\frac{\partial\rho}{\partial r}\right)_{\text{eff}} = \left(\frac{d\rho}{dr}\right) + \left(\frac{\rho^0}{\bar{v}_{s,0}}\right)\left(\frac{d\bar{v}_2}{dr}\right) = \frac{\omega^2 r}{\beta_{\text{eff}}}$$

takes into account variations in hydration and pressure across the band of DNA and appropriate values for a number of salts are available (Schmid and Hearst, 1971). The effective density gradient is measured by determining the distance between bands of normal DNA and ^{15}N-DNA, Δr, due to the difference, $\Delta\rho$, in density as a result of ^{15}N-substitution. The observed value of $\Delta\rho/\Delta r$ is a good measure of the effective density gradient. If, as in the original treatment, thermodynamic ideality is assumed, that is

$$\left(\frac{\partial\mu_2}{\partial m_2}\right)_{\mu_1} = \frac{RT}{m_2}$$

then substitution into eqn (8.8) and integration gives

$$m_2 = m_{2,0}\exp[-M_{s,0}\bar{v}_{s,0}\omega^4 r_0^2\delta^2/2\beta_{\text{eff}}RT]$$

where $m_{2,0}$ is the concentration of DNA at band centre. This is a gaussian distribution of the form

$$m_2 = m_{2,0}\exp\left[-\frac{\delta^2}{2\sigma^2}\right]$$

where the standard deviation is given by

$$\sigma^2 = \frac{RT\beta_{\text{eff}}}{M_{s,0}\bar{v}_{s,0}\omega^4 r_0^2}$$

This earlier analysis omitted the effect of non-ideality which is very large for

DNA in solution (Schmid and Hearst, 1969). So, in order to obtain equations of practical value, the expansion

$$\left(\frac{\partial \mu_2}{\partial m_2}\right)_{\mu_1} = \frac{RT}{m_2}\left[1 + 2B_{m_2} + \ldots\right]$$

must be substituted into eqn (8.8). Further rearrangement and applications in practice led to the most satisfactory form for practical use:

$$\ln M_{app} = \ln M_{s,0} - B\langle m_2 \rangle \tag{8.9}$$

where

$$M_{app} = \frac{RT\beta_{eff}}{\langle \delta^2 \rangle \bar{v}_{s,0} r_0^2 \omega^4}$$

The moments

$$\langle m_2 \rangle = \frac{\displaystyle\int_{-\infty}^{+\infty} m_2^2 \, d\delta}{\displaystyle\int_{-\infty}^{+\infty} m_2 \, d\delta}$$

and

$$\langle \delta^2 \rangle = \frac{\displaystyle\int_{-\infty}^{+\infty} \delta^2 m_2 \, d\delta}{\displaystyle\int_{-\infty}^{+\infty} m_2 \, d\delta}$$

which are derived from numerical integration of the scanner data, are used for accuracy instead of a single measure of the bandwidth of the DNA distribution. Thus, experiments are carried out at a number of DNA concentrations, very accurate scanner data are obtained and extrapolation using eqn (8.9) allows unambiguous evaluation of the molecular weight of the DNA (*Figure 8.11*).

The thermodynamic treatment of Cohen and Eisenberg (1968) for this system sidesteps the evaluation of the interaction parameters and partial volumes required above. However, a knowledge of the variation of the term $(\partial \rho/\partial c_2)_\mu^0$, which accounts for the preferential interactions in the system, with radius, is required, see eqn (8.6). The quantity χ is defined by

$$\chi = \left[\frac{d}{dr}\left(\frac{\partial \rho}{\partial c_2}\right)_\mu^0\right]_{r_0}$$

and values can be extracted from isotope substitution experiments as for effective density-gradient determination. Thus integration of eqn (8.6) gives a gaussian distribution around the band centre where the standard deviation is obtained from

$$\sigma^2 = -\frac{RT}{M_2 \omega^2 r_0 \chi}$$

Non-ideal effects can be corrected for as before and the molecular weight obtained. These independent theoretical approaches of Hearst and co-workers and Cohen and Eisenberg (1968) are in agreement, so that both the physical and practical basis of measuring the molecular weight of homogeneous DNA preparations is well established.

DNA analysis by means of sedimentation equilibrium methods is not restricted to three-component systems and a discussion of four-component systems obtained by adding further chemicals, such as ethidium bromide, to buoyant density gradients is available (Bauer and Vinograd, 1969).

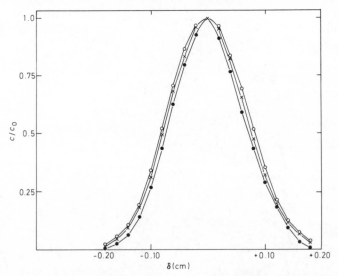

Figure 8.11 Equilibrium concentration distributions of SV40 DNA in buoyant density gradients of CsCl. Different amounts of DNA, 0.5 µg (●—●), 3.0 µg (×—×) and 18.2 µg (○—○), were centrifuged in CsCl, density 1.7 g/cm³, for 72 h at 30 000 rev/min at 22 °C, and scanned at either 265 nm or 290 nm

Sedimentation analysis of DNA in buoyant density gradients often yields data with overlapping bands of DNA and, in many instances, sloping baselines, due to the presence of low-molecular-weight solutes, are present (*Figure 8.12*; Bauer and Vinograd, 1968). Data from the photoelectric scanner may readily be quantified by subtracting the baseline, estimated by appropriate curve-fitting procedures, and the bands of DNA can be resolved using a gaussian, or similar, numerical resolving procedure (Bevington, 1969; Fraser and Suzuki, 1973). Buoyant density-gradient analysis in the analytical ultracentrifuge has been extended to macromolecules other than DNA, although specific practical difficulties, such as those with RNA or protein, can arise (Ifft, 1969; Williams and Vinograd, 1971).

SEDIMENTATION VELOCITY ANALYSIS

Areas of Application

Sedimentation velocity analysis, in common with other flow processes, is an experimental system in which the movement of a boundary or band of solute in

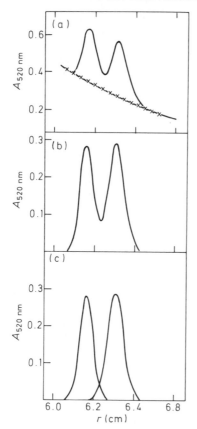

Figure 8.12 Resolution of overlapping bands of DNA at equilibrium in solutions of CsCl containing ethidium bromide. (a) Tracing of scanner output from an experiment in which a mixture of M. luteus DNA and SV40 form II DNA was centrifuged in CsCl, density 1.6 g/cm³, 100 μg/ml ethidium bromide at 44 000 rev/min. The sloping baseline, due to the free dye, has been fitted by least squares to a polynomial. (b) The baseline has been removed from the data leaving the overlapping distributions of the dye–DNA complexes. (c) The overlapping bands have been resolved by fitting two gaussian distributions (Fraser and Suzuki, 1973)

response to some driving force is measured. The differences between sedimentation velocity analysis and, say, gel chromatography are related to the nature of the driving force and the properties of the flow medium. Thus, in sedimentation velocity analysis the driving force is the gradient in total potential (see p. 259) and is subject to pressure effects from the large centrifugal fields. In gel chromatography, the flow depends on solute partitioning in the gel matrix, and on axial dispersion due to diffusion and flow perturbation by the gel beads. These phenomena affect the shapes of the boundaries or zones obtained but the same basic information is given by all such transport systems.

Sedimentation velocity experiments are usually regarded as two-component systems, as the macromolecule is sedimented at an ionic strength high enough to minimize drag by small non-sedimenting counter-ions, thus avoiding this primary salt effect. However, if sedimentation is performed at too high a concentration of salt, then the system will constitute a three-component system and attendant precautions in data analysis must be taken (see pp. 265–270). A clear and rigorous treatment of flow processes in the ultracentrifuge, incorporating the thermodynamics of irreversible processes, has been developed by Williams *et al.* (1958). Changes of concentration with time in a sector-shaped cell are described by the differential equation of the ultracentrifuge, or the Lamm equation (Lamm, 1929). Thus

$$\left(\frac{\partial c}{\partial t}\right)_r = \frac{1}{r}\frac{d}{dr}\left[Dr\left(\frac{\partial c}{\partial r}\right)_t - s\omega^2 r^2 c\right]$$

where D is the diffusion coefficient and s is the sedimentation coefficient of the sedimenting species. The various solutions of this equation which are available have all introduced some simplifying assumptions such as zero concentration dependence of s and D or a rectangular rather than sectorial cell shape (Mason and Weaver, 1924). Integration of the Lamm equation to yield

$$\frac{d}{dt}\int_{r_1}^{r_2} cr dr = \left.Dr\frac{dc}{dr} - s\omega^2 r^2 c\right|_{r_1}^{r_2}$$

allows the concentration change between radial positions r_1 and r_2 to be related to s and D. If two pairs of r_1 and r_2 are chosen then, in principle, the two simultaneous equations may be solved for s and D, and concentration-dependent effects can be included (Bethune, 1970). Alternative solutions to the Lamm equation require complex numerical manipulations (Dishon, Wiss and Yphantis, 1967).

A conceptually much simpler approach yielding far more tractable equations that describe the behaviour of sedimenting particles is the kinetic approach. A particle sedimenting with constant velocity in a centrifugal field will experience a sedimentation force, determined by its buoyant mass and the centrifugal field, opposed by a frictional force, reflecting the frictional coefficient of the molecule and its velocity. Hence,

$$\frac{M_2}{N}(1 - \bar{v}_2\rho)\omega^2 r = f\frac{dr}{dt} \tag{8.10}$$

where f is the molar frictional coefficient of the molecule, dependent on its size, shape and hydration. The sedimentation coefficient, s, is defined by the velocity of the particle per unit centrifugal field; thus,

$$s = \frac{dr/dt}{\omega^2 r} = \frac{d \ln r}{dt}\frac{1}{\omega^2} \tag{8.10a}$$

and, from eqn (8.10),

$$s = \frac{M_2(1 - \bar{v}_2\rho)}{Nf} \tag{8.11}$$

Sedimentation coefficients are usually expressed in terms of Svedberg units, one Svedberg unit being 10^{-13} sec.

The sedimenting molecules tend to diffuse. Diffusion is itself a transport process when the driving force is the gradient in chemical potential. For ideal solutions,

$$D = \frac{RT}{Nf} \qquad (8.12)$$

and the diffusion is also dependent on the molar frictional coefficient. Substituting eqn (8.12) in eqn (8.11), we obtain the Svedberg equation

$$M_2 = \frac{RT_s}{D(1 - \bar{v}_2 \rho)} \qquad (8.13)$$

From eqn (8.13) it can be seen that, if D is known, the particle molecular weight can be obtained. The equilibrium method is more frequently used to determine particle molecular weight and is the more accurate, but sedimentation velocity analysis has been used quite extensively, especially for very large molecules. The diffusion coefficient must usually be measured as well but many empirical equations based on absolute techniques give a direct relationship between sedimentation coefficient and molecular weight. Equation (8.11) shows the other use of sedimentation velocity analysis. The frictional coefficient can be determined if the molecular weight is known and, from this, deductions about molecular shape can be made. The sensitivity of frictional coefficient to shape can lead to conformational resolution of two species of the same molecular weight by sedimentation velocity experiments. A third application of sedimentation velocity analysis is in the field of molecular interactions. Here, resolution frequently depends on both molecular weight and frictional coefficient differences between the associating species.

Measurement of Sedimentation Coefficient

The two methods for measuring the sedimentation coefficient of a macromolecule in the analytical ultracentrifuge are boundary sedimentation and band sedimentation (*Figure 8.13*). In boundary sedimentation the solution contains an even distribution of macromolecules at the start of the experiment. The solute then sediments away from the centre of rotation during the course of the experiment so that the concentration of the macromolecule at the meniscus falls to zero. In the centre of the cell there is a plateau region of solute concentration and at the bottom of the cell the concentration of the macromolecule rises sharply as the sedimenting solute accumulates. In a simple system, the point on the scanner or densitometer trace where the boundary curve is at its steepest, usually at 50% of the plateau concentration, is taken as the representative point on the boundary and is used to determine the distance of the moving boundary from the centre of rotation (*Figure 8.14*). In a mixture of macromolecules which do not interact extensively and which have similar sedimentation coefficients, two boundaries are observed.

The boundary method is experimentally simple to operate and is probably the most accurate method for obtaining the sedimentation coefficient of a purified macromolecule. One main defect is that it requires a large quantity of experimental material compared with the band method. Another problem in boundary

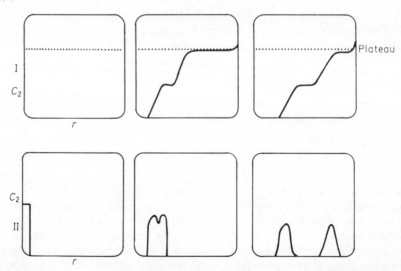

Figure 8.13 Boundary and band sedimentation: the separation of two macromolecules by boundary (I) and band (II) sedimentation analysis. The effect of radial dilution is shown in (I). The apparent concentration drops during centrifugation due to sedimentation in a sector-shaped cell. Diffusion effects lead to boundary and band broadening with time

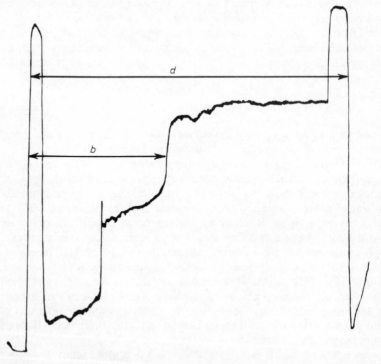

Figure 8.14 A typical densitometer trace showing reference edges and boundaries. The distance from the inner to outer reference edges is 1.6 cm. Hence, $r = 5.7 + (b/d \times 1.6)$. The measurement of b should be from the inner reference edge to the second moment of the boundary

sedimentation occurs when more than one macromolecular species is present. Because of hydrodynamic interactions with the non-sedimenting solvent, sedimentation rate decreases with increase in concentration in most systems. Consequently, in mixed systems, all components except the slowest sedimenting ones are in the presence of other solutes which may affect their sedimentation rate. The slowest sedimenting material travels faster on the meniscus side of the first boundary because the total concentration is lower in that part of the cell, and there is a build-up of material on the meniscal side of the first boundary which masks the true size of that boundary. The estimates of both the sedimentation coefficient and the relative proportions of the sedimenting species may therefore be in error.

The band sedimentation system is a variation of zone sedimentation in which a preformed gradient is used. In band sedimentation, a small amount of material is layered on top of a solvent system which generates its own density gradient during the run, thus stabilizing the solute band. The solvent is typically a concentrated solution of an alkali halide, or, if ionic effects are to be avoided, D_2O. A variety of analytical cells which allow the introduction of the solution layer as centrifugation starts is now available. The main advantages of the band system are (a) very small amounts of material are required, and (b) in mixed solutions the sedimentation of the two solutes is independent. However, the dense solvent medium and the small molecular-weight solution components interact in a complex manner to give an absolute value for the sedimentation coefficient which is not easily corrected for solvent effects (Belli, 1973), and the band method is not advisable for accurate determinations of $s^0_{20,w}$.

Once a method has been selected, the correct cell assembly, speed and solution conditions must be chosen. For photoelectric scanner experiments a double-sector cell is used, whereas for the ultraviolet photographic system a single-sector cell is more suitable. The speed of rotation depends on the expected sedimentation coefficient but for sedimentation coefficients of up to about 20 S it is customary to use a speed of between 40 000 and 60 000 rev/min. The buffer should have a low ultraviolet absorbance for single-sector work and an ionic strength of about 0.1 will minimize salt effects which can be caused by retardation of highly charged macromolecules by low-molecular-weight counter-ions.

The scanner or densitometer traces obtained are used to determine the radial distance of the boundary or band at the selected time intervals. A graph of log r against t (in sec) can be expected to yield a straight line of slope $\omega^2 s$ (*Figure 8.15*). Measurement of t presents no problems but the measurement of r for accurate determinations of s is more complex. In boundary sedimentation, the effect of radial dilution in the sector-shaped cell means that the true maximum gradient of the boundary is not the correct position of measurement of r. The flow equation for movement of the plateau concentration is best satisfied by taking r at the square root of the second moment of the concentration gradient curve, a point slightly further down the cell than the maximum gradient (Goldberg, 1953). In a sharp boundary, the difference is very slight. In band sedimentation in a sector-shaped cell, the mass averaged value of ln r, given by

$$\langle \ln r \rangle = \int_{r_a}^{r_b} \ln r \, dm \bigg/ \int_{r_a}^{r_b} dm \qquad (8.14)$$

where dm is the mass increment of solute in the zone, should be used rather than

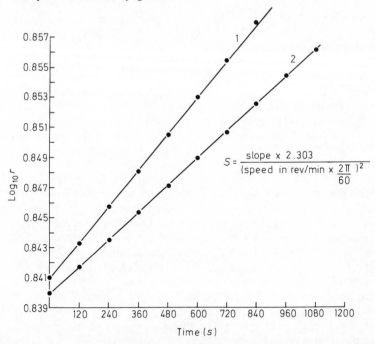

Figure 8.15 Sedimentation coefficient determinations of (1) T7 DNA and (2) SV40 DNA at 39 000 rev/min and 20 °C. Note that at longer time intervals the experimental points tend to lie above the line in the case of T7 DNA, showing the effect of radial dilution in concentration-dependent sedimentation

ln r (Schumaker and Rosenbloom, 1965). However, again the two are very close and in a sharp band the centre of the band where the concentration is maximal is usually selected for the measurement of r.

The value of the sedimentation coefficient obtained from one experiment must be corrected for both solvent and concentration. The standard solvent condition is water at 20 °C, so that

$$s_{20,w} = s \frac{\eta}{\eta_{20,w}} \frac{(1 - \bar{v}\rho_{20,w})}{(1 - \bar{v}\rho)} \tag{8.15}$$

where η and ρ represent the viscosity and density of the solvent at the temperature of the experiment, and $\eta_{20,w}$ and $\rho_{20,w}$ represent the viscosity and density of water at 20 °C.

The use of absorption optical systems minimizes the effect of concentration on sedimentation coefficient as the concentrations used are usually so low that virial effects are not effective. However, the sedimentation coefficient can vary either linearly with respect to concentration as in

$$s = s^0(1 - kc) \tag{8.16}$$

or non-linearly, in which case it is better described by

$$s = \frac{s^0}{1 + kc} \tag{8.17}$$

Concentration dependence of the form given by eqn (8.16) has yielded approximate solutions of the Lamm equation, but with concentration dependence given by eqn (8.17), no exact solution of the Lamm equation has been described (Fujita, 1956). Numerical solutions of the Lamm equation have allowed the two forms of concentration dependence of the sedimentation coefficient in velocity experiments to be examined (Dishon, Wiss and Yphantis, 1967). Concentration dependence of sedimentation coefficients always represents the operation of a retarding force due to hydrodynamic effects. Any increase in s-value with concentration can be attributed to self-association (see pp. 282–283). Consequently, in a concentration-dependent boundary system, the components at lower concentration will be accelerated and the components at higher concentration will be retarded. In the usual circumstance of the boundary-sharpening effect of concentration dependence being exactly balanced by the boundary-spreading effect of diffusion, it is possible to estimate s/D from the data (Creeth, 1964; Dishon, Wiss and Yphantis, 1967). One effect of concentration dependence in the boundary method is that radial dilution, due to the sector-shaped cell, causes an increase in sedimentation rate with time and the graph of $\ln r$ against t is not linear. Extrapolation to obtain the s-value at $t = 0$ then becomes necessary. The effect of radial dilution is usually small unless the concentration dependence is high, and is expressed by the equation

$$C_{\text{plateau}} = C_0 \exp\left(-2s\omega^2 t\right) \tag{8.18}$$

In band centrifugation, strong concentration dependence produces skew bands as the molecules on either side of the band, where the concentration is low,

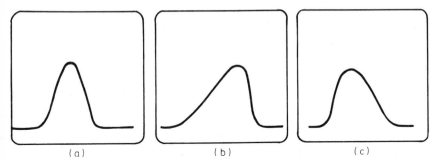

(a)   (b)   (c)

Figure 8.16 The effect of concentration on band shape in band sedimentation analysis: (a) no concentration dependence of s; (b) positive concentration dependence of s. This type of band is characteristic of self-association. The multimers with higher sedimentation coefficient are located at the front of the band and the monomers trail at the back. (c) Negative concentration dependence of s. The molecules accelerate on either side of the band where the concentration is low. Direction of sedimentation, left to right

accelerate (*Figure 8.16*). In the case of large molecules, it is possible to determine the concentration effect in a single experiment from consideration of the shape of the bands (Vinograd *et al.*, 1963).

Use of Sedimentation Coefficients in Molecular-weight Determination

For most macromolecules, molecular weight cannot be directly determined from

s^0 without the measurement of D^0, the diffusion coefficient extrapolated to zero concentration. The Svedberg equation

$$M_2 = \frac{s^0}{D^0} \frac{RT}{(1 - \bar{v}_2 \rho)}$$

is consequently not frequently used in accurate molecular-weight determinations. Several methods for the measurements of diffusion coefficients are available, but few have been considered accurate enough, especially for large DNA molecules and viruses which have very small diffusion coefficients. The analytical ultracentrifuge itself has often been used to determine diffusion coefficients of the smaller macromolecules, as the optical systems can be employed to observe the spreading of a boundary or band with time (Gosting, 1956). The boundary is formed in a synthetic boundary cell (*Figure 8.5a*) or by sedimentation, and spreading is observed during the experiment. The accuracy of such methods is limited, especially with slowly-diffusing solutes, and correction for concentration dependence may be required. Consequently, the determination of molecular weights using sedimentation velocity analysis is less accurate and frequently more time consuming than by sedimentation equilibrium.

However, large DNA molecules and viruses which are less amenable to equilibrium treatment can be studied by sedimentation velocity analysis if the diffusion coefficient is determined by optical mixing spectroscopy. This technique requires the analysis of the spectral distribution of scattered laser light. From the Doppler shift in the Rayleigh beam, the translational and rotational diffusion coefficients of large and small molecules can be obtained with a high degree of accuracy. Extremely large DNA molecules, however, have an anomalous sedimentation coefficient which depends on rotor speed and pressure effects (Rosenbloom and Schumaker, 1967; Schumaker and Zimm, 1973). Together with sedimentation velocity analysis, optical mixing spectroscopy has provided an extension to the range of molecular-weight determinations possible (Dubin *et al.*, 1970).

There are many empirical equations which have been used to relate molecular weight to sedimentation coefficient. *Table 8.1* shows the most recent equations published for the main types of macromolecule. However, it must be remembered

Table 8.1 EMPIRICAL EQUATIONS RELATING SEDIMENTATION COEFFICIENT AND MOLECULAR WEIGHT

Equation	Molecular type	Mol. wt range	Ref.
$s^0_{20,w} = 0.002\,42\ M^{0.67}$	proteins	1.7×10^4 to 4.9×10^7	Halsall (1967)
$s^0_{20,w} = 2.8 + 0.008\,34\ M^{0.479}$	linear-DNA, double-stranded	10^6 to 10^8	Freifelder (1970)
$s^0_{20,w} = 0.052\,8\ M^{0.4}$	alkaline DNA, single-stranded	1.5×10^6 to 7×10^7	Studier (1965)
$s^0_{20,w} = 0.010\,5\ M^{0.549}$	neutral DNA, single-stranded	1.5×10^6 to 7×10^7	Studier (1965)
$s^0 = 2.7 + 0.175\,9\ M^{0.445}$	circular DNA, double-stranded	3×10^6 to 3×10^7	Gray *et al.* (1967)
$s^0 = 7.44 + 0.002\,43\ M^{0.58}$	supercoiled DNA, double-stranded	10^6 to 3×10^7	Hudson *et al.* (1968)

that the sedimentation coefficient of a macromolecule depends both on the size and shape of the particle. When molecular-weight markers are used to construct an empirical equation, the assumption is made that the unknown macromolecule being studied has the same hydrodynamic properties. Thus, in the case of a protein, if the assumption is made that the macromolecule behaves as a hydrated sphere, then the empirical equation will give a molecular weight within the limitations of that assumption. In the case of DNA the conformation of the molecule has a profound effect on the sedimentation coefficient, and different equations apply to each form of DNA. If the conformation of the DNA under study is not known (for example, superhelical, circular, etc.), empirical equations cannot be used. Similarly, if an additional component, such as ethidium bromide, is present in the DNA solution, the binding which occurs changes the effective size and shape of the macromolecule so that the empirical equations cannot be used.

In heterogeneous or polydisperse systems a weight-average sedimentation coefficient is obtained if the boundary measurement is correctly made to the square root of the second moment. In such situations a total analysis of the distribution, $g(s)$ is sometimes made (Schumaker and Schachman, 1957; Halsall, 1967). This type of analysis is particularly suitable for polynucleotides but has also been used for lipoproteins (Oncley, 1969). However, many assumptions are involved in the derivations, the concentration dependence of the components may be complex, and the diffusion coefficient distribution may be hard to determine if a molecular-weight distribution is required.

Sedimentation Velocity Analysis and Molecular Shape

Although the value of s is quite sensitive to molecular shape, it is seldom possible to predict the conformation of a macromolecule from the sedimentation coefficient alone. In general, the more compact a molecule the higher its sedimentation rate, so that a spherical molecule can be expected to sediment faster than a rod-like molecule of the same molecular weight since the latter will experience frictional resistance from the solvent for the statistically larger proportion of the time that it will spend perpendicular to the direction of sedimentation. If the molecular weight is known, the frictional coefficient can be obtained from s by the equation

$$f = \frac{M_2(1 - \bar{v}_2\rho)}{Ns^0} \tag{8.19}$$

The frictional coefficient can be related to the radius, r, of a spherical molecule by Stokes' law

$$f = 6\pi\eta r \tag{8.20}$$

but in the case of asymmetric molecules which have a frictional coefficient f greater than the frictional coefficient f_0 of a sphere of equivalent volume, sedimentation alone gives insufficient information on shape, and this must be combined with data from other hydrodynamic measurements (Scheraga and Mandelkern, 1953). Few molecules are close enough to a spherical shape and, consequently, sedimentation velocity analysis alone is little used in detailed conformational analysis of proteins.

In the case of stiff chains, an attempt can be made to make a direct correlation between sedimentation coefficient and molecular shape using the Kirkwood formulation for polymer chains (Kirkwood, 1954; Hearst and Stockmayer, 1962). This allows the effect of individual frictional elements of the chain on each other to be calculated and summated to give a hypothetical sedimentation coefficient for a variety of conformations. This can be compared to the experimental value. The fundamental equation is

$$s^0 = \frac{M_2(1 - \bar{v}_2\rho)}{f} \left[1 + \left(\frac{a}{2bN}\right) \sum_{i=1}^{N} \sum_{j=1}^{N} (1 - \delta_{ij})(R_{ij})^{-1} \right] \quad (8.21)$$

where f is the frictional coefficient of each element, b is the spacing of elements along the contour, a is the effective Stokes diameter of an element, R_{ij} is the average reciprocal displacement of elements i and j, and δ_{ij} is the Kronecker delta function. The method has been used to investigate various possible DNA conformations but has been little used for other biological polymers (Hearst and Stockmayer, 1962; Gray, 1967).

Although it is difficult to predict absolute conformation from sedimentation velocity experiments alone, they are frequently used to analyse the proportion of various conformations of molecules of the same molecular weight in a mixture. Thus, compact superhelical DNA of molecular weight 3×10^6 sediments at 21 S, the relaxed circular form of the same molecule sediments at 17 S, and the linear form, which is the least compact, sediments at 14.5 S (Vinograd and Lebowitz, 1966). This resolution is very useful in assaying the proportion of molecules of each type during the progress of an enzyme-catalysed interconversion.

Large Molecule–Small Molecule Interactions

There are two totally different situations in which large molecule–small molecule interactions are studied in the analytical ultracentrifuge. In the first case, the interaction is used to study the sedimentation characteristics of very small quantities of enzyme (less than 1 μg/ml) in a band sedimentation experiment. The enzyme is sedimented through a solution containing substrate and the appearance of the product is monitored at a suitable wavelength (Cohen, Giraud and Messiah, 1967; Cohen and Mire, 1971). In this way, the progress of the active form of the enzyme down the cell can be observed. Dehydrogenases are particularly suitable for this type of experiment as the NADH produced can be monitored at 340 nm. Two advantages apart from the obvious economy of enzyme are claimed for this technique. The first is that the hydrodynamic properties of the active enzyme are being observed rather than those of the enzyme protein which may have different activities in different stages of aggregation, and the second that the hydrodynamic properties of a single enzyme in an impure mixture can be determined if one assumes that its sedimentation characteristics are not altered by combination with any other component of the mixture.

The main use of the ultracentrifuge for the study of large molecule–small molecule interactions is in situations in which the binding of the small molecule drastically alters the hydrodynamic properties of the large molecule. A type of experiment which illustrates this situation particularly well is the study of DNA–drug interactions. Supercoiled DNA is circular DNA whose ends have been sealed when the double helix was slightly unwound, so that it takes on the appearance of a twisted circle and sediments at 21 S at a molecular weight of 3×10^6 (see p. 280). If a number of dye molecules intercalate in between the bases of the double helix and unwind it, then twists are progressively removed until the DNA molecule is a 'relaxed' circle. The dye has then unwound the DNA to the extent to which it was originally unwound when the circles were first sealed (Crawford and Waring, 1967; Bauer and Vinograd, 1968). The sedimentation coefficient drops to that of the open circle and, if further dye is bound, rises again as the molecule takes on supertwists in the opposite direction (*Figure 8.17*). This type of experiment

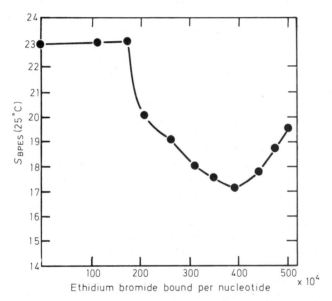

Figure 8.17 Ethidium bromide titration of supercoiled DNA. The minimum of the curve represents the point at which the DNA molecule is unwound to a circular form. The DNA in this experiment is φX 174 replicative form

has been used extensively to classify drugs according to whether or not they intercalate between base-pairs. From the number of drug molecules bound at the minimum sedimentation value the extent of unwinding induced by each drug can be determined (Waring, 1970).

An elegant use of the scanner has been made in some cases where differences in shape are very small between, say, an enzyme and the enzyme substrate complex, so that the conformational change induced by the substrate seems almost impossible to measure. The reference sector has enzyme alone and the sample sector the enzyme–substrate complex. In this way, a difference sedimentation pattern is obtained (Schachman, 1963). Differential sedimentation of two conformations of DNA at a pH difference of 0.05 units has demonstrated changes in shape which would be impossible to detect in separate experiments (Ostrander and Gray, 1973).

Large Molecule–Large Molecule Interactions

The majority of proteins have the potential to self-associate given the appropriate conditions (Klotz, Langerman and Darnall, 1970; Nichol *et al.*, 1964). The phenomenon is of functional interest because of the variable catalytic activity of oligomers of the same protein. Full analysis of associating systems is best achieved by equilibrium sedimentation, but the effect is often first detected in transport experiments and a helpful preliminary analysis can then be made. Association of this type is most easily observed by analysis of the shape of the moving boundary or band or, alternatively, the concentration dependence of the sedimentation coefficient (Rao and Kegeles, 1958; Gilbert and Jenkins, 1959; Gilbert, 1960; Nichol and Ogston, 1965; Cox, 1969; Cann, 1970).

The sedimentation coefficient in non-self-associating systems shows a negative concentration dependence (see pp. 273–277). An increase in sedimentation coefficient with concentration suggests a self-associating system as the weight-average sedimentation coefficient reflects the increased proportion of oligomer at the higher concentration. Frequently, the curve of *s* against concentration shows a maximum as association becomes complete and then decreases as the customary negative concentration dependence of *s* takes over.

The boundary shape and its variation with concentration also give information about self-association, although they give little detailed information about the type of association occurring. The two types of polymerization can be described as follows: discrete, which follows the equation $nM \rightleftharpoons M_n$, and continuous, which follows the equations $M + M \rightleftharpoons M_2 + M \rightleftharpoons M_3 + M \rightleftharpoons M_4 + M \rightleftharpoons M_5 \ldots$. In the latter case, a unimodal boundary is always observed. It may have a skew shape indicative of association but it is possible for extensively associating protein systems to show completely gaussian boundaries (Cox, 1969). In the case of discrete association where *n* is three or greater, two phases are detectable in the boundary and the graph of dc/dr against *r* shows two peaks. The slower component remains constant at all concentrations but the faster moving one increases in amount with concentration (Gilbert, 1963). The best example of this behaviour is to be found in the case of β-lactoglobulin which self-associates in a discrete manner to form tetramers (Timasheff and Townend, 1960).

Other flow processes, such as gel chromatography, provide very similar data to velocity sedimentation. However, such processes are not subject to pressure effects which can lead to convectional disturbances in associating systems (Harrington and Kegeles, 1973; Johnson, Yphantis and Weiss, 1973). The trailing peak in sedimentation is often more widely spread due to the higher diffusion coefficient of the monomer. The dispersion of the trailing peak of solute in gel chromatography is lessened with the smaller molecules, leading to a sharpening of the trailing edge of the boundary (Zimmerman, Cox and Ackers, 1971).

Theoretical equations to analyse the experimental information from systems of the type $A + B \rightleftharpoons AB$ have been developed by Gilbert (1960). In the case where the velocities are in the order $v_c > v_a > v_b$, it is possible to predict a slow-moving boundary corresponding to the unreacted component and a faster moving boundary composed of the other components of the mixture. A series of experiments with varying proportions of reactants *A* and *B* then allow the determination of the kinetics of the association. Experimentally, it has been difficult to find relevant biological systems which correspond to such idealized behaviour, but the reaction of a univalent protein antigen (component B) with antibody corresponds quite closely, being a system of the type $A + 2B \rightleftharpoons AB_2$ and which follows the predicted behaviour (Pepe and Singer, 1959; Singer, Pepe and Ilten, 1959).

Many of the more relevant interactions among dissimilar macromolecules have been difficult to interpret owing to the heterogeneity of the nucleic acid samples available. However, sedimentation velocity analysis has been used most effectively to demonstrate the co-operativity of histone binding to homogeneous SV40 DNA (Barclay and Eason, 1972). In the presence of excess DNA only two species were found to sediment, one representing DNA fully saturated with histone and the other naked DNA. In this type of experiment, as in many involving interactions between nucleic acids and proteins, the association constant is sufficiently great that kinetic analysis is unnecessary and indeed impossible.

CONCLUSIONS

Analytical ultracentrifugation has had a considerable effect on the study of biological macromolecules, in particular proteins and nucleic acids. Theoretical and practical advances have been made in recent years both in the equilibrium and velocity techniques, and data from analytical ultracentrifuge experiments are highly accurate due to the introduction of automatic data-processing equipment and the photoelectric scanner.

Sedimentation equilibrium analysis continues to supplement other absolute methods, such as osmotic pressure and light-scattering methods, which have their basis in equilibrium thermodynamics. The equilibrium method is very powerful in determining molecular weights in all types of solution conditions. In complex solutions where preferential interactions with the macromolecule occur, solute–solvent interactions at dialysis equilibrium are accounted for by determining the appropriate density increments using precision densitometry. The development of three-component theory which applies to such conditions has allowed a clear understanding of the meaning of molecular-weight data obtained from centrifuge experiments. A most important area of progress has been the study of associating systems at chemical and sedimentation equilibrium, and alternative theoretical approaches are now available to confirm the assignment of particular types of model.

Highly accurate procedures for s-value determination and for the detection of very small differences in s-value in similar solutions have been developed in the case of sedimentation velocity analysis. The analysis and simulation of boundary and band shapes in interacting systems has undergone important development, and the dangers of misinterpretation have been emphasized in recent years. Pressure-dependent effects and their influence in associating systems have been defined for velocity analysis. The introduction of boundary analysis in active enzyme–substrate complexes represents a most interesting extension of the velocity technique.

Parallel developments in transport processes, such as gel chromatography and gel electrophoresis, have more than kept pace with developments in analytical ultracentrifugation. Thus, in studying associating systems, for example, gel chromatography has probably replaced velocity analysis due to the absence of complicating pressure effects in the former. However, useful preliminary information continues to be available from the velocity method in this area of research. These developments have greatly expanded the potential for analysing biological macromolecules in solution, and it is clearly desirable to use as many of these as possible in order to define completely the properties of the system under examination.

REFERENCES

ACKERS, G.K. (1970). In *Advances in Protein Chemistry*, vol. 24, p. 343. Eds C.B. Anfinsen, J.T. Edsall and F.M. Richards. London; Academic Press

ADAMS, E.T. (1967). *Fractions No. 3*. Palo Alto; Beckman Instruments Inc.

ADAMS, E.T. (1969). *Ann. N.Y. Acad. Sci.*, **164**, 226

BALDWIN, J.O., BOSELEY, P.G., BRADBURY, E.M. and IBEL, K. (1975). *Nature, Lond.*, **253**, 245

BARCLAY, A.B. and EASON, R. (1972). *Biochim. biophys. Acta.*, **269**, 37

BAUER, J.H. and PICKELS, E.G. (1936). *J. exp. Med.*, **64**, 503

BAUER, W. and VINOGRAD, J. (1968). *J. molec. Biol.*, **33**, 141

BAUER, W. and VINOGRAD, J. (1969). *Ann. N.Y. Acad. Sci.*, **164**, 192

BEAMS, J.W., DIXON, H.M., ROBESON, A. and SNIDOW, N. (1955). *J. phys. Chem.*, **59**, 915

BELLI, M. (1973). *Biopolymers*, **12**, 1853

BETHUNE, J.L. (1970). *Biochemistry*, **9**, 2737

BEVINGTON, P.R. (1969). *Data Reduction and Error Analysis in the Physical Sciences*. New York; McGraw-Hill

BOWEN, T.J. (1970). *An Introduction to Ultracentrifugation*. Aberdeen Univ. Press; Wiley Interscience

BREWER, J.M., PESCE, A.J., SPENCER, T.E. (1974). In *Experimental Techniques in Biochemistry*, p. 161. Eds J.M. Brewer, A.J. Pesce and R.B. Ashworth. New Jersey; Prentice Hall

CAMPBELL, A.M. and JOLLY, D.J. (1973). *Biochem. J.*, **133**, 209

CANN, J.R. (1970). *Interacting Macromolecules*. London; Academic Press

CASSASSA, E.F. (1969). *Ann. N.Y. Acad. Sci.*, **164**, 13

CASSASSA, E.F. and EISENBERG, H. (1961). *J. phys. Chem.*, **65**, 427

CASSASSA, E.F. and EISENBERG, H. (1964). *Adv. Prot. Chem.*, **19**, 287

CHERVENKA, C.H. (1971). *Fractions No. 1*. Palo Alto; Beckman Instruments Inc.

CHUN, P.W., KIM, S.J., WILLIAMS, J.D., COPE, W.J., TANG, L.H. and ADAMS, E.T. (1972). *Biopolymers*, **11**, 197

COATES, J.H. (1970). In *Physical Principles and Technique of Protein Chemistry*, pt B, ch. 10, p. 2. Ed. S.J. Leach. London; Academic Press

COHEN, G. and EISENBERG, H. (1968). *Biopolymers*, **6**, 1077

COHEN, R., GIRAUD, B. and MESSIAH, A. (1967). *Biopolymers*, **5**, 203

COHEN, R. and MIRE, M. (1971). *Eur. J. Biochem.*, **23**, 267

COX, D.J. (1969). *Archs Biochem. Biophys.*, **129**, 106

CRAWFORD, L.V. and WARING, M.J. (1967). *J. molec. Biol.*, **25**, 23

CREETH, J.M. (1964). *Proc. R. Soc.*, **A282**, 403

CREETH, J.M. and PAIN, R.H. (1967). In *Progress Biophysics molec. Biol.*, **17**, 217

DISHON, M., WISS, G.H. and YPHANTIS, D.A. (1967). *Biopolymers*, **5**, 697

DUBIN, S.B., BENEDEK, G.B., BANCROFT, F.C. and FRIEFELDER, D. (1970). *J. molec. Biol.*, **54**, 547

DYSON, R.D. and ISENBERG, I. (1971). *Biochemistry*, **10**, 3233

EDELSTEIN, S.J. and SCHACHMAN, H.K. (1967). *J. biol. Chem.*, **242**, 306

EISENBERG, H. (1962). *J. phys. Chem.*, **36**, 1837

FISH, W.W., REYNOLDS, J.A. and TANFORD, C. (1970). *J. biol. Chem.*, **245**, 5166

FRASER, R.D.B. and SUZUKI, E. (1973). In *Physical Principles and Techniques of Protein Chemistry*, pt C, p. 301. Ed. S.J. Leach. London; Academic Press

FREIFELDER, D. (1970). *J. molec. Biol.*, **54**, 567

FUJITA, H. (1956). *J. chem. Phys.*, **24**, 1084

GILBERT, G.A. (1960). *Nature, Lond.*, **186**, 882

GILBERT, G.A. (1963). *Proc. R. Soc.*, **A276**, 354

GILBERT, G.A. and JENKINS, R.C. (1959). *Proc. R. Soc.*, **A253**, 420

GOLDBERG, R.J. (1953). *J. phys. Chem.*, **57**, 194

GOSTING, L.J. (1956). *Adv. Prot. Chem.*, **11**, 429

GRAY, H.B. (1967). *Biopolymers*, **5**, 1009

GRAY, H.B., BLOOMFIELD, V.A. and HEARST, J.E. (1967). *J. chem. Phys.*, **46**, 1493

HALSALL, H.B. (1967). *Nature, Lond.*, **215**, 880

HARRINGTON, W.F. and KEGELES, G. (1973). In *Methods in Enzymology*, vol. XXVII, pt D, ch. 13, p. 306. Eds C.H.W. Hirs and S.N. Timasheff. London; Academic Press

HASCHEMEYER, R.H. and BOWERS, W.F. (1970). *Biochemistry*, **9**, 435

HEARST, J.E. and STOCKMAYER, W. (1962). *J. chem. Phys.*, **37**, 1425

HEARST, J.E. and VINOGRAD, J. (1961a). *Proc. natn. Acad. Sci., U.S.A.*, **47**, 999

HEARST, J.E. and VINOGRAD, J. (1961b). *Proc. natn. Acad. Sci., U.S.A.*, **47**, 1005

HUDSON, B., CLAYTON, R.A. and VINOGRAD, J. (1968). *Cold Spring Harb. Symp. quant. Biol.*, **33**, 435

HUSTON, J.D., FISH, W.W., MANN, K.G. and TANFORD, C. (1972). *Biochemistry*, **11**, 1609

IFFT, J.B. (1969). In *Laboratory Manual of Analytical Methods of Protein Chemistry*, vol. 5, p. 151. Eds P. Alexander and H.P. Lindgren. New York; Pergamon

IFFT, J.B., MARTIN, W.R. and KINZIE, K. (1970). *Biopolymers*, **9**, 597

IFFT, J.B., VOET, D.H. and VINOGRAD, J. (1961). *J. phys. Chem.*, **65**, 1138

JOHNSON, M., YPHANTIS, D.A. and WEISS, G.H. (1973). *Biopolymers*, **12**, 2477

KIRBY-HADE, E.P. and TANFORD, C. (1967). *J. Am. chem. Soc.*, **89**, 5034

KIRKWOOD, J. (1954). *J. Polym. Sci.*, **12**, 1

KLOTZ, I.M., LANGERMAN, N.R. and DARNALL, D.W. (1970). *A. Rev. Biochem.*, **39**, 25

KUPKE, D.W. (1973). In *Physical Principles and Techniques of Protein Chemistry*, pt C, p. 1. Ed. S.J. Leach. London; Academic Press

LAMM, O. (1929). *Ark. Nat. Astron. Fys.*, **21B**, No. 2

LEE, J.C. and TIMASHEFF, S.N. (1974). *Biochemistry*, **13**, 257

MASON, M. and WEAVER, W. (1924). *Phys. Rev.*, **23**, 412

MESELSON, M., STAHL, F.W. and VINOGRAD, J. (1957). *Proc. natn. Acad. Sci., U.S.A.*, **43**, 581

MUNK, P. and COX, D.J. (1972). *Biochemistry*, **11**, 687

NICHOL, L.W. and OGSTON, A.G. (1965). *Proc. R. Soc.*, **B163**, 343

NICHOL, L.W., BETHUNE, J.L., KEGELES, G. and HESS, E.L. (1964). In *The Proteins*, ch. 9, p. 305. Ed. H. Neurath. London; Academic Press

NOELKEN, M.E. and TIMASHEFF, S.N. (1967). *J. biol. Chem.*, **242**, 5080
ONCLEY, J.L. (1969). *Biopolymers*, **7**, 119
OSTRANDER, D.A. and GRAY, H.B. (1973). *Biopolymers*, **12**, 1387
PEPE, F.A. and SINGER, S.J. (1959). *J. Am. chem. Soc.*, **81**, 3878
RAO, M.S.N. and KEGELES, G. (1958). *J. Am. chem. Soc.*, **80**, 5724
REISLER, E. and EISENBERG, H. (1969). *Biochemistry*, **8**, 4572
RICHARDS, E.G., TELLER, D.C. and SCHACHMAN, H.K. (1968). *Biochemistry*, **7**, 1054
ROARK, D.E. and YPHANTIS, D.A. (1969). *Ann. N.Y. Acad. Sci.*, **164**, 245
ROSENBLOOM, J. and SCHUMAKER, V. (1967). *Biochemistry*, **6**, 276
ROWE, H.J. and ROWE, A.J. (1970). *Biochim. biophys. Acta*, **222**, 647
SAVITZKY, A. and GOLAY, M.J.E. (1964). *Analyt. Chem.*, **36**, 1629
SCHACHMAN, H.K. (1959). *Ultracentrifugation in Biochemistry*, p. 32. London; Academic Press
SCHACHMAN, H.K. (1963). *Biochemistry*, **2**, 887
SCHACHMAN, H.K. and EDELSTEIN, S.J. (1966). *Biochemistry*, **5**, 2681
SCHERAGA, H.A. and MANDELKERN, L. (1953). *J. Am. chem. Soc.*, **75**, 179
SCHMID, C.W. and HEARST, J.E. (1969). *J. molec. Biol.*, **44**, 143
SCHMID, C.W. and HEARST, J.E. (1971). *Biopolymers*, **10**, 1901
SCHUMAKER, V.N. and ROSENBLOOM, J. (1965). *Biochemistry*, **4**, 1005
SCHUMAKER, V.N. and SCHACHMAN, H.K. (1957). *Biochim. biophys. Acta*, **23**, 628
SCHUMAKER, V.N. and ZIMM, B. (1973). *Biopolymers*, **12**, 869
SINGER, S.J., PEPE, F.A. and ILTEN, D. (1959). *J. Am. chem. Soc.*, **81**, 3887
STABINGER, H., LEOPOLD, H. and KRATKY, O. (1967). *Mh. Chem.*, **98**, 436
STEINBERG, I.Z. and SCHACHMAN, H.K. (1966). *Biochemistry*, **5**, 3728
STUDIER, F.W. (1965). *J. molec. Biol.*, **11**, 373
SVEDBERG, T. and CHIRNOAGA, E. (1928). *J. Am. chem. Soc.*, **50**, 1399
SVEDBERG, T. and FAHRAEUS, R. (1926). *J. Am. chem. Soc.*, **48**, 430
SVEDBERG, T. and PEDERSON, K.O. (1940). *The Ultracentrifuge*. Oxford; Oxford University Press
SVEDBERG, T. and RINDE, H. (1924). *J. Am. chem. Soc.*, **46**, 2677
TANFORD, C., KAWAHARA, K. and LAPANJE, S. (1967). *J. Am. chem. Soc.*, **89**, 729
TANFORD, C., NOZAKI, Y., REYNOLDS, J.A. and MAKINO, S. (1974). *Biochemistry*, **13**, 2369
TANG, L.H. and ADAMS, E.T. (1973). *Archs biochem. Biophys.*, **157**, 520
TELLER, D.C., HORBETT, T.A., RICHARDS, E.G. and SCHACHMAN, H.K. (1969). *Ann. N.Y. Acad. Sci.*, **164**, 66
TIMASHEFF, S.N. and TOWNEND, R. (1960). *J. Am. chem. Soc.*, **82**, 3157
TUNG, M.S. and STEINER, R.F. (1974). *Eur. J. Biochem.*, **44**, 49
UPHOLT, W.B., GRAY, H.B. and VINOGRAD, J. (1971). *J. molec. Biol.*, **62**, 21
VAN HOLDE, K.E. (1967). *Fractions No. 1*. Palo Alto; Beckman Instruments Inc.
VAN HOLDE, K.E. and BALDWIN, R.L. (1958). *J. phys. Chem.*, **62**, 734
VINOGRAD, J. and LEBOWITZ, J. (1966). *J. gen. Physiol.*, **49**, 103
VINOGRAD, J., RADLOFF, R. and BRUNER, R. (1965). *Biopolymers*, **3**, 481
VINOGRAD, J., BRUNER, R., KENT, R. and WEIGLE, J. (1963). *Proc. natn. Acad. Sci., U.S.A.*, **49**, 902
VISSER, J., DEONIER, R.C., ADAMS, E.T. and WILLIAMS, J.W. (1972). *Biochemistry*, **11**, 2634

WARING, M.J. (1970). *J. molec. Biol.,* **55,** 247

WEBER, K., PRINGLE, J.R. and OSBERN, M. (1972). In *Methods in Enzymology,* vol. XXVI, pt C, p. 3. Eds C.H.W. Hirs and S.N. Timasheff. London; Academic Press

WILLIAMS, A.E. and VINOGRAD, J. (1971). *Biochim. biophys. Acta,* **228,** 423

WILLIAMS, J.W. (1972). *Ultracentrifugation of Macromolecules.* London; Academic Press

WILLIAMS, J.W., VAN HOLDE, K.E., BALDWIN, R.L. and FUJITA, H. (1958). *Chem. Rev.,* **58,** 715

WINZOR, D.J. (1970). In *Physical Principles and Techniques of Protein Chemistry,* pt A, ch. 9, p. 451. Ed. S.J. Leach. London; Academic Press

YPHANTIS, D.A. (1964). *Biochemistry,* **3,** 297

ZIMMERMAN, J.K., COX, D.J. and ACKERS, G.K. (1971). *J. biol. Chem.,* **246,** 4242

9 Characteristics of Ultracentrifuge Rotors and Tubes

J. MOLLOY
MSE Scientific Instruments Ltd, Crawley, Sussex
D. RICKWOOD
Department of Biology, The University of Essex, Colchester

The rotors and tubes of preparative ultracentrifuges stand at the point of contact between the user, the instrument and the sample. Obviously, it is important that users are aware of the capabilities and limitations of these instruments since such factors determine the procedures to be used for particular separations and enable one to predict the likely result of the experiment. In the case of rotors speed, capacity and geometry are the main variable parameters, whereas tubes may vary in their ability to withstand centrifugal forces and their resistance to solvents.

This chapter discusses the general and, more specifically, the practical aspects of the design, use and care of rotors, tubes and tube caps. The Appendix presents a detailed analysis of the physical characteristics of rotors made by the main centrifuge manufacturers and should help the reader to judge which one might best suit his requirements.

DESIGNS AND TYPES OF ROTOR

Materials

Almost all of the rotors currently available are manufactured from either aluminium or titanium alloys by a process involving forging followed by annealing and machining. The rotors are then given a durable finish either by anodizing in the case of aluminium rotors or by applying black epoxy paint. The external finish of the rotor is particularly important for the even transfer of heat from the bowl and, since most systems employ infrared detectors, a uniform surface is also necessary for good temperature control.

Originally, only aluminium rotors were available for ultracentrifuges and indeed some of the older models will only accommodate these rotors. The composition of the aluminium alloys used for rotors varies not only among different manufacturers but also among different rotors, depending on the speed and capacity of each. The chief disadvantage of aluminium rotors is the reactive nature of the alloy. Normally, the surface is protected by a film of oxide; however, this film is very brittle and tends to craze during centrifugation when the rotor stretches. It is also destroyed by salt solutions (see p. 302), the metal being rapidly corroded and thus weakening the rotor. Unfortunately, the most usual area of corrosion is the most difficult to inspect, namely, the bottom of the rotor pockets. Even in a short time rotors can be extensively corroded and hence rendered unsafe to use. Aluminium rotors also exhibit both metal fatigue and stress corrosion. The latter effect is particularly serious because the centrifugal force not only strains the crystalline structure of the alloy but it also forces water into the grain boundaries. Once formed, the rate at which cracks propagate increases almost exponentially. Aluminium rotors can be autoclaved, although they should not be maintained at temperatures in excess of 120 °C. However, rotors manufactured from two or more metals whose coefficients of expansion differ significantly should not be heated.

Rotors manufactured from titanium alloys are much heavier, but also much more durable than their aluminium counterparts and, while they do exhibit metal fatigue, they are particularly resistant to stress corrosion. In addition to their resistance to corrosion by salt solutions, the high strength of these alloys means that they can be centrifuged at much higher speeds. For example, whereas small aluminium rotors have a top speed of 60 000 rev/min, an equivalent titanium rotor can be used at speeds up to 80 000 rev/min. Titanium rotors are less subject to fatigue than aluminium rotors.

Other types of material have been investigated and indeed Beckman's elutriator rotor (JE6) is made of plastic, although its speed is restricted to 6000 rev/ min. The possibility of using composite compounds, for example glass fibre, carbon fibre, etc., has also been investigated but the results so far have not been encouraging. The chief problem is that the strength of these materials is extremely dependent on the orientation of the fibres, and with present technology it is not possible to get the correct alignment of the fibres homogeneously throughout the rotor. If this problem is solved then it may be possible to make light-weight rotors capable of very high speeds indeed.

Types of Rotor

Centrifuge rotors can be broadly divided into three classes: (*a*) fixed-angle (including vertical rotors), (*b*) swing-out, and (*c*) zonal.

Fixed-angle Rotors

These rotors are very widely used and there is a wide diversity of them (see Appendix). The tube size of such rotors can vary from 0.2 ml to 500 ml and they are generally used for fractionating material by differential pelleting (see Chapter 1). Fixed-angle rotors are the easiest to design and it is possible to manufacture rotors of this type which are capable of being used up to 80 000 rev/min. The design of these rotors was originally fairly empirical in that the number and volume of tubes was first chosen and then, on a basis of trial and error, a lump of metal of the correct size was obtained which could reach the desired speed without disintegrating. More recently, it has been possible to optimize the shape of rotors by carrying out stress analysis of the rotor shape using a computer. The basis of this is that every gram of metal that does not directly strengthen the rotor should be removed, because during high-speed centrifugation the presence of the excess metal actually weakens the rotor. Indeed, it can be calculated that most of the rotor strength is required not to retain the sample tube but rather to hold the rotor together. Thus, now it is possible to determine which parts of the originally round rotor can be machined away to give maximum strength and this explains the angular appearance of the more recently designed fixed-angle rotors.

The angle of the sample tube can vary from 14 to 40 degrees away from the vertical (see Appendix). As a general rule, the wider the angle the more compact is the pellet obtained. On the other hand, particles pellet more rapidly in narrow-angle rotors because of the shorter column of liquid through which they have to migrate; moreover, narrow-angle rotors give particularly good resolution of material which is banded isopycnically in self-forming gradients (see Chapter 6). Centrifuge manufacturers have now capitalized on this observation and introduced vertical tube rotors (zero-angle rotors) in which, as the name suggests, the tubes are vertical in the rotor. The manufacturers claim that such rotors should prove especially useful for both rate-zonal and isopycnic separations. However, they would seem to be much less suitable for separations involving differential pelleting and it is likely that the resolution obtained in rate-zonal separations will be adversely affected by wall effects and the short path length of the particles.

Swing-out Rotors

'Swing-out' or 'swinging-bucket' rotors are much more difficult to design than fixed-angle rotors and the high-performance rotors, for example the MSE 6 × 4.2 or Beckman SW60 Ti types, represent the absolute limits of present-day stress engineering technology. These rotors can be used for pelleting material, particularly if it is in a small volume (say, 5 ml or less), although they are less efficient than fixed-angle rotors. The main use of these rotors is for separating material by density-gradient centrifugation (see Chapters 3 and 5).

There is some variation in rotor design in the way that the individual buckets are supported. The original design (*Figure 9.1a*) uses a hinge pin threaded into the rotor and passing through two eyes on the bucket. When the rotor spins the buckets move out to a horizontal position where they are supported by the central portion or yoke of the rotor. From the user's point of view this design is undoubtedly the safest in that it is not possible to dislodge the buckets accidentally and loaded rotors can be carried from one place to another. However, its major drawback is that the geometry of the design dictates that there is only room to attach three buckets.

In an attempt to overcome this limitation two other designs have been introduced. The first of these is the hook-on bucket system (*Figure 9.1b*) which is used for six-place rotors and which exists in two forms, depending on whether the hook is on the bucket or on the yoke of the rotor. There is considerable variation in designs and in some versions the buckets are more secure than others; the main problem is that frequently the hook is too shallow to ensure the security of the bucket. The general rules that should be followed are, (*a*) avoid carrying the rotor with its buckets attached, and (*b*) be extremely careful not to jar the rotor when putting it into the ultracentrifuge. The other design that has been used for six-place rotors is the ball-and-socket arrangement (*Figure 9.1c*) which

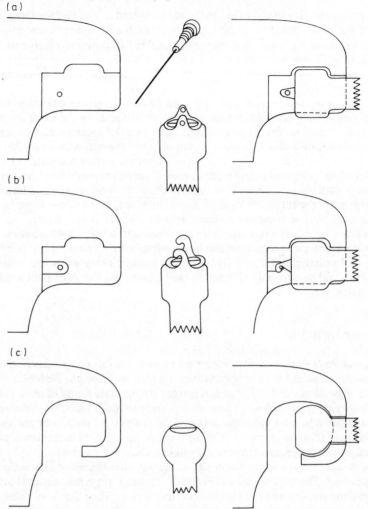

Figure 9.1 The design of swing-out rotors: (a) hinge pin; (b) hook-on bucket; (c) ball and socket

has been used for a number of different rotors. Again, the security of the buckets depends on the design; the sockets of some rotors are so shallow that the buckets are easily dislodged. As with the hook-on bucket design, in general it is best to attach the buckets on to the rotor just prior to loading the rotor into the ultra-centrifuge. In addition, it is very important to keep the surfaces of the ball and socket clean and polished. IEC also recommend lubricating the two surfaces to ensure a smooth transition from the vertical to horizontal position.

The real advantage of the six-place rotors is the flexibility in the number of samples that can be run at any one time, that is, 2, 3, 4 or 6 samples, although of course all six buckets *must* be present during the run. One significant cost-saving advance by Beckman has been the introduction of more than one set of buckets for a single rotor. The SW27 rotor can be used as a large-volume swing-out rotor (6 × 38.5 ml) or as an analytical long-bucket rotor (6 × 17 ml).

Zonal Rotors

These rotors were developed at the Oak Ridge National Laboratory by a team led by Dr N.G. Anderson. Zonal rotors are designed to fractionate large amounts of material using rate-zonal or isopycnic techniques (in some cases, they are also used for differential pelleting), whereas continuous-flow zonal rotors can be used to process tens of litres of liquid. The designs of commercially available rotors are based on those developed by Anderson and consist of essentially four different types, designated series A, B, J and K. A detailed description of the design and applications of zonal rotors is given in Chapter 4.

Stability of Rotors

One of the main considerations in both designing and using rotors is stability during centrifugation. At low speed even perfectly balanced centrifuge rotors precess, that is the centre of the rotor is not stationary but rather it describes a circular motion. Precession, which is also termed 'synchronous whirl', occurs with all rotors, although the speed at which this occurs and the magnitude of the effect depends upon the design of the rotor. The degree of precession can be minimized by designing the drive assembly so that the movement of the drive shaft is damped by the bearing housing. An alternative, generally less satisfactory, solution is the incorporation of a stabilizer at the top of the rotor, as used in the Beckman centrifuges prior to the introduction of the L5 model. However, whereas the amplitude of precession may be quite large, there is essentially no evidence to suggest that it spoils separations or affects the resolution of particles.

A more serious problem is that of asynchronous whirl which occurs at higher speeds and which can affect the resolution of rotors. Asynchronous whirl is caused by mechanical imperfections or imbalance of the rotor and drive assembly and such defects are usually detected when the centrifuges and rotors are tested at the factory. However, instability of a rotor resulting in asynchronous whirl can, and does, occur for a number of other reasons. The most likely cause is incorrect balance of the sample tubes. The magnitude of the effect of a small imbalance depends on the weight, design and rotational speed of the rotor; at high speeds it can be very large indeed. Correct balancing is particularly important when centrifuging density gradients, especially when using swing-out rotors; in this case the sample tubes should be balanced to within 0.1 g.

Correct capping of tubes is of paramount importance also, particularly in swing-out rotors. High-speed rotors are spun in a high vacuum so that a small leakage can cause a tube to lose several millilitres of liquid in just a few hours. If the leakage is severe the rotor can become so unstable that the drive-shaft is bent; the rotor may even be thrown off the spindle. It must also be remembered that not only should the tubes be balanced so far as the total weight is concerned but also balanced with respect to their centre of gravity. This is particularly important in density-gradient centrifugation; in such experiments the tubes are only correctly balanced if they contain identical gradients. Finally, it must be stressed that the empty buckets and caps together form an integral part of high-speed swing-out rotors so far as balance is concerned. Consequently, each bucket must be hung at its correct position and in the correct orientation on the yoke of the rotor and, moreover, the tubes and their contents must be balanced by themselves, *not* in the buckets.

A second common cause of asynchronous whirl is running a rotor in a centrifuge with a bent drive-shaft. As stated previously, unbalanced rotors can bend the drive-shaft, although more usually the damage occurs when rotors are removed from the spindle without due care. Rotors should always be lifted off using the correct extractor key when one is provided, particularly when the rotors are stuck on the spindle. Rotors should always be removed by applying a vertical lifting force and *never* by wrenching or twisting the rotor at an angle to the spindle. If it is suspected that a spindle is damaged it can be checked by placing a rotor in position and spinning it slowly by hand; if the spindle is bent the centre of the rotor will describe a small but discernible circle. Once the spindle is bent it must be replaced before the centrifuge is used again. A bent spindle not only ruins the resolution normally obtainable, but also causes serious (and expensive) damage to the bearing housing and its vacuum seal and the gearbox.

Finally, it should be noted that any object which touches a spinning rotor can also cause asynchronous whirl; consequently, any attempt to restrain the low-speed synchronous whirl movements of a rotor can actually do more harm than good so far as rotor stabilization is concerned. Also, it is clear that any contact between a spinning rotor and, for example, a temperature sensor will not only damage the sensor but also destabilize the rotor.

CENTRIFUGE TUBES AND CAPS

Tubes

Glass centrifuge tubes, even when made from borosilicate or Corex glass, are unable to withstand the stress of high-speed ultracentrifugation, and so a number of other materials have been developed for centrifuge tubes. However, even the most durable of these only has a finite lifetime. The properties of the materials commonly used for centrifuge tubes can be summarized as follows:

Special glasses (Corex, Pyrex, Kimax, etc.). These materials have the great advantage that they are transparent and easy to clean and sterilize. They are much stronger and more resistant to alkalis than conventional glass. Their durability makes it possible to re-use them many times. Glass is generally regarded as a fragile material, so it is notable that Corex tubes, for example, can be used up to 40 000g in the correct adaptors.

Polycarbonate. This is a transparent material, one of the strongest plastics currently in use and it is autoclavable. It is less chemically inert than polypropylene, polyethylene or polyallomer but it is resistant to dilute aqueous acids and neutral salt solutions. It is slowly attacked by dilute solutions of alkalis and diethylpyrocarbonate (Baycovin), the latter of which has been used as a sterilization agent and nuclease inhibitor. The solvent in certain rotor polishes also dissolves polycarbonate tubes and so it is important to remove excess polish from the bottom of rotor pockets. Some reagents are listed 'satisfactory' in *Table 9.1*; however, if they have a pH greater than 7, the tubes may crack during centrifugation.

Polypropylene. This material has a wide chemical resistance; it is autoclavable and is more transparent than polyethylene. It also offers higher temperature limits than polyethylene, but is less satisfactory for work at low temperatures. Although polypropylene is affected by organic solvents and oxidizing agents, an unsatisfactory rating in the chemical resistance chart (*Table 9.1*) for either poly-ethylene or polypropylene may not disqualify these plastics for short-term centrifugation. Especially at ambient temperatures or lower, both plastics have excellent short-term chemical resistance (particularly to solvents) and would be affected only after long-term exposure.

Polyethylene. A milky white translucent material, low-density polyethylene is *not* autoclavable but high-density polyethylene is. Polyethylene is useful for centrifuging at low temperatures and has a wide chemical resistance; it may be used with acids, alkalis, salt and aqueous solutions. For applications involving organic acids, halogens, strong oxidizing agents and aromatic, aliphatic and chlorinated hydrocarbons, polyethylene tubes and bottles should be tested before general use.

Polyallomer. A copolymer of propylene and ethylene monomers, polyallomer exhibits many of the desirable properties of both high-density polyethylene and crystalline polypropylene. It is autoclavable at 120 °C for 30 min. The chemical resistance characteristics are similar to polypropylene but it is more transparent than either polypropylene or polyethylene.

Cellulose acetate butyrate. A strong, transparent thermoplastic material. Tubes made of it are ideal for gradient work in that this plastic is water-wettable and the tubes can be easily sliced and pierced. Tubes can be used satisfactorily with dilute salt solutions, weak acids and very weak bases, as well as aromatic and chlorinated hydrocarbons. In general, concentrated solutions of acids and bases attack cellulose acetate butyrate, as will most solvents such as alcohols, esters, aldehydes, ethers and ketones. Solvent materials attack most aggressively. Because of its chemical structure, it cannot be used near heat or autoclaved.

Cellulose nitrate. This has similar properties to cellulose acetate butyrate but it is both highly inflammable and explosive! The tubes age on storage and should not be kept for longer than one year. A wide variety of tubes, with either thick or thin walls, are available for Beckman rotors.

Nylon. Translucent and autoclavable but, on the other hand, it does absorb large amounts of water. Nylon is substantially inert to hydrocarbons, ketones, alcohols, organic acids [with the exception of formic and carbolic (phenol) acids, meta-cresol, cresylic acid and xylenol], oxidizing agents and mineral acids. Nylon tubes will withstand high speeds without failure.

Kynar. A high-molecular-weight homopolymer of vinylidene fluoride. It is suitable for use with halogens and strong oxidizing agents. It is very inert to a

Table 9.1 CHEMICAL RESISTANCE CHART FOR TUBES AND ZONAL ROTORS

S, satisfactory; M, marginal, test before using; U, unsatisfactory, not recommended; –, not tested.

Reagent	Polycarbonate	Polypropylene	Polyethylene	Polyallomer	Cell. nitrate	Cell. acetate butyrate	Nylon	Kynar	Noryl	S. steel
Acetaldehyde (100%)	U	–	–	–	U	U	–	–	–	–
Acetic acid (5%)	S	S	S	S	S	S	S	S	S	U
Acetic acid (60%)	U	S	U	–	U	U	S	S	S	U
Acetic acid (glacial)	U	U	U	S	U	U	–	S	–	U
Acetone	U	M	S	M	U	U	S	M	–	–
Allyl alcohol	S	S	S	S	–	U	S	–	–	–
Alum, concentrated	–	S	S	S	–	–	S	–	–	–
Aluminium chloride	S	S	S	S	S	S	S	S	–	U
Aluminium fluoride	U	S	S	S	–	–	S	S	–	–
Ammonium acetate	S	S	S	S	–	–	–	–	–	–
Ammonium carbonate	U	S	S	S	S	S	S	S	–	–
Ammonium hydroxide (10%)	U	S	S	S	U	U	S	S	–	–
Ammonium hydroxide (conc.)	U	S	S	S	U	U	S	–	–	–
Ammonium persulphate (sat'd)	–	S	–	S	–	–	–	–	–	–
Ammonium sulphide	U	S	–	S	–	–	S	S	–	–
Amyl alcohol	S	M	S	M	U	U	S	S	S	–
Aniline	–	U	S	U	–	–	–	S	–	–
Aqua regia	U	U	–	U	U	U	–	S	–	U
Benzene	U	M	U	M	S	S	S	S	–	–
Benzyl alcohol	U	U	–	U	S	U	S	S	–	–
Boric acid	S	S	S	S	S	S	S	S	S	–
N-Butyl alcohol	S	S	M	S	U	U	S	S	S	–
Calcium chloride	M	S	S	S	S	S	S	S	S	–
Calcium hypochlorite	S	S	S	S	–	–	S	S	–	U
Carbon tetrachloride	U	U	U	U	S	S	S	S	–	U
Cetyl alcohol	–	–	–	–	–	U	S	–	–	–
Chlorine water	S	S	M	S	–	S	–	–	–	–
Chloroacetic	–	–	–	–	–	–	–	–	S	U
Chlorobenzene	U	U	U	U	U	U	–	S	–	–
Chloroform	U	M	U	M	S	M	S	S	–	–
Chromic acid (10%)	S	S	S	S	U	U	–	S	S	–
Chromic acid (50%)	U	S	S	S	S	U	–	S	–	U
Citric acid (10%)	S	S	S	S	–	S	S	S	S	–
Cresol	U	S	S	S	–	–	U	S	–	–
Cyclohexyl alcohol	M	S	S	S	–	U	S	–	–	–
Diacetone	–	S	S	S	–	U	–	–	–	–
Diazo salts	–	S	S	S	–	–	–	–	–	–
Diethyl ketone	U	M	–	U	U	U	S	–	–	–
Dimethylformamide	U	S	S	S	–	–	–	–	–	–
Diethylpyrocarbonate	U	S	S	S	–	–	–	–	U	–
Dioxane	U	M	M	M	–	U	–	S	–	–
Ether diethyl	U	M	M	M	U	U	–	–	–	–
Ethyl acetate	U	S	S	M	U	U	S	S	–	–
Ethyl alcohol (50%)	U	S	S	S	S	S	S	S	S	–
Ethyl alcohol (95%)	U	S	S	S	U	U	S	S	S	–
Ethylene dichloride	U	U	U	M	U	U	S	–	S	–
Ethylene glycol	S	S	S	S	S	S	S	S	–	–
Ferric chloride	–	S	S	S	–	–	S	S	S	U
Fluoboric acid	–	S	S	S	–	–	–	–	–	–
Formaldehyde (40%)	S	S	S	S	S	–	S	S	S	–
Formic acid (100%)	M	S	S	S	–	U	U	S	S	–

Table 9.1 *continued*

Reagent	Polycarbonate	Polypropylene	Polyethylene	Polyallomer	Cell. nitrate	Cell. acetate butyrate	Nylon	Kynar	Noryl	S. steel
Gallic acid	—	S	S	S	—	—	S	S	—	—
Glycerol	—	S	S	S	S	—	—	S	S	—
2-Heptyl	—	S	S	S	—	U	—	—	—	—
Hydrochloric acid (10%)	S	S	S	S	S	S	S	S	S	U
Hydrochloric acid (50%)	S	M	S	M	U	U	—	S	S	U
Hydrochloric acid (conc.)	M	—	—	—	U	U	—	S	S	U
Hydrofluoric acid (10%)	M	S	S	S	M	M	S	S	—	U
Hydrofluoric acid (100%)	U	S	S	S	U	U	—	S	—	U
Hydroformic acid (100%)	—	S	S	S	—	—	—	—	—	U
Hydrogen peroxide (3%)	S	S	S	S	S	S	S	S	S	—
Hydrogen peroxide (100%)	S	S	S	S	S	S	—	—	S	—
Isobutyl alcohol	—	S	S	S	—	U	S	—	—	—
Isopropyl alcohol	U	S	S	S	U	U	S	—	S	—
Lactic acid (20%)	S	S	M	S	—	—	—	—	S	U
Lactic acid (100%)	—	—	—	—	—	—	—	—	S	U
Lauryl alcohol	—	S	S	S	—	U	S	—	—	—
Lead acetate	—	S	S	S	—	S	—	S	S	—
Linseed oil	S	S	S	S	S	S	—	S	—	—
Magnesium hydroxide	U	S	S	S	—	U	—	S	S	—
Maleic acid	—	S	S	S	—	—	—	S	—	—
Manganese salts	—	S	S	S	—	S	—	—	—	—
Mercury	S	S	S	S	—	S	—	S	—	—
Methyl alcohol	U	S	S	S	U	U	S	S	—	—
Methyl ethyl ketone	U	S	S	S	U	U	S	M	—	—
Methyl salicylate	U	—	—	—	—	U	—	—	—	—
Methylene chloride	U	S	S	S	U	U	—	S	—	—
Nickel salts	S	S	S	S	S	S	—	—	—	—
Nitric acid (10%)	S	S	S	S	S	S	M	S	S	U
Nitric acid (50%)	M	S	M	S	M	M	M	S	S	U
Nitric acid (95%)	U	M	U	M	U	U	U	S	—	U
Oleic acid	S	S	S	S	S	S	S	S	—	—
Oxalic acid	S	S	S	S	S	S	S	S	—	U
Perchloric acid (10%)	—	S	S	S	—	—	—	S	—	U
Phenol	U	U	S	U	—	—	U	S	—	—
Phenyl ethyl alcohol	—	S	S	S	—	U	S	—	—	—
Phosphoric acid (10%)	S	S	S	S	S	S	—	S	S	—
Phosphoric acid (conc.)	S	M	S	M	M	M	—	S	—	U
Phosphorus trichloride	U	S	S	S	—	—	—	S	—	—
Potassium acetate	M	S	S	S	—	—	—	—	—	—
Potassium carbonate	—	S	S	S	S	S	S	S	S	—
Potassium chlorate	S	S	S	S	S	S	S	S	S	—
Potassium chloride	S	S	S	S	S	S	S	S	S	—
Potassium hydroxide (5%)	U	S	S	S	S	S	S	S	—	—
Potassium hydroxide (conc.)	U	S	S	S	U	U	—	—	—	—
Potassium permanganate	—	S	S	S	—	—	S	S	—	—
Silicic acid	—	S	S	S	—	—	—	—	—	—
Silicone fluids	—	M	M	M	—	—	—	—	—	—
Silver cyanide	—	S	S	S	—	—	—	S	—	—
Sodium bisulphate	S	S	S	S	S	S	S	S	S	—
Sodium borate	S	S	S	S	S	S	S	—	—	—
Sodium carbonate	M	S	S	S	S	S	S	S	S	—
Sodium chloride (10%)	S	S	S	S	S	S	S	S	—	—
Sodium chloride (sat'd)	—	S	S	S	S	—	S	—	—	—
Sodium dichromate	—	S	S	S	—	—	S	—	—	—

Table 9.1 *continued*

Reagent	Polycarbonate	Polypropylene	Polyethylene	Polyallomer	Cell. nitrate	Cell. acetate butyrate	Nylon	Kynar	Noryl	S. steel
Sodium hydroxide (1%)	U	S	S	S	S	S	S	S	—	—
Sodium hydroxide (10%)	U	S	S	S	U	U	S	S	—	—
Sodium hydroxide (conc.)	U	M	S	M	U	U	S	—	—	—
Sodium hypochlorite	S	S	S	S	S	S	S	S	—	U
Sodium nitrate (10%)	U	S	S	S	—	—	S	S	S	—
Sodium peroxide	—	S	S	S	—	—	S	S	S	—
Sodium sulphide	U	S	S	S	—	S	S	S	—	—
Sodium thiosulphate	S	S	S	S	—	—	S	S	S	—
Sulphuric acid (10%)	S	S	S	S	S	S	S	S	S	U
Sulphuric acid (50%)	S	S	S	S	U	U	M	S	S	U
Sulphuric acid (75%)	S	S	U	S	U	U	U	S	S	U
Sulphuric acid (conc.)	S	S	U	S	U	U	U	S	—	U
Tannic acid	—	S	S	S	—	—	S	S	—	—
Toluene	U	U	S	U	S	S	S	S	—	—
Trichlorethylene	U	U	U	U	—	—	S	S	—	U
Trichloroacetic acid	U	S	S	S	—	—	—	—	—	—
Trichloroethane	U	U	U	U	—	S	S	S	—	—
Trisodium phosphate	M	S	S	S	—	—	—	S	—	—
Turpentine	U	M	U	M	—	U	S	S	—	—
Urea	S	S	S	S	S	S	S	S	—	—
Urine	S	S	S	S	S	S	S	S	—	—
Xylene	U	U	U	U	—	S	U	S	—	—
Zinc chloride	S	S	S	S	S	S	S	S	—	U

wide variety of chemicals but is not recommended for use with highly polar solvents such as ketones and esters. It is resistant to high temperatures and has low absorption qualities. Kynar has been used as a coating agent because of the balance of physical, thermal and chemical properties not found in other polymers. The greater mechanical strength, superior resistance to abrasion and lesser cold flow properties are retained over a useful temperature range of −60 °C to 150 °C. This material is used as a spray-coated lining on stainless steel tubes and bottles manufactured by Dupont-Sorvall.

Noryl. A modified polyphenylene oxide, used for zonal rotor cores, Noryl is non-toxic and has high-strength properties, although it may weaken slightly after repeated autoclaving over a long period. It is unaffected by aqueous media, including weak and strong acids and bases. It will, however, soften or dissolve in many halogenated and aromatic hydrocarbons. Dilute solutions of diethyl-pyrocarbonate drastically weaken Noryl, and after such treatment the septa of zonal rotor cores can disintegrate during centrifugation.

Stainless steel. Used for holders, caps and tubes, this material offers advantages in particular cases where specific chemical resistance and high-speed capabilities are required. Some tubes and bottles are also available with Kynar lining for special applications.

The choice as to which type of centrifuge tube should be used is extremely dependent on the composition of the solution put into it. *Table 9.1* gives an extensive list of chemicals many of which are frequently used in one or other types of preparative technique involving centrifugation. However, an additional factor to be considered is that the stress of centrifugation can enhance the susceptibility of tubes to chemical attack.

Although thin-walled tubes can be used in swing-out rotors, in fixed-angle and vertical rotors there is a strong force acting across the tube trying to squash it; thus, thicker walled tubes must be used. Even so, as a general rule centrifuge tubes can only withstand the stress of centrifugation at high speed if they are completely full, the only exceptions to this rule being when stainless steel tubes, thick-walled tubes, or Oak Ridge type polycarbonate tubes are used, although in no case can the rotors be run at maximum speed (see p. 301). Besides strength, flexibility is also important, particularly when centrifuging density gradients. The difference in density between the top and bottom of the gradient can crack and even shatter the more brittle tubes (for example, those made from polycarbonate and polypropylene) when they are centrifuged in fixed-angle rotors (see p. 000). On the other hand, the additional strength of polycarbonate tubes is necessary when centrifuging at speeds in excess of 50 000 rev/min.

Sealing Caps for Tubes

All tubes used in fixed-angle and vertical rotors must be securely capped before they are centrifuged. The function of the cap is threefold. First, it ensures that the sample is contained within the tube, which is particularly important when working with biohazardous, radioactive or corrosive samples. Secondly, it prevents the sample being exposed to the high vacuum surrounding the rotor when the seal on the lid of the rotor leaks. Thirdly, and perhaps its most important role, it supports the top of the tube and prevents deformation during centrifugation.

The sealing caps which were originally developed, and which are still widely used, are quite complex; the components of a typical cap are shown in *Figure 9.2*. A common feature of all caps is the use of compressed rubber 'O'-rings to make a liquid-tight seal. Caps are made in plastic, aluminium and stainless steel, although in the case of the last of these the top speed of the rotor must be derated by 25% to compensate for the extra weight. The rigid nature of polycarbonate has enabled other, simpler, types of caps to be devised. Threaded polycarbonate tubes either of the parallel-sided or Oak Ridge types can be sealed with aluminium screw caps. However, these simpler types of cap do have a tendency to leak and so, whereas they strengthen the top of the tube, they do not always contribute significantly to the containment of the sample.

One problem that does arise with caps is cleaning them after use. Aluminium caps are particularly susceptible to corrosion and they must never be left to soak in the usual laboratory detergents, most of which are extremely alkaline. Caps should be disassembled, briefly rinsed in a neutral detergent, thoroughly rinsed with tap and distilled water and finally wiped dry. When reassembling caps, care should be taken to ensure that the correct individual components are combined with each other, and some manufacturers recommend that the rubber

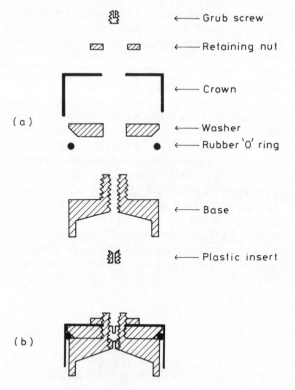

Figure 9.2 *Components of tube sealing caps: (a) components; (b) assembled cap*

'O'-rings are lightly smeared with silicone grease, although it is better to use no grease than too much.

DERATING ROTORS

The time needed to complete any centrifugal separation is inversely proportional to the square of the speed, and so increasing the speed of the rotor markedly reduces the time required for any separation. However, in some cases it is not possible to utilize the maximum speed of the rotor, because either the high centrifugal force destroys the integrity of the sample (for example, cells) or it adversely affects the resolution by, for example, increasing the slope of the gradient (see p. 174). More usually it is the physical properties of the sample, tubes, caps, adaptors and rotors which place restraints on the maximum speed of each particular rotor. Even when full, cellulose nitrate and polypropylene tubes cannot be centrifuged above 50 000 rev/min. At higher speeds it is necessary to use thick-walled polycarbonate or polyallomer tubes. On the other hand, some polycarbonate tubes will not withstand centrifugation at very high speeds if they are only partly filled. Similar problems arise in the use of Delrin adaptors in high-speed rotors, since it is the strength of the Delrin adaptors themselves that limit the speed of the rotor.

All ultracentrifuge rotors are designed and tested assuming that the density of the samples will be close to unity and that they will be contained in lightweight tubes sealed, where necessary, with similarly lightweight caps. However, tubes are also made from denser materials (for example, stainless steel) and, when using these tubes, it is necessary to reduce the maximum speed of the rotor by up to 25% (derating factor 0.75). Similarly, when using stainless steel caps for plastic tubes the top speed of the rotors must be reduced by 25%.

When the sample to be centrifuged has a density greater than 1.2 g/cm³ (for example, CsCl solutions), the maximum speed of the rotor must be reduced. Failure to observe this rule can lead to the failure of the rotor during centrifugation. The extent to which any rotor must be derated can be calculated from the equation

$$V' = V \left(\frac{1.2}{\rho}\right)^{\frac{1}{2}}$$

where V is the manufacturers' quoted maximum speed and V' is the maximum speed for a sample with a density of ρ g/cm³. *Table 9.2* encompasses the most common range of densities encountered. One serious problem when working

Table 9.2 EFFECT OF SPECIFIC GRAVITY OF SAMPLES ON THE MAXIMUM SPEED OF ROTORS*

Specific gravity	Derating factor
1.2	1.00
1.4	0.93
1.6	0.87
1.8	0.82
2.0	0.77
2.2	0.74
2.4	0.71
2.6	0.68
2.8	0.65
3.0	0.63

*To obtain the maximum speed of any rotor multiply the top speed of the rotor by the derating factor corresponding to the density of the sample

with concentrated salt solutions is the possibility that crystallization of the solute may occur if, as a result of centrifugation, the limit of solubility is exceeded. For example, in the case of CsCl, its solutions do not exceed 1.9 g/cm³, whereas the crystals that can form during high-speed centrifugation at low temperatures have a density of 4 g/cm³, far beyond the design limits of any rotor. Great care should therefore be taken to ensure that crystallization does not occur during centrifugation.

Some manufacturers also recommend that older rotors should be derated after either a fixed period or a given number of runs. The basis of this policy is that, since the stress level of a rotor is proportional to the square of its speed, derating will help to slow down the rate of stress-corrosion mediated propagation of cracks. However, the rate at which cracks propagate increases almost exponentially and so, even after derating, a rapid failure of corroded rotors can occur. If the user is in any doubt as to the age or extent of corrosion of any rotor, it should be immediately returned to the manufacturer for a detailed examination.

ROUTINE MAINTENANCE OF ROTORS

Detailed descriptions of the routine maintenance necessary for individual rotors of all types are given in the manufacturers' instruction manuals. It cannot be overemphasized that the instructions given in these manuals must be rigorously followed, otherwise there is a real danger that rotors may disintegrate during high-speed centrifugation. Here, some of the main points are reiterated to stress their importance.

The main aim of maintenance is to prevent corrosion of the rotors. As stated previously, aluminium rotors are particularly susceptible to corrosion, but prolonged exposure to adverse conditions can also seriously affect the performance of titanium rotors and the procedures described here should always be followed. First, rotors should always be rinsed with tap water and then distilled water *immediately* after use. The rotors can be left inverted to drain dry or they may be dried with tissue but on no account should moisture be left at the bottom of the rotor pockets. The rotors should be stored in a dry atmosphere, preferably at room temperature. They should not be kept in the coldroom for prolonged periods otherwise condensation accumulates in the pockets of the rotor and the moisture can combine with salt deposits, leading to corrosion. While stored in the coldroom, rotors should be left inverted to prevent any accumulation of condensation.

Aluminium rotors require special care and they should be regularly and carefully polished inside and out with a suitable wax which both displaces and repels moisture, but special care must also be taken not to trap moisture or salt deposits below the wax film. The theory is that during centrifugation the wax flows into any cracks that form in the surface aluminium oxide layer, thus preventing moisture coming into contact with the aluminium alloy. Great care should be taken not to scratch or mark the inside or outside of any rotor, since scratches particularly can act as nucleation centres for cracks. Aluminium rotors are also weakened by impact forces, for example by dropping the rotor even a few inches on to a hard surface, and by heating above $140\,^{\circ}C$. Finally, aluminium rotors are so susceptible to corrosion that salt solutions [for example, $CsCl$, Cs_2SO_4, $(NH_4)_2SO_4$ and mercury salts] should be centrifuged in titanium rotors whenever possible.

In addition to combating corrosion it is also important to keep the exterior of the rotors clean and undamaged. Most temperature-control systems rely on monitoring the infrared emission from the rotors, and at the low temperatures usually employed changes in the rotor surface can result in erratic temperature control. A number of rotors also have a disc on the base which defines the speed at which they can run. These discs must be kept clean and care taken that they do not become scratched.

ACKNOWLEDGEMENTS

The writing of this chapter and its Appendix has been greatly facilitated by the extensive help given by Mr Meddings (Beckman RIIC Ltd), Mr Hopkin (Damon–IEC Ltd), Dr Robson (Dupont-Sorvall Ltd) and Mr Fogg (Kontron Instruments Ltd).

APPENDIX: PHYSICAL DIMENSIONS AND CHARACTERISTICS OF ROTORS

Centrifuge rotors have been divided into four main classes: swing-out, fixed-angle, vertical and zonal (*Tables 9.3–9.6*). In turn, the swing-out and fixed-angle rotors have been set out in three tables, corresponding to small, intermediate and large volume rotors – *Tables 9.3(I)* and *9.4(I–III)*. Continuous-flow rotors have been excluded from the table of zonal rotors, since each rotor, unlike all other types of rotor, is generally very individual and it is difficult to relate one rotor to another.

Notes

1. *Bucket sizes.* The bucket sizes given by the manufacturers and included in the tables are nominal. The volume given is frequently the volume of the rotor pocket, rather than the volume that the tube will hold. The exact volume a tube holds is dependent on its thickness.

   2. *Efficiency and resolution of rotors.* The efficiency of a rotor can be determined from its k-factor, which is a constant relating the time in hours required to pellet a given particle with a sedimentation coefficient, S. Thus

$$k = S \times t$$

where k is related to the dimensions and speed of the rotor by the equation

$$k = \frac{(\ln R_{max} - \ln R_{min})2.78 \times 10^9}{\omega^2}$$

As can be seen from the first equation, *the value of* k *decreases as the efficiency of the rotor increases.* The k-factor of rotors can vary from as little as ten to a thousand or more. It is important to remember that the value of k is dependent on the speed of the rotor. All values given in the tables are calculated using the top speed of each rotor; lower speeds mean higher k-factors. Also, these k-factors have been calculated from bucket sizes; they make no allowance for the shape of the tube or bottle (see Note 1). In practice, k-factors will be lower than those quoted.

   An alternative method for assessing the capability of swing-out rotors has been devised by A. Fritsch [*Analyt. Biochem.* (1973) **55**, 57]. Using his procedure, the resolving power of rotors, Λ, has been calculated for all swing-out rotors. It compares the resolution of the various rotors that can be achieved in rate-zonal separations using an approximately isokinetic 5–20% sucrose gradient. In this case, Λ increases as the resolving power of the rotor increases. However, as reference to *Table 9.3 (I–III)* shows, the comparative values obtained by the two methods are similar, although they do not agree exactly.

   3. *Relative centrifugal force.* The following equation is a convenient way to calculate RCF for any radial position in a rotor; it is accurate to within 0.1%:

$$RCF = 11.18 \times r \times \left(\frac{rev/min}{1000}\right)^2$$

where r is the distance from the centre of rotation, in centimetres.

Table 9.3(I) SWING-OUT ROTORS: SMALL VOLUME

Manufacturer*	Rotor designation	Bucket No. and volume ml	Rotor material	Maximum speed rev/min	R_{min} cm	R_{max} cm	Max RCF g-force	k-factor	Λ
Beckman	SW65 Ti	3 × 5.0	Ti	65 000	4.12	8.90	420 000	46	1.14
	SW60 Ti	6 × 4.4	Ti	60 000	6.30	12.00	485 000	45	1.37
	SW56 Ti	6 × 4.4	Ti	56 000	5.60	11.63	408 000	59	1.28
	SW50 L	3 × 5.0	Al	50 000	4.70	9.80	274 000	74	0.96
	SW50.1	6 × 5.0	Al	50 000	5.97	10.73	300 000	59	1.00
	SW39 L	3 × 5.0	Al	39 000	4.70	9.80	175 000	122	0.75
Dupont-Sorvall	AH650	6 × 5.0	Al	50 000	5.97	10.73	300 000	59	1.00
Heraeus Christ	S52/61	6 × 5.0	Ti	52 000	5.15	9.88	299 000	61	0.98
	S40/105	3 × 5.0	Al	40 000	5.01	9.74	174 000	105	0.75
IEC	SB405	6 × 4.2	Ti	60 000	5.10	10.10	405 900	48	1.17
Kontron	TST60	6 × 4.4	Ti	60 000	6.40	12.13	489 000	45	1.39
	TST54	6 × 5.0	Ti	54 000	6.40	11.14	363 500	48	1.10
MSE	6 × 4.2	6 × 4.2	Ti	60 000	5.18	12.42	500 000	62	1.51
	3 × 6.5	3 × 6.5	Ti	60 000	4.27	10.43	420 000	63	1.27
	3 × 5	3 × 5.0	Al	50 000	4.69	10.73	300 000	84	1.07
	6 × 5.5	6 × 5.5	Al	45 000	5.64	10.62	242 700	80	0.92

*The manufacturers in *Tables 9.3–9.6* are listed alphabetically.

Table 9.3(II) SWING-OUT ROTORS: INTERMEDIATE VOLUME

Manufacturer	Rotor designation	Bucket No. and volume ml	Rotor material	Maximum speed rev/min	R_{min} cm	R_{max} cm	Max RCF g-force	k-factor	Λ
Beckman	SW41 Ti	6 × 13.2	Ti	41 000	6.63	15.23	286 500	125	1.25
	SW40 Ti	6 × 14.0	Ti	40 000	6.67	15.87	284 000	137	1.28
	SW36	4 × 13.5	Al	36 000	5.99	13.30	193 000	156	0.96
	SW27.1	6 × 17.0	Al	27 000	6.80	16.60	135 000	310	0.91
Dupont-Sorvall	AH627	6 × 17.0	Al	27 000	6.79	16.56	135 000	310	0.91
Heraeus Christ	S40/135	6 × 15.0	Ti	40 000	6.81	16.02	287 000	135	1.29
IEC	SB283	6 × 14.0	Ti	41 000	5.55	15.15	283 200	151	1.28
Kontron	TST41	6 × 14.0	Ti	41 000	6.80	16.00	301 000	128	1.32
	TST28.17	6 × 17.0	Al+Ti*	28 000	7.02	16.87	148 000	283	0.96
MSE	6 × 14	6 × 14.0	Ti	40 000	6.34	15.86	284 000	145	1.29
	6 × 16.5	6 × 16.5	Al	30 000	6.12	15.80	159 000	267	0.97

*Buckets only are titanium.

Table 9.3(III) SWING-OUT ROTORS: LARGE VOLUME

Manufacturer	Rotor designation	Bucket No. and volume ml	Rotor material	Maximum speed rev/min	R_{min} cm	R_{max} cm	Max RCF g-force	k-factor	Λ
Beckman	SW27	6 × 38.5	Al	27 000	7.50	16.10	131 000	265	0.86
	SW25.1	3 × 34	Al	25 000	5.60	12.90	90 000	338	0.65
	SW25.2	3 × 60	Al	25 000	6.70	15.23	107 000	335	0.76
Dupont-Sorvall	AH627	6 × 38.5	Al	27 000	7.50	16.07	131 000	265	0.86
Heraeus Christ	S30/230	3 × 30	Al	30 000	5.68	12.85	129 000	230	0.77
	S20/584	3 × 75	Al	20 000	6.32	15.88	71 000	584	0.65
IEC	SB110	6 × 40	Al	25 000	6.32	15.27	110 000	355	0.77
Kontron	TST28.38	6 × 38	Al+Ti*	28 000	7.08	15.67	137 500	256	0.87
MSE	3 × 25	3 × 25	Al	30 000	5.88	12.87	129 500	220	0.77
	6 × 38	6 × 38	Al	25 000	8.10	16.50	115 300	288	0.80
	3 × 70	3 × 70	Al	23 500	6.41	16.27	100 000	427	0.78

*Buckets only are titanium.

Table 9.4(I) FIXED-ANGLE ROTORS: SMALL VOLUME

Manufacturer	Rotor designation	Bucket No. and volume ml	Rotor material	Tube angle deg	Maximum speed rev/min	R_{min} cm	R_{max} cm	Max RCF g-force	k-factor
Beckman	A100	6 × 0.2	Al	18.0	100 000	0.95	1.47	165 000	11
	75 Ti	8 × 13.5	Ti	25.5	75 000	3.7	8.0	503 500	35
	65	8 × 13.5	Al	23.5	65 000	3.7	7.8	368 800	45
	50 Ti	12 × 13.5	Ti	26.0	50 000	3.8	8.1	226 600	77
	50	8 × 13.5	Al	20.0	50 000	3.7	7.1	198 700	66
	40	12 × 13.5	Al	26.0	40 000	3.8	8.1	145 000	120
	40.2	12 × 6.5	Al	40.0	40 000	3.5	8.0	143 100	131
	40.3	18 × 6.5	Al	20.0	40 000	4.8	8.0	143 200	81
	30.2	20 × 10.5	Al	14.0	30 000	6.3	9.4	94 500	113
	25*	100 × 1.0	Al	25.0	25 000	8.1	10.0	70 100	84
			Al	25.0	25 000	9.7	11.6	81 200	71
			Al	25.0	25 000	11.3	13.4	92 300	62
Dupont-Sorvall	T865.1	8 × 13.5	Ti	23.5	65 000	4.6	8.7	410 000	38
Heraeus Christ	W70/38 Ti	8 × 12	Ti	25.0	70 000	3.6	7.6	416 300	38
	W65/40 Ti	12 × 12	Ti	25.0	65 000	4.2	8.2	385 000	40
	W60/44	8 × 12	Al	20.0	60 000	4.0	7.5	301 500	44
	W40/77	20 × 12	Al	25.0	40 000	6.3	10.3	183 500	77
	W40/41*	40 × 5	Al	20.0	40 000	6.8	9.2	164 700	48
			Al	20.0	40 000	8.2	10.6	188 900	41
IEC	A321	8 × 12	Al	35.0	60 000	2.7	8.0	321 400	76
	A269	10 × 12	Al	20.0	55 000	4.2	8.0	269 100	53
	A168	20 × 12	Al	14.0	40 000	6.5	9.7	168 600	60
Kontron	TFT65.13	12 × 13	Ti	26.0	65 000	3.7	8.1	**384 300**	47
	TFT50.13	12 × 13	Ti	26.0	50 000	3.7	8.1	227 400	80
MSE	8 × 14	8 × 14	Ti	29.0	75 000	3.5	8.1	511 100	38
	10 × 10	10 × 10	Ti	35.0	72 000	4.3	8.6	499 400	34
	8 × 14	8 × 14	Al	25.5	65 000	3.5	7.8	368 490	46
	10 × 10	10 × 10	Ti	35.0	65 000	3.5	8.1	407 000	42
	10 × 10	10 × 10	Al	20.0	50 000	3.9	7.1	199 217	61
	8 × 10	8 × 10	Al	35.0	40 000	3.7	8.0	142 363	127

*The tubes of these rotors are arranged in multiple rows.

Table 9.4(II) FIXED-ANGLE ROTORS: INTERMEDIATE VOLUME

Manufacturer	Rotor designation	Bucket No. and volume ml	Rotor material	Tube angle deg	Maximum speed rev/min	R_{min} cm	R_{max} cm	Max RCF g-force	k-factor
Beckman	70 Ti	8 × 38.5	Ti	23.0	70 000	3.9	9.2	504 000	44
	60 Ti	8 × 38.5	Ti	23.5	60 000	3.7	9.0	326 000	63
	50.2 Ti	12 × 38.5	Ti	24.0	50 000	5.3	10.8	303 000	72
	42.1	8 × 38.5	Al	30.0	42 000	3.9	9.9	195 000	134
	30	12 × 38.5	Al	26.0	30 000	5.0	10.5	105 000	209
Dupont-Sorvall	T865	8 × 36	Ti	23.5	65 000	3.8	9.1	429 275	52
	A841	8 × 36	Al	23.5	41 000	3.8	9.1	170 000	130
Heraeus Christ	W40/128	8 × 35	Al	25.0	40 000	4.1	9.3	166 700	128
	W60/58	8 × 35	Ti	23.5	60 000	3.9	9.0	361 000	58
IEC	A237	8 × 40	Al	23.0	50 000	3.4	8.6	237 200	92
	A211	8 × 40	Al	30.0	45 000	3.1	9.4	211 200	137
	A192	8 × 50	Al	33.0	40 000	3.7	10.7	192 100	168
	A110 (109)	12 × 40	Al	26.0	30 000	5.4	11.0	109 800	199
Kontron	TFT70	8 × 38	Ti	20.0	70 000	4.3	9.2	504 000	40
	TFT65.38	8 × 38	Ti	20.0	65 000	4.3	9.2	434 500	46
	TFT50.38	12 × 38	Ti	23.0	50 000	5.5	10.8	303 300	70
MSE	8 × 35	8 × 35	Ti	21.0	60 000	3.6	9.4	379 200	68
	8 × 25	8 × 25	Ti	30.0	60 000	4.4	9.4	379 200	54
	8 × 35	8 × 35	Al	21.0	50 000	3.6	9.4	263 000	98
	8 × 25	8 × 25	Al	30.0	50 000	4.4	9.4	263 000	78
	8 × 50	8 × 50	Al	30.0	40 000	4.6	10.9	195 000	137

Table 9.4(III) FIXED-ANGLE ROTORS: LARGE VOLUME

Manufacturer	Rotor designation	Bucket No. and volume ml	Rotor material	Tube angle deg	Maximum speed rev/min	R_{min} cm	R_{max} cm	Max RCF g-force	k-factor
Beckman	45 Ti	6 × 94	Ti	24.0	45 000	3.7	10.4	235 800	129
	35	6 × 94	Al	25.0	35 000	3.5	10.4	142 800	225
	21	10 × 94	Al	18.0	21 000	6.0	12.0	59 200	398
	19	6 × 250	Al	25.0	19 000	4.4	13.3	53 700	776
	15	4 × 500	Al	15.0	15 000	4.1	14.2	35 500	1405
Dupont-Sorvall	A641	6 × 98	Al	22.5	41 000	3.4	9.9	186 466	162
Heraeus Christ	W30/281	6 × 90	Al	20.0	30 000	3.5	9.6	96 800	281
	W23/534	4 × 250	Al	20.0	23 500	3.7	12.0	74 000	534
	W15/1460	4 × 500	Al	20.0	15 000	4.2	15.2	38 400	1460
IEC	A170	6 × 75	Al	20.0	40 000	3.2	9.7	170 300	176
	A54	6 × 250	Al	20.0	20 000	3.7	12.6	54 800	767
	A28	4 × 500	Al	20.0	14 000	1.5	13.2	28 500	2854
Kontron	TFT45	6 × 94	Ti	23.0	45 000	3.9	10.4	236.600	125
	TFA20	6 × 250	Al	23.0	20 000	5.1	13.5	60 400	616
MSE	6 × 100	6 × 100	Al	25.0	35 000	4.0	11.2	154 000	213
	10 × 100	10 × 100	Al	18.0	25 000	6.3	12.7	88 500	284
	6 × 300	6 × 300	Al	25.0	21 000	5.9	15.3	75 000	547
	4 × 500	4 × 500	Al	18.0	14 000	4.1	14.7	32 230	1650

Table 9.5 VERTICAL TUBE ROTORS

Manufacturer	Rotor designation	Bucket No. and volume ml	Rotor material	Maximum speed rev/min	R_{min} cm	R_{max} cm	Max RCF g-force	k-factor
Beckman	VTi-65	8 X 5.0	Ti	65 000	7.2	8.5	401 700	10
	VTi-50	8 X 38.5	Ti	50 000	6.1	8.6	241 200	35
	VAl-26	8 X 38.5	Al	26 000	7.1	9.6	70 000	120
Dupont-Sorvall	TV865	8 X 5.0	Ti	65 000	7.2	8.5	400 000	10
	TV850	8 X 36.0	Ti	50 000	5.9	8.5	236 547	36
	TV865B	8 X 17.0	Ti	65 000	6.0	8.5	399 765	21
MSE	VWR65	8 X 5.0	Ti	65 000	7.4	8.5	400 600	8
	VWR50	8 X 35.0	Ti	50 000	6.2	8.6	240 600	34

Table 9.6 ZONAL ROTORS

Manufacturer	Rotor designation	Rotor volume ml	Rotor material	Maximum speed rev/min	Particle pathlength cm	Depth of rotor chamber cm	Max RCF g-force
Beckman	Z60	330	Ti	60 000	5.2	3.1	256 000
	Ti-14	650	Ti	48 000	5.4	5.4	171 800
	Al-14	650	Al	35 000	5.4	5.4	91 300
	Ti-15	1665	Ti	32 000	7.6	7.6	102 000
	Al-15	1665	Al	22 000	7.6	7.6	48 100
	CF-32 Ti	430	Ti	32 000	2.4	6.8	102 000
	B4	1700	Al	40 000	3.4	25.6	91 000
	B16	750	Al	40 000	1.2	25.6	91 000
	JCF-Z						
	zonal	1900	Ti	20 000	6.9	8.9	40 000
	reorienting	1750	Ti	20 000	6.9	8.9	40 000
	continuous flow	680	Ti	20 000	2.8	8.4	40 000
Dupont-Sorvall	TZ28	1350	Ti	28 000	5.9	—	83 400
	SZ14	1380	Al	19 500	5.3	—	40 500
Electro-nucleonics	K3	3200	Ti	35 000	1.1	76.2	90 000
	K5	8390	Ti	35 000	4.5	76.2	90 000
	K6	3200	Ti	35 000	1.1	76.2	90 000
	K10	8000	Ti	35 000	3.9	76.2	90 000
	K11	380	Ti	35 000	0.1	76.2	90 000
	RK3	1600	Ti	35 000	1.1	38.1	90 000
	RK6	1600	Ti	35 000	1.1	38.1	90 000
	RK10	3980	Ti	35 000	3.9	38.1	90 000
	RK11	900	Ti	35 000	0.1	38.1	90 000
	J1	780	Ti	35 000	0.9	38.1	150 000
Heraeus Christ	B14	650	Ti	45 000	5.2	5.4	151 000
	B15	1650	Ti	30 000	7.4	6.4	89 500
	B29	1400	Ti	30 000	7.1	6.4	86 000
IEC	Z15	780	Perspex/Al	8 000	9.5	2.6	9 100
	B30	659	Ti	50 000	4.8	5.4	186 400
	B30I*	570	Ti	50 000	4.5	5.4	176 900
	B29	1674	Ti	35 000	6.9	7.6	121 800
	B29I*	1480	Ti	35 000	6.5	7.6	117 000

*Dimensions with inserts in rotors B30 and B29, respectively.

Table 9.6 *continued*

Manufacturer	Rotor designation	Rotor volume ml	Rotor material	Maximum speed rev/min	Particle pathlength cm	Depth of rotor chamber cm	Max RCF g-force
Kontron	TZT48	650	Ti	48 000	5.2	5.4	171 800
	TZT32	1650	Ti	32 000	7.4	7.6	102 000
MSE	B14	650	Ti	47 000	5.4	5.4	165 000
	B14	650	Al	35 000	5.4	5.4	91 000
	B15	1670	Ti	35 000	7.6	7.6	121 800
	B15	1670	Al	25 000	7.6	7.6	62 000
	B20	325	Ti	35 000	1.6	6.6	122 000
	continuous action	–	s. steel	21 000		7.0	28 300
	HS	695	Perspex/Al	10 000	7.6	1.9	12 800
	A-Type (AXII)	1300	Perspex/Al	5 000	13.6	1.5	5 000

The Relationship Between Radius and Volume of Zonal Rotors (*Figures 9.3–9.9*)

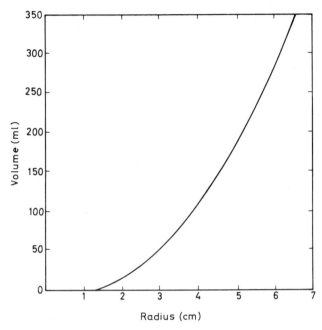

Figure 9.3 *The relationship between radius and volume of the Beckman Z60 rotor*

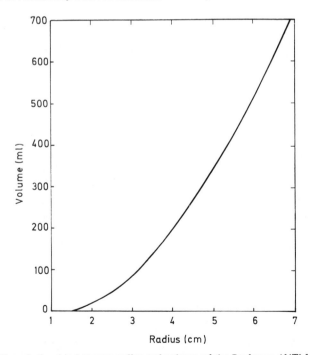

Figure 9.4 *The relationship between radius and volume of the Beckman Al/Ti-14 rotor*

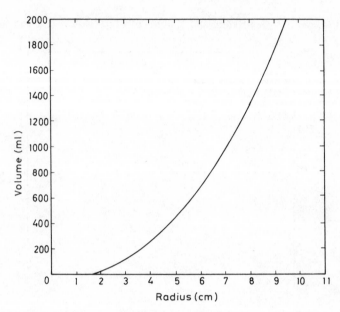

Figure 9.5 The relationship between radius and volume of the Beckman Al/Ti-15 rotor

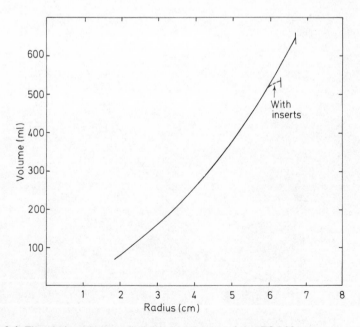

Figure 9.6 The relationship between radius and volume of the IEC B30 rotor

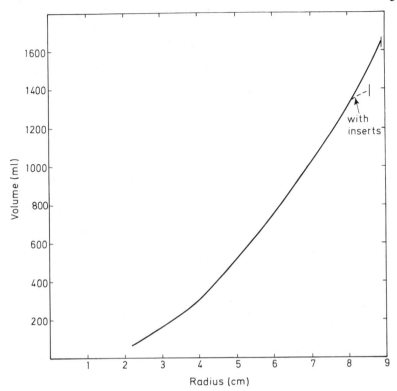

Figure 9.7 The relationship between radius and volume of the IEC B29 rotor

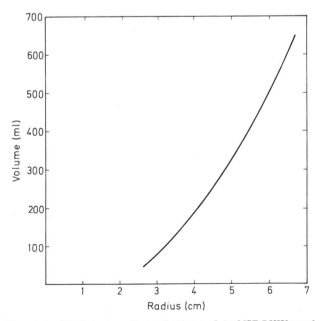

Figure 9.8 The relationship between radius and volume of the MSE BXIV zonal rotor

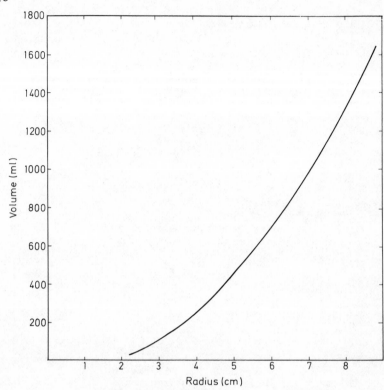

Figure 9.9 The relationship between the radius and volume of the MSE BXV zonal rotor

Index

317